Birds of Aruba, Bonaire, and Curaçao

Birds of Aruba, Bonaire, and Curaçao

A SITE AND FIELD GUIDE

JEFFREY V. WELLS
ALLISON CHILDS WELLS

Illustrated by

ROBERT DEAN

A Zona Tropical Publication
FROM
COMSTOCK PUBLISHING ASSOCIATES
a division of
Cornell University Press
Ithaca and London

First published 2017 by Cornell University Press

First printing, Cornell Paperbacks, 2017
Printed in China

Library of Congress Cataloging-in-Publication Data

Names: Wells, Jeffrey V. (Jeffrey Vance), 1964– author. | Wells, Allison Childs, author. |
    Dean, Robert, 1955– illustrator.
Title: Birds of Aruba, Bonaire, and Curaçao : a site and field guide / Jeffrey V. Wells,
    Allison Childs Wells ; illustrated by Robert Dean.
Description: Ithaca : Comstock Publishing Associates, a division of Cornell University
    Press, 2017. | "A Zona Tropical publication." | Includes bibliographical references and
    index.
Identifiers: LCCN 2016050673 | ISBN 9781501701078 (pbk. : alk. paper)
Subjects: LCSH: Birds—Aruba—Identification. | Birds—Bonaire—Identification. | Birds—
    Curaçao—Identification. | Bird watching—Aruba—Guidebooks. | Bird watching—
    Bonaire—Guidebooks. | Bird watching—Curaçao—Guidebooks.
Classification: LCC QL688.A78 W44 2017 | DDC 598.0972986—dc23
LC record available at https://lccn.loc.gov/2016050673

Book design: Gabriela Wattson
Cover design: David Rotstein

We dedicate this book to Professor Doctor K. H. Voous, a pioneer of Caribbean ornithology who established the baseline for ornithology in the "Netherlands Antilles," as the islands were known during his time. His *Birds of the Netherlands Antilles* was indispensable during our early years on the islands and continues to be an important resource and inspiration.

This book is also dedicated to the warm, welcoming people of Aruba, Bonaire, and Curaçao. We are grateful for the opportunity to have spent so much time on the islands that are their home, and it is their hands that hold the future of the islands' birds and other wildlife.

Finally, we dedicate this book to our son, Evan, who made his first visit to Aruba when Allison was four months pregnant with him, then again at five months old, and many times since. Evan has come to love the islands—their birds, culture, people, and special places—as much as we do, and we trust that he will continue to answer the call of the islands throughout his life.

Contents

Preface

We made our first trip to Aruba in 1993, as members of the twenty-piece Ithaca Ageless Jazz Band. As lead trumpeter (Jeff) and vocalist (Allison), we performed music ranging from swing numbers of the Big Band Era to contemporary jazz, at resorts throughout the island. Following our gigs, our bandmates often headed off to casinos or clubs, but we hit the sack as early as possible to prepare for birding at dawn the next day.

None of the field guides we carried with us in those early years was exactly what we needed. Several locally produced bird guides, including *Birds of Aruba* (Reuter) and *Our Birds* (de Boer), served a very important purpose. The latter, written for a local audience, is a valuable educational tool, with text provided in three languages. However, both books covered very few species and provided only limited information about those species. Field guides covering the Caribbean or regions in South America did not include all of the birds of Aruba, Bonaire, and Curaçao (ABCs).

As a guide for our birding trips—and as a reference work for our research—*Birds of the Netherlands Antilles* by K. H. Voous was indispensable, providing detailed information about the life histories of most of the birds that occur on the ABCs, along with critical ecological information for each of the islands. Packing this book became as routine to us as packing our toothbrushes and binoculars. In his tome, Voous mentions several of the most important birding spots on each of the islands—Bubali Bird Sanctuary and Arikok National Park on Aruba; the Pekelmeer and Washington-Slagbaai National Park on Bonaire; and Malpais and Christoffel Park on Curaçao. His book, which was published in 1983, was never intended to be either a site guide or a bird-identification guide, however, and much of the careful research he conducted for it had become outdated before we first set foot on the ABCs.

We eventually left the band and moved from Ithaca, New York, back to our home state of Maine, but our trips to the ABCs continued. The more birding we did on the islands, and the more data we collected about the resident species and migratory visitors, the more it became apparent that a new, modern field guide was in order. We started by creating two websites, one for the birds of Aruba (www.arubabirds.com) and the other for the birds of Bonaire (www.bonairebirds. com). The websites allowed us to share information and, at the same time, to collect information about sightings and other data from our fellow birders. With each visit to the islands, as we gathered more data about the birds and where to find them, our vision for this book—a bird identification and site guide—came into clearer focus, and we recognized the importance of also including a section on conservation.

Over the years we built many important relationships with the staff at Arikok National Park on Aruba and with the researchers and managers at STINAPA and CARMABI who run the national parks on Bonaire and Curaçao, respectively. We learned a great deal from them about the conservation issues on their islands and the struggles relating thereto, including the lack of ornithological training for park rangers. With them and with the support of concerned members of the islands' private sector, we helped send a few park rangers to the United States, where they learned bird study techniques, bird identification and biology, and bird conservation practices—knowledge that they were then able to share with other park rangers. We also learned that a book of this kind would be extremely useful as a reference for those earning their livelihoods protecting their natural resources. We hope, too, that the book will be useful for young people from Aruba, Bonaire, and Curaçao who have an interest in learning more about the birds that are part of their natural heritage.

We know from our own experience that the tourists unloading their bags from the taxi in front of their resort hotel are often the very same people who want to know the name of those little black-and-yellow birds that hop onto the rim of their glasses for a sip of mango juice at breakfast. We hope that birders from all parts of the world—including the ABCs—will find this book useful and enjoyable. But we also hope to reach the beachgoers, golfers, shoppers, windsurfers, and scuba divers—indeed anyone wishing to learn more about the birds of Aruba, Bonaire, and Curaçao and where to find them.

Best in birding,
Allison and Jeff Wells

Acknowledgments

During our more than two decades of studying the birds of Aruba, Bonaire, and Curaçao, we have benefitted from the assistance, advice, and good company of many people and institutions. We want to thank and acknowledge as many of them as possible and apologize in advance to anyone we might have left off the list.

First and foremost we would like to extend our utmost thanks to the Ellsworth Kelly Foundation, whose crucial support for the book's artwork finally moved this project toward completion. Sadly, Ellsworth himself passed away before this book was published, but we gratefully acknowledge his long-held interest in birds and his accomplishments as a world renowned artist.

The nonprofit organizations by whom we have been employed were also crucial partners in the multiyear process of researching and writing this book. Our early years of researching the birds of Aruba, Bonaire, and Curaçao took place while we were at the Cornell University Lab of Ornithology (Allison) and the National Audubon Society (Jeff). Both institutions were great places to work and were supportive of our interests in the birds of these islands and their conservation. Over the last decade, the Boreal Songbird Initiative (Jeff) and the Natural Resources Council of Maine (Allison) have similarly encouraged our work on the book. In particular we would like to thank Lane Nothman and Fritz Reid from the Boreal Songbird Initiative, and Lisa Pohlman, Kathy Thompson, and Gretta Wark from the Natural Resources Council of Maine.

Although we never had the opportunity to meet world renowned ornithologist K.H. Voous, who wrote *Birds of the Netherlands Antilles*, we greatly value the opportunities we had to communicate with him in the first years that we visited Aruba. The world owes him tremendous gratitude for the years of thorough ornithological research that he compiled in not only several editions of the aforementioned book but also numerous journal publications. Even after the English version of *Birds of the Netherlands Antilles* was published in 1983, Voous continued to collect bird records from the islands. After his death in 2002, these reports were compiled under the leadership of Tineke Prins and Hans Reuter of the Zoological Museum Amsterdam and published as an annotated checklist in 2009; coauthors Adolphe Debrot, Jan Wattel, and Vincent Nijman made important contributions. Their work has been a valuable resource for people studying the avifauna of Aruba, Bonaire, and Curaçao.

The Caribbean Research and Management of Biodiversity (CARMABI) foundation, headquartered on the island of Curaçao, provided all kinds of help to us over the years, including the use of a vehicle and lodging in their field station, GIS support, and advice and encouragement. Former director Adolphe Debrot provided key support and encouragement for our efforts. Dolfi, as he is affectionately known by many, has been one of the most prolific scientists ever to have focused attention on the islands, which were once known as the Netherlands Antilles. His massive list of publications covers topics as diverse as coral reef ecology, marine mammals, birds, insects, restoration ecology, and much more. We gratefully acknowledge his contributions to our work and to furthering the understanding of the ecology of Aruba, Bonaire, and Curaçao. Leon Pors, formerly of CARMABI, and John De Freitas, still with CARMABI, have provided information and advice that have been helpful.

The staff at Arikok National Park on Aruba, Washington-Slagbaai National Park on Bonaire, and Christoffel Park on Curaçao deserve praise for their continuing efforts over many decades to ensure the ecological integrity of the lands under their stewardship. Many of them have been helpful to us in a variety of ways. Longtime Arikok National Park ranger Julio Beaujon has continued to be a great source of knowledge over many years. Dilma Arends, formerly one of the park leaders, was a strong supporter of research and education efforts in earlier years and was instrumental in putting together financial support to bring park ranger Everaldo Raffini to the famed Hog Island Audubon Ornithology Camp in Maine. Greg Peterson, chair of the Arikok National Park Foundation, founder and president of Aruba Birdlife Conservation, and board member of the Dutch Caribbean Nature Alliance, has been a champion of bird and ecological education and conservation on Aruba. His tireless efforts on behalf of the birds and people of Aruba deserve recognition and support.

We have been assisted in large and small ways by a number of staff of STINAPA Bonaire, including Sabine Engel of the Bonaire National Marine Park, who arranged a visit to Klein Bonaire for conducting bird surveys; former manager of Washington-Slagbaai National Park, Fernando Simal, for arranging access to the park for bird surveys and sound recording; George "Kultura" Thode, for providing background on status of bird species in the park, as well as his thoughts on the natural history of White-tailed Nightjars; and park ranger Clifford Cicilia, for his interest in attending the Audubon Camp at Hog Island, Maine, to learn more about birds and birding. We thank Seth Benz, Sabine Engel, Janneke van Gerwen, Rene Hakkenberg, George de Salvo, Fernando Simal, and Larry Theilgard for helping to put together sponsorships that allowed Clifford to attend the camp.

Many other people not associated directly with the national parks have also been instrumental in supporting our work. Gerard Phillips, an Irish birder and artist now living in Canada, accompanied Jeff on a bird survey trip on Curaçao. His birding skills and his enthusiasm and good humor, despite a grueling survey schedule, were much appreciated. We also greatly appreciate the accommodations provided by the Buddy Dive Resort on one of our trips to Bonaire, through the support of Carol Bradovchak, Ruud van Baal, and Marcel Westerhoff. The late Jerry Ligon, a birding expert and bird guide from Bonaire, provided a valuable service by compiling bird sightings and documentation; we remember him for his enthusiasm for teaching people about birds and sharing information. Seabird expert Ruud van Halewijn was one of the earliest to document the offshore marine birds around the ABCs. Ruud continues to be involved in bird research and conservation on the islands.

Thriving birding communities now exist on all three islands, and these resident birders have been wonderfully supportive in providing information and photographs. They include Aruba birders Julio Beaujon and Tyrone de Kort. It was a pleasure to spend time birding on Aruba with Ferdinand Kelkboom, Michiel Oversteegen, Sven Oversteegen, Peter Sprockel, and Ross Wauben. Curaçao birders we wish to thank include Bart de Boer, Michelle Da Costa Gomez, Carel de Haseth, Eric Newton, Chris Richards, Cisca Rusch, and Rob Wellens; and on Bonaire, Elly Albers, the late Jerry Ligon, and Sipke Stapert.

Many scientists and professionals responded to our queries. These include Kalli De Meyer, Emeray Martha-Neuman, and Alice Ramsay (Dutch Caribbean Nature Alliance), Adrián Azpiroz (University of Missouri-St. Louis), Gerard van Buurt (herpetology expert from Curaçao), Jessica Eberhard (Louisiana State University), Kyle Harms (Louisiana State University), Jean-Claude Hippolyte (Centre Européen de Recherche et d'Enseignement des Géosciences de l'Environnement), Pepijn Kamminga (Naturalis Biodiversity Center, Netherlands), Jolanda Luksenburg (George Mason University), Milton Ponson, Jonathan Putnam (US National Park Service), Howard Reinert (The College of New Jersey), Mark Robbins (University of Kansas), Gordon Taylor (Stony Brook University), and Matt Whelchel (Florida Aqua-store).

For assistance with our work documenting and archiving the bird sounds of the ABCs, we thank the current and former staff of the Cornell Lab of Ornithology's Macaulay Library of Natural Sounds, especially Greg Budney, Jack Bradbury, Annette Finney, Martha Fischer, Mark Reaves, and Matt Medler.

Kimberly Bostwick, Kevin McGowan, Charles Dardia, and Irby Lovette of the Cornell Museum of Vertebrates provided support, including loans of bird specimens from other institutions.

For discussion of various bird identification questions, we are grateful to Louis Bevier, Martin Frost, Jeff Gerbracht, Floyd Hayes, Marshall Iliff, Jay McGowan, and Steve Mlodinow. Over the years, we have had the pleasure of corresponding with many people who traveled to the islands and provided their bird reports and other data. We very much appreciate their filling out checklists and/or writing up reports that were helpful as we worked on the book and also provided guidance and encouragement to others interested in traveling to these islands. These contributors and correspondents include the following:

Kathy and Steve Abbott, Bob Abraham, Shanti Aiyer, Tony Ambrose, Niels Peter Andreasen, Alison Aun, Patrick Baglee, Tony Baker, Carl Ball, Neal Baltz, Manny Barrera, Mark Barrett, Jon Bartol, Juliana Bastidas, Fred Baumgarten, Patricia Beitzinger, Leigh Anne Bell, Pete Bengtson, Michele and Justin Berger, Elsmarie Beukenboom, Adrian Binns, Brian Bockhahn, Ross Bonander, Holly Booker, Sam Bordovsky, Lindsay Bosch, Terry Boyd, Julie Bowen, Paul Bowley, Eelco Brandenburg, Hilke Breder, Bill Brooking, Kay Brown, Bob Bryant, Paul Buckley, Susannah Buhrman, Annette Burdges, Ken Burgener, Malcolm Calvert, David Campbell, Lew Candura, Joe and Kathy Canzano, Russ Carr, Kimberley Casey, John Cecil, Ron Cedrone, Kathryn Chandler, Tim and Linda Clos, Martin and Shelagh Coates, Alan Collier, Anne Cooke, Donna Cooper, Candace Cornell, Leon Corrall, Norm Cote, Marilyn Cote-Miller, Robin Cox-Laird, August Croes, Ivar Croes, Bruce Cryder, Tim and Kathy Cybulski, Brian Daly, Harry Davies, Adam Davison, Marijke de Boer, Han de Bruijne, Theo de Kool, Tyron de Kort, Ben de Kruijff, Adrian del Nevo, Mathias Deming, Walter and Marian de Mooy, Tim den Outer, Brian Dering, Susan and Johan de Roos, Vladimir Dinets, Emile Dirks, Mike Dougherty, Bruno Giorno Eberhardt, Robert Einhorn, Mark Eising, Dave Eshbaugh, Lynda Eunson, Franchelle Everon, Gian Fabbri, Doug Faulder, Art Feagles, Rik Feije, Bob Feldberg, Cheryl Ferguson, Juan Carlos Fernández-Ordóñez, Kevin Finley, Stokes Fishburne Wayne Fisher, Erin Foley, Mary Frey, Tom Froman, S.R. Fopma, Linda Fuller, Steve Hey, Heather Gallant, John Galluzzo, Larry Gardella, Gehan Gehale, Kathy Genaw, Chris and Sandra Gibson,

Isabel Gibson, John Gibson, Martin Gottschling, Agnieszka Götz, Paul Goudriaan, Nancy Governali, Aaron Gwin, Anne Marie Hartman, Thannee Hassell, Chuck Hay, Bill Hedberg, Lolly Hedeen, Dick Heintz, Kathe and John Hendrickson, Carolyn Hernandez, Stephen Hey, Natalie Hodges, Patrick Holian, Kim Hubbard, Dave Hubler, Barbara Hunsberger, Brian Hobbs, Muriel Horacek, John Hoogerheide, Barb Houston, Arlette Hunnakko, Gina Jie-Sam-Foek, Graham Jones, Joy Joy, Jim and Jo Ellen Kalat, Wim Kamphuis, Rob Kelder, Mary Kersh, Alan Key, Alf King, Dennis and Alice Kirschbaum, Kees Klaij, Sarah Knutie, Erik Kramshøj, Margaret Krolick, J. Kroll, Jane Labun, Jessica Lajoie, Wim Lange, Selma Lampe, Craig Lanken, Steve Lebrun, Andrea Leistra, Peter Lensvelt, Erwin Lenting, Ted Lenz, John Lewis, Paul Ligorski, Eleanor Linkkila, Marion Lippmann, Kris Littlefield, Mary Lohse, Debra Lovley, Jacque Lowery, Cynthia Mailman, Sheila Maloney, Beth Mangia, Laura Manske, Neil Markowitz, Ad Martens, Rowan Martin, Becky Marvil, Barbara Masey, Andre Mazeron, Sean McMahon, Eileen McVey, David McWeeney, John Paul Mereen, Linsey Miller, Steve Mlodinow, Robert Mocko, Pete Mooney, Gary Morris, Rick Morris, Nelson Mostow, Dan Mudge, T.L. Murphy, Erik Neuteboom, Eric Newton, Hugo Nüssler, Ted O'Callahan, Sergio Ocampo, Daniel O'Malley, Hob Osterlund, Doris Parry, Annette Pasek, Ted Pearlman, Wayne and Carol Pembroke, Terry Peters, Roy and Marie Peterson, Sander Pieterse, Terry Piggott, Jeff Pippen, Jeannie Pitcher, Joan Pirogjo, Ted Post, Tim Potuyt, Jay and Jen Powell, Joe Prochaska, Raju Raman, Karen Rapp, Knud Rasmussen, Kerry Redding, Antoinette and Gary Renz, John Reynolds, Jan Hein Ribot, Jason Riggio, Elaine Robarge, Magnus Robb, Laura Robinson, Lyn and Dave Robinson, Bruce Ruppel, Jeffrey Ryan, Jaye Rykunyk, Jan Schaafsma, Diane Schellack, Lisa Schipper, Marjorie Schrader, Andy and Patricia Sheldrick, Steve Shultz, Beverly Schwartz-Katsh, Eileen Schwinn, Connie and Robert Shertz, Laurie Shrimpton, Antonio Silveira, Terry Sisson, Bill Sloan, Kathy Smith, Josh Southern, Gunter Speckmann, David Spence, Leo Spoormakers, Taco Spanbroek, Roberta Stemp, Liz Steppe, Lisbeth Stockman, Gary Stone, Skyler Streich, Alexandra and Detlef Stremke, Christie Johnson Stuber, Julie Suchecki, Ladislau Suli, Sven-Erik Sundberg, Nate Swick, Marcia Taylor, Kris Terrillion, Abha Tilokani, Alison Thompson, Mark Thompson, Sue Tichy, Jonnie Tilma, James Toledano, Lisa Tromp, Jim Trotter, Bob Tull, George and Lorna Tuthill, Susan Van Clieaf, Tineke Van Den Hoven, Bob van der Ree, John van der Woude, Jan van der Winden, Monika and Hans van Wijk-Ritsche, Hans Verdaat, Matt Victoria, Gerrit Vink, Sarah Viviano, Linda Walfield, Kathy Walker, Rob Walker, Sally Walters, Wendy Ward, Alicia Weitzel, Benjamin Whitcomb, Jan Wierda, Jan Harm Wiers, Roy Weaver, Diane Webster, Pollyanna West, Enny Wever, David Whiteley, Wayne Wilkinson, William Wind, Rich Wolfert, Mark Worsey, Denise Young, Sue Youngs, and Willow Zuchowski.

Finally a very special thanks to Jim Miller, Inez Boyd, and Peter Vickery, all of whom provided a solid foundation early in our birding life and fostered our excitement for birds and birding over the years.

With love and gratitude, we acknowledge our grandmothers, Audrey Giles Chase and Ida Moore Cuthbertson, for sparking our interest in birds at an early age. Our parents, Arthur Wells and Konni Chase Wells and Dana and Jewell Childs, nurtured our adventurous spirits, curiosity about the natural world, and a deep love of writing. For these gifts and so much more, we cannot thank them enough.

Introduction

Birds of Aruba, Bonaire, and Curaçao: A Site and Field Guide is intended both for visitors to the islands and the people who live there. This is the first book ever that is both a site guide and a field guide to the birds of the ABCs. So, whether you are a visitor thumbing through it in the gift shop of your hotel or a resident who wants to learn more about the birds that inhabit your home, this book should serve you well. We have tried to make it accessible to beginners and experts alike. Our hope is that it will inspire readers—regardless of their nationality—to contribute in whatever fashion to efforts that will ensure that these birds remain a part of the islands' natural heritage for generations to come.

The book opens with a series of introductory chapters, the first a general description of the islands themselves, including notes on history and culture; second, information about the fascinating ecology of the ABCs; and finally an introduction to the varied avifauna of the islands.

The site guide is organized by island. For each of the three main islands—Aruba, Bonaire, and Curaçao—we have chosen sites that experience has taught us are the best, most accessible locations for birding. Of course, there are always new places to explore, and we hope birders will let us know of other locations that they think deserve attention in future editions of the book. For each site, there is a map, directions to get there, and a summary of information about what birds are likely to be found there, as well as helpful hints about when and how to look for birds there.

The bulk of this book is devoted to the field guide, which is divided into two parts. A series of beautiful color plates by artist Robert Dean illustrate the species we considered appropriate for inclusion in this book as of August 2016. We have purposely not included illustrations for a handful of birds that were likely escaped birds or for which there is only limited supporting documentation. The color plates appear on right-hand pages; the facing page gives a short description of the key identification features of the bird and its length. Generally, you will be able to indentify most birds seen on the islands simply by perusing the plates and the accompanying short descriptions. If you encounter a bird whose identity is still a puzzle to you, or if you just want to learn more about the bird's status, there is a more in-depth written account describing plumage, similar species, vocalizations (when relevant), status, and range.

This book ends with a section on bird conservation that includes a proposed list of endangered, threatened, and special concern species on each island, a history of the major factors impacting bird populations on the islands, a summary of conservation organizations and projects, and recommendations for each island. It also includes a summary of species conservation initiatives and needs.

Bird Names

For English and scientific bird names, we have followed the eBird/Clements Checklist of Birds of the World, version 2016, with three exceptions: 1) we use

Troupial rather than Venezuelan Troupial, as it seems misleading to describe a native species as being from another country; 2) we continue to separate Caribbean Coot and American Coot and; 3) we leave Bananaquit in its own family, Coerebidae, rather than include it in the Thraupidae family. We have included common names in both Papiamento (including island-specific variations) and Dutch bird names in each species account. These are taken from Prins et al. (2009), with some name variations also included from Voous (1983) and a few new recent names from Peterson (2016).

Species Included

We have included illustrations for 294 species in the book (including three non-native parrot species that evidence suggests are now established in the suburbs of Kralendijk on Curaçao). There are a number of species kept in captivity that have escaped or been released, some seen for years in certain locations, that we do not include in the book. A good example is a record of a single Tricolored Munia, a bird native to parts of Asia, which was seen over a period of several months in 2013 at the Divi Links on Aruba.

Several of the species that Prins et al. (2009) thought did not have established populations—Chestnut-fronted Macaw, Blue-crowned Parakeet, and Scarlet-fronted Parakeet—we now consider to be established (or possibly established), and so include them.

Among the 280 species that Prins et al. (2009) thought had been appropriately documented, we take issue in several instances, which is to be expected as authors often disagree on criteria for accepting bird species. Many regions of the world have established bird record committees that follow fairly strict rules as to what constitutes acceptable documentation. No such committee exists for Aruba, Bonaire, or Curaçao, either individually or collectively.

On reviewing purported photographic evidence, we can cite several cases in which birds that would have been new to the islands were in fact erroneously identified. Other reported species that we considered at the time of the book's publication to be insufficiently documented for inclusion on the list include Great Frigatebird, Northern Jacana, Mouse-colored Tyrannulet, Swainson's Flycatcher, Yellow-throated Warbler, Swainson's Warbler, Greater Antillean Grackle, Boat-tailed Grackle, and Oriole-Blackbird; these species are all very similar to resident species and/or are generally difficult to identify. For a number of birds included on a recent checklist of birds recorded from Aruba (Peterson 2016), we are not aware of any published or publically available documentation; therefore we do not include them on our official list. These species include some that are very likely captive escaped birds, including Greylag Goose, Muscovy Duck, and Java Sparrow. Others, such as White-winged Dove, Smoky-brown Woodpecker, Northern Rough-winged Swallow, and Bronzed Cowbird, appear on the list without any supporting information. We also do not include some species currently appearing on eBird.org because, again, they are either very likely escaped caged birds or we feel that they are not substantiated by sufficient evidence.

There are a number of misidentified species on Observado.org, some so improbable that they should charitably be considered data entry errors; in other cases, the photographs provided as supporting evidence show a different species. These include California Quail, Abdim's Stork (photo shows a Tricolored Heron), Long-billed Curlew, Zenaida Dove, Campo Troupial, Common Grackle, and Yellow-headed Blackbird (photo shows a Yellow-hooded Blackbird).

Species that have been recorded on one or more of the ABCs since Prins et al. (2009)—and that we consider well documented—include Whistling Heron, Long-winged Harrier, Turkey Vulture, American Pygmy Kingfisher, Smooth-billed Ani, Pied-water Tyrant, and Lined Seedeater (as the book was going to press the occurrence of a new species, Great Kiskadee, was confirmed by photographs on Curaçao). We recognize that some of the species that we have not included on our list may in fact have been correctly identified, but we believe it is important to err on the side of prudence. Proper documentation is important because it establishes scientific credibility about the true status and distribution of species, information vitally important for conservation efforts. Archived documentation (photos, videos, audio recordings) also allows future researchers to evaluate new information that might provide further insight into the biology and status of a species.

Notes on Status

While we have endeavored to summarize all known published records of uncommon and rare birds from the ABCs, it is virtually impossible to search all possible places now available on the Internet where sightings, photos, and video of birds may be posted. Some sources, like the trip reports posted to our websites Birds of Aruba (www.arubabirds.com) and Birds of Bonaire (www.bonairebirds.com), are readily available to us and to others. Sightings input into eBird.org and Observado.org are also generally available, although sometimes details are lacking and it can be time consuming to track down each individual observer for more information. There are other trip report websites and photo websites where some sightings and documentation can occasionally be found but there is no systematic way to track down such records. Many Facebook sites (including our own Birds of Aruba and Birds of Bonaire Facebook pages) invite people to post photos and sightings of birds from the ABCs. These can be very helpful for learning about the status of birds, although many of these posts lack specific information about when and where the birds in the photos were seen, making it difficult to use the information to understand the status of various species. We have tried our best, given those caveats, to describe the status of each species. We apologize in advance for records we may have missed. If you have records you would like us to consider for future editions of this book, please visit ArubaBirds.com and BonaireBirds.com. Also be sure to input your records in eBird at www.ebird.org. Note that we cite our own observations in the status sections of the book with our initials as AJW.

The Islands

For first-time visitors, especially those from cold climates where winter colors are primarily a range of grays, stepping onto any of these islands may be a wonderful shock to the senses. It's not just hype—the Caribbean islands, including the ABCs, are surely as close as it gets to a colorful paradise for travelers escaping an icy winter.

Aruba, Bonaire, and Curaçao are located in the southwestern Caribbean, not far from the northern coast of Venezuela (on a clear day, you can see that country from parts of Aruba), within the Atlantic Time Zone. Although after-effects of strong storms elsewhere in the Caribbean have on occasion caused considerable damage along the coasts of the islands, such weather is so infrequent that the phrase "out of the hurricane zone" is used as a major marketing point in promotional literature, enticing travelers to visit the islands at any time of year.

The close proximity of the ABC islands to one another means that they share many features. The water temperature is a soothing 78 to 82 °F (26 to 28 °C). The air temperature hovers at about 80 °F (27 °C), with a warm, almost constant breeze. The sky is a delightful clear blue most days. Although there is a designated rainy season, October through January, showers can come on quickly at any time of year. If you're out birding at any distance from shelter, accept your fate: you will be drenched. Rain showers on the ABC islands may last just minutes, but the water sometimes falls in torrents; nevertheless, the rain here is never accompanied by cold weather and you will dry off nearly as quickly as you get wet.

All of the ABC islands have lovely white sand beaches (especially Aruba) and coral reefs (especially Bonaire). Much of the native thorn-scrub habitat has been widely disturbed or destroyed by feral populations of goats and donkeys brought to the islands centuries ago by European settlers. The resort industry has also taken a toll on the land, particularly on Aruba, which receives more than a million tourists each year.

Along the coasts, primarily on the north and northwestern sides of the islands, lie ancient coral reef beds, sharp, craggy, and largely uninhabited by plant and animal life; near the water, however, skulk beautiful white or blue crabs, and, toward the interior, you may see the occasional lizard. The ocean is wild and unpredictable here; more than a few swimmers have lost their lives to the ruthless combination of strong waves and jagged rocks. For the visitor, nevertheless, a trip to the craggy north side is a must, if for no other reason than to appreciate the contrast between these austere coasts and the exquisite white-sand beaches on other parts of the island.

The ABCs have a combined human population of about 275,000, with approximately 39% living on Aruba, 6% on Bonaire, and 54% on Curaçao. The official language is Dutch, but the native tongue is Papiamento, a rich combination of Spanish, Portuguese, Dutch, French, and English, with some indigenous Arawakan names, along with an occasional word of African origin. On most parts of the islands, particularly in the more populated areas, English is widely spoken. Most housekeeping and general custodial help in the resort industry primarily speak Spanish (they are often from mainland South America) and understand little English.

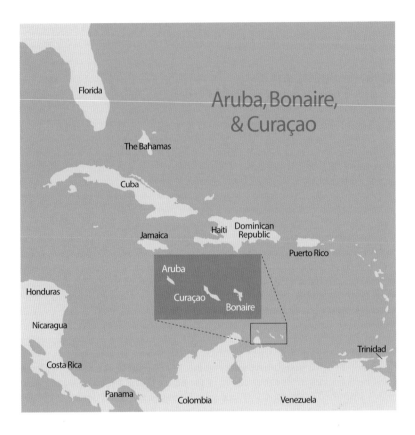

The crime rate is low on all of the islands, and during our many visits we have never been the victims of petty theft. Needless to say, however, criminal acts occur in every country in the world, and petty theft is often a problem in places with lots of tourists. A family member of ours who scuba dives often in Bonaire reports that rental car agencies have asked her to leave her car unlocked, with no valuables in it, as the island has recently seen an uptick in thieves breaking into cars. Always use caution, listen to the advice of hotel staff about when and where it is safe to go, and, ideally, travel with others.

The first inhabitants of the islands were indigenous peoples that, according to archeological evidence, lived there starting thousands of years ago. Europeans arrived in the early 1500s, and so began the modern colonial history of Aruba, Bonaire, and Curaçao. The earliest European visitors were Spaniards, who laid claim to the islands by the early 1500s. They captured many of the indigenous Caquetio and took them away as slaves. A small number of Spaniards lived on the islands throughout the sixteenth century; they came under Dutch control in the 1600s and remained administrative units of the Netherlands until very recently and with only a short interlude (the British held the islands for a brief period in the early 1800s). The islands today reflect their rich historical and cultural origins.

Aruba, Bonaire, and Curaçao were, until recent decades, all considered part of the Netherlands Antilles—sometimes called, unofficially, the Dutch Caribbean. Aruba and Curaçao are now independent, while Bonaire is considered a special municipality of the Netherlands. The islands have also often been referred to collectively as the ABC islands. The authors use that shorthand designation in the book at times but recognize that the islanders and the island governments generally prefer not to consider themselves in a collective sense, as each island now governs itself independently.

Although the islands have much in common, each has notable differences, something worth keeping in mind when you are deciding on which of the ABC islands to visit.

Aruba

Aruba is the smallest of the ABC islands, a mere 70 mi² (180 km²) in area; it is 19 mi (30 km) long and, at its widest point, 5 mi (8 km) wide, with about 43 mi (68 km) of coastline. The westernmost of the three islands, it lies 48 mi (77 km) to the west of Curaçao (and 98 mi [158 km] to the west of Bonaire), making Venezuela, which is just 17 mi (27 km) away, its closest neighbor. Like all of the ABCs, Aruba has the good fortune to be located outside of the Caribbean hurricane zone.

The average annual temperature on Aruba is 82 °F (28 °C) with lows averaging around 78 °F (25.9 °C) and highs of 89 °F (31.5 °C). There is very little seasonal variation, but the highest temperatures are generally experienced June through October. The climate is very dry, with near constant tradewinds. Rainfall averages about 20 in (500 mm) annually; although showers can occur throughout the year, the heaviest rains occur from October through January.

Aruba is famous for the white sand beaches that extend along its southwestern coast. This is where most of its resorts are located, along with Oranjestad, Aruba's capital city. Just offshore are several coral reefs that are popular snorkeling spots, especially Baby Beach at the southeastern tip and Malmok on the southwestern side. Dense mangroves cover reef islands off Oranjestad, while wetlands at Bubali, Spanish Lagoon, and Savaneta are lined with mangroves and/or buttonbush. A water treatment facility at Bubali (sometimes called Bubaliplas), located behind the high-rise hotels, is a haven for birds. Intermittent wetlands (salinas) occur in a number of areas on the southwestern side of the island, notably near Malmok, north of Bubali, and at Tierra del Sol.

The interior of the island is characterized by semi-arid thorn scrub. Although Aruba is primarily flat, about 18% of the land falls within Arikok National Park, in the northeast; its rolling hills, covered with thorn scrub, offer breathtaking views of sparkling turquoise Caribbean waters. Here you will find the divi-divi tree (*Caesalpinia coriaria*), for which the island is known; these grow pointed sharply south as a result of the near constant northeasterly tradewinds. Various species of cacti are prevalent in some areas of the island, especially those in which goats have grazed for hundreds of

The ABC islands are a popular destination for cruise ships, luring hundreds of thousands of visitors annually.

years, resulting in the loss of other forms of vegetation. The desolate north side is virtually devoid of vegetation, and few people live there. It consists of a series of ancient coral reef steps that are sharp and craggy, in many places taking the form of steep cliffs.

Aruba is a major tourist destination. As you pass through the airport, you will find yourself bombarded with a flurry of coupons and brochures from employees of local restaurants and time shares. Taxis line the street in front of the airport, and booths representing car rental companies dot the sidewalk. The stream of energetic tourists coming and going seems at times to contrast with the laid-back attitude of some of the locals, resulting in a curious "hurry up and relax" atmosphere. While for many tourists the main attractions on Aruba are the beaches and the shopping at the bustling boutiques, many do visit Arikok National Park and other nature sites.

There are two main cities: Oranjestad, the capital, with a population of about 20,000, and San Nicolas, home to some 17,000 people. Oranjestad is the tourism hub, located as it is within an easy walk or ride to the resorts. It's also the port city for the cruise ships. By day, it bustles with government employees and shoppers; at night, the clubs, restaurants, and casinos are hopping. San Nicolas, by contrast, currently receives relatively few tourists (efforts are underway to change that), although many tour and party buses stop at the famous Charlies' Bar and Restaurant, where mementos left by fishermen, oil workers, scuba divers, and others line the walls. It used to be the larger of the two cities, flourishing because of a boom in phosphate mining and the construction in 1924 of an oil refinery built to handle Venezuelan oil extracted from its Maracaibo Basin oil deposits. The refinery was being dismantled and converted to an oil storage facility but recently plans have been put forth to reopen it as a refinery.

In 1986, Aruba seceded from the Netherlands Antilles to become an independent member of the Kingdom of the Netherlands. Although Aruba is a member of the Kingdom of the Netherlands, its government is autonomous, with a prime minister who is elected every four years and who oversees a Council of Ministers; parliament has 21 members. The island's official language is Dutch, though Papiemento is the local language. English is now *almost* universally spoken throughout the island. Spanish is also common, and is the primary (sometimes only) language spoken by many employees of the resort industry. As of 2014, the total Aruba population was 107,394 (Central Bureau of Statistics Aruba 2015).

Aruba's currency is the Aruban florin (sometimes also called the guilder), but the US dollar is accepted virtually everywhere—shops in rural areas may not be able to break large bills or give change in US currency. Other currencies can be easily exchanged at island banks. ATM machines are located at hotels and throughout the island in urban areas but your best bet is to use machines associated with banks. Traveler's checks and credit cards are readily accepted; personal checks typically are not.

For decades Aruba's slogan was "One Happy Island," and do be prepared to encounter many warm, friendly people during your stay. The first human inhabitants of Aruba were indigenous peoples from mainland South America. Later came Caquetio peoples (from the broad Arawakan language group), who were already living on Aruba when European's first arrived. By the mid-1600s, the Dutch had taken over and remained in control for several hundred years, with the English assuming control briefly during the Napoleonic wars. Add to this mix people who have arrived from South America in recent times to work in the tourist industry, immigrants from Europe, and visitors who end up staying forever, and you have what today is an island composed of people with wonderfully diverse cultural heritages.

Bonaire

Bonaire is the easternmost of the three ABC islands, located about 54 mi (87 km) from the coast of Venezuela and 30 mi (50 km) from the nearest ABC island of Curaçao. It is 37 mi (60 km) east of the uninhabited atolls of Islas Las Aves, Venezuela. Bonaire has a land mass of approximately 111 mi² (272 km²); 21 mi (35 km) long, it is 9 mi (15 km) at its widest point.

The average annual temperature on Bonaire is 82 °F (28 °C) with lows averaging around 78 °F (25.9 °C) and highs reaching 89 °F (31.5 °C). The hottest months are generally June through October. The climate is very dry, with near constant tradewinds. Rainfall averages about 19 in (490 mm) annually; there is generally

The ruins of Malmok Lighthouse in Washington-Slagbaai National Park, Bonaire.

more rainfall October through January, though showers can occur throughout the year. Like the other two ABCs, Bonaire is located outside the hurricane zone.

The population of Bonaire was estimated at 17,400 in 2013. While Dutch is the official language of the island, the local language is Papiamento. Spanish and English are also widely spoken.

Some 70,000 tourists visit Bonaire annually, but their presence is less noticeable than on faster paced Aruba. It is an easy island to get around on, with far less traffic than Aruba, and it tends to attract more tourists interested in nature rather than in sunbathing on the beach. Lacking the great number of white sand beaches that characterize Aruba, Bonaire's biggest draw is its marine park—and its birdlife. Bonaire is internationally known for its pristine coral reefs that provide excellent scuba diving and snorkeling; the island's government has gone to great lengths to protect these reefs, recognizing that they are an important natural resource.

The interior of the island is primarily dry, semi-arid, thorn scrub. Much of the wooded hills and valleys of the northwestern part of the island have been preserved as the Washington-Slagbaai National Park, which totals 9,300 acres (3,800 hectares) and is managed by STINAPA, the Netherlands Antilles National Parks Foundation. The park is dominated by Mount Brandaris, which, at 797 ft (243 m), is the island's highest point. Also within the park—and bordering it—are rocky coasts, sandy beaches, inland bays, lagoons, and salinas.

On the southern end of the island, where the land is flat and approaches sea level, you will find the area known as the Pekelmeer, a series of shallow lagoons that over the course of hundreds of years have been modified for salt production. After the middle of the twentieth century, the lagoons were further modified for large-scale industrial salt production; water levels are tightly controlled within a series of condenser lagoons to maximize salt production. Much of the salt ends up in US homes, in water softener systems. The Pekelmeer is also where American Flamingos gather by the hundreds and shorebirds flock for feeding in the winter.

About a half-mile (800 m) off the western side of Bonaire lies a small uninhabited islet, Klein Bonaire (in English, Little Bonaire). The flat, coral-limestone area that composes the islet is a mere two meters (at most) above the surface of the surrounding waters; in size, it is about 1,500 acres (607 hectares). The islet is fringed with beaches, and the vegetation consists only of low-growing scrub. Klein Bonaire is a popular spot for scuba diving and snorkeling, as a pristine coral reef surrounds the islet. It was privately owned from 1968 until 1999, when it was purchased by the government of Bonaire, the World Nature Fund, and the Foundation to Preserve Klein Bonaire, and is now part of the Bonaire National Marine Park. There are plans to reintroduce native vegetation.

Spain occupied Bonaire—and claimed it as its own—in 1499. When the Spanish arrived, they overlooked the fact that the island was already home to indigenous peoples, including the Caquetio (from the Arawak language group). In fact the island had been inhabited for thousands of years. The Dutch took possession in 1636 and, despite brief periods of occupation by the British during the early part of the 1800s, the island has remained under Dutch influence (Hartog 1978). Slavery forms a tragic chapter of the history of Bonaire, and one can still see huts that were built for the slaves who labored in the salt pans in the southern end of the island. Slavery was abolished on Bonaire in 1863 (Hartog 1978).

In 1954 the Netherlands granted self-rule to all the Caribbean islands then under its possession. These islands, then forming the new nation of Netherlands Antilles, were nominally granted autonomy, although they still belonged to the Netherlands. The capital of the Netherlands Antilles was located in Curaçao, with each island electing a lieutenant governor and representatives to parliament. In 2010, The Netherlands Antilles was dissolved, and Bonaire became a "special municipality" of the Netherlands, its people now officially Dutch citizens.

The capital of Bonaire is Kralendijk. It has some souvenir shops, but unlike Aruba's Oranjestad it is not a tourist enclave. Overall, the town has something of a European feel. Life moves a little slower here. Rincon, located in the interior on the way to Washington-Slagbaai National Park, is the island's oldest settlement. Hardly a metropolis, it is a quaint settlement with many century-old buildings.

The island's currency, formerly the guilder, is now the US dollar. Traveler's checks and credit cards are readily accepted within the tourist industry and in Kralendijk, and you can normally find access to ATMs. The island economy is largely dependent on ecotourism, with the main focus on scuba diving in the

Colorful colonial buildings are a striking feature of downtown Willemstad, Curaçao. The city, rich with history, has been designated a UNESCO World Heritage Site.

waters of the marine park that surrounds the island. The salt production industry, a small oil transfer facility, government offices, banking, and fishing also provide employment. Unemployment on Bonaire is quite low, and so is the crime rate.

Curaçao

Curaçao is the largest of the three islands, with a land area of 285 mi² (444 km²). It is 30 mi long (60 km) and 7 mi (11 km) wide at its thickest point. Sandwiched between Aruba and Bonaire—48 mi (77 km) east of Aruba and 30 mi (50 km) west of Bonaire—and just 43 mi (70 km) off mainland Venezuela—Curaçao, like Aruba and Bonaire, lies outside of the Caribbean hurricane zone.

The average annual temperature on Curaçao is 82 °F (28 °C). Low temperatures average around 78 °F (25.9 °C); the highest temperatures (89 °F [31.5 °C]) are generally experienced June through October. The seemingly perpetual tradewinds keep the climate very dry. Average rainfall of about 22 in (550 mm) slightly exceeds that on Aruba and Bonaire; while you shouldn't be surprised by rains at virtually any time of year, the rainiest months of the year tend to be October through January.

Curaçao's geological features include eroded volcanic hills, ancient limestone terraces, small pocket beaches, and semi-enclosed inland bays that are surrounded by dense mangroves. Several shallow, enclosed hyper-saline lagoons provide important habitat for waterbirds. There are also a number of freshwater catchment dams—many built originally for the use of the oil industry—that may hold a large amount of water even during the dry season. The natural landscape is largely characterized by rugged, rocky terrain covered with scrub woodlands; some areas are dominated by columnar cacti. The generally flat land is punctuated by limestone terrace cliffs. While most of the island is surrounded by coral reefs teeming with life, the island's northern coast is exposed to wild, rough surf that rolls in from the open sea. This contrasts sharply with the sheltered coves of the southern coast, which boast calm, sandy beaches with turquoise waters. On the northwestern end of the island is St. Christoffelberg, the island's highest peak (1,230 ft [375 m]), which lies within Christoffel National Park. It and nearby Shete Boka National Park are characterized by thorn scrub vegetation and windswept bluffs. The parks provide important habitat for birds and other wildlife as well as protection for a high concentration of rare plants.

Spain claimed Curaçao in 1499, and it remained in Spanish hands until the Dutch seized control in 1634. In 1807, Curaçao was taken by the British, but the Dutch regained control in 1815. It was during the seventeenth century that the Dutch began trading slaves; they took Africans from their homelands and brought them to countries like Curaçao, which was one of the largest slave depots in the Caribbean. Slavery was abolished in 1863, and today the island's slave huts serve as a museum. Historically, Curaçao was the capital of the Netherlands Antilles, which—until that entity was dissolved in 2010—comprised Aruba and Bonaire as well as the islands of Saba, St. Eustatius, and St. Maarten. Curaçao has a long history as a trading and commercial hub of the region, and that history is reflected in its people, whose cultures and ancestries are fascinatingly diverse.

Curaçao is an autonomous country within the Kingdom of the Netherlands. Its people are primarily of Dutch and African ancestry, and the island has much more of a European feel to it compared to Aruba and Bonaire. Dutch is the official language, though island residents speak Papiamento. English and Spanish are also widely spoken.

A 2011 census estimated the population of Curaçao at 150,563, which makes it the most populated of the three islands. Indeed, many parts of the island are highly urbanized, with about 30% of the land said to be occupied by housing and industry (Debrot and Wells 2008). In bygone eras, the cultivation of seasonal crops, charcoal production, and livestock grazing altered the land to a considerable degree. Today, however, secondary woodlands are gradually recuperating, especially on the western half of the island. Tourism, real estate development, contamination by pollutants, and disturbance due to recreational activities are among the principal threats to remaining habitat areas.

The capital city of Curaçao is Willemstad. Full of historic buildings, the city retains its colonial Dutch charm; in fact the city center has been designated a UNESCO World Heritage Site. It is very walkable and full of shops and vendors, making it a popular destination for tourists.

Curaçao's official currency is the guilder (called the florin in Papiemento), although the US dollar is widely accepted. The Euro is also accepted—to a lesser extent—at many hotels and restaurants. You will be able to use major credit cards and find ATMs in larger population centers throughout the island—withdrawals can usually be made in either local or US currency.

Tourism is one of the island's main economic drivers. The majority of visitors come from the Netherlands, but there are also tourists from the rest of Europe, the US, and South America. Justly famous for its rich cultural history, the island's coral reefs also afford great scuba diving and snorkeling. It is also known for its financial services industry. At one time, Curaçao was home to the world's largest oil refinery. Oil refining continues to take place on the island, and remains a significant employer and major part of the island economy.

Ecology

This section attempts to give a general description of the ecology of the ABCs by considering five factors:

1. Geology
2. Geography
3. Climate
4. The surrounding ocean
5. Land use

Ancient coral reef escarpments show remnants of changes in sea level over millions of years. They provide important specialized habitats for bats, owls, parrots, and other birds.

Geology

The history of the geology of the islands is a fascinating one, as here you will find volcanic rocks, fossilized coral reefs, and even boulder fields that are thought to have been caused by ancient tsunamis (Alexander 1961; Engel et al. 2010; Hippolyte and Mann 2010; Muhs et al. 2012). Aruba, Bonaire, Curaçao, and some tiny Venezuelan islands to the west of Bonaire are the tips of the mostly submerged Leeward Antilles Ridge, which roughly parallels the South American coast (Hippolyte and Mann 2011). This ridge runs along the northern edge of the South American tectonic plate and just south of the point where the Caribbean tectonic plate pushes against and under the South American tectonic plate.

The oldest geological regions of the islands, volcanic and sedimentary rock found in the interior of each island, were formed in the Cretaceous period (145.5 to 65.5 million years ago). The islands sat above sea level through the Oligocene (33.9 to 23 million years ago) and then submerged during the Miocene (23 to 5.3 million years ago), after which they slowly reemerged from the sea. The process of submerging and reemerging has left the islands with a layered, "wedding-cake" landscape. The higher elevation Cretaceous volcanic cores of each island are encircled by a relatively narrow band of limestone deposits formed by sea creatures from the Miocene, when the islands were submerged. These deposits are themselves bordered by a series of 3 to 5 terraces, some stepped and some gradually sloped, that each represent an ancient coral reef laid down over the last 2.5 million years (Quaternary Period); as ocean levels rose and subsided through the ice ages (Alexander 1961; Schellmann et al. 2004; Muhs et al. 2012), the reefs were sometimes submerged and at other periods exposed. When exposed, the reef terraces would be eroded by wave action, forming cliffs in some areas. These ancient reef terrace cliffs are oriented generally parallel to the shoreline, hundreds of meters inland from the sea. They are not only a striking feature of the landscape but are also incredibly important specialized habitats for animals (bats, owls, parrots, and parakeets, among others) and plants (Lace and Mylroie 2013).

Geography

The plants and animals that occur on the ABCs are a fascinating mix; although most species are of South American origin, some come from other regions of the West Indies, and yet others have been introduced from remote parts of the world. Even the oldest inhabitants of the islands arrived relatively recently in geological terms (van Buurt 2005); as sea levels subsided, more land area was exposed, providing more terrain for colonizing plants to establish themselves, which in turn provided more habitat for colonizing animals.

In general, the number and the identity of species that colonize an island are a function of the degree of difficulty of reaching that island from other land masses; the size of the island; the types of habitats and food—and their abundance; and the amount of freshwater that is available. The degree of difficulty of reaching an island is in part a function of distance but also, of course, relates to the ability of a given species to travel long distances. Given that the ABCs are much closer to the South American mainland than to West Indian islands, it is not surprising that there are many more species of animals of South American affiliation than of West Indian affiliation. In fact, when sea levels were low during the Quaternary Period (2.6 million years ago to present day), Aruba may have been connected by land to mainland South America.

There are, nonetheless, a number of species on the ABCs that either bear a family relation to species on other West Indian islands or that in fact have migrated from those islands. For example, the closest relatives of *Anolis bonairensis*, a small lizard that is endemic to Bonaire, occur in the Lesser Antilles (van Buurt 2005). Among the bird species found on the ABCs that are of West Indian origin, note the following:

- **Black-whiskered Vireo.** Breeds from coastal southern Florida through the Greater and Lesser Antilles and on the ABCs but not (as far as is known) on the South American mainland (although most populations are thought to winter in northern South America).
- **Caribbean Elaenia.** Resident along the Caribbean coast of the Yucatan Peninsula; also occurs from Puerto Rico south through the Lesser Antilles, to Grenada, and on the ABCs. It is not known from the South American mainland.
- **Scaly-naped Pigeon.** Resident in Cuba, Hispaniola, and Puerto Rico. Also occurs south through the Lesser Antilles, to Grenada, and then on Curaçao and Bonaire, but not on South American mainland.
- **Pearly-eyed Thrasher.** Resident in southern Bahamas, Puerto Rico, and south through Lesser Antilles to St. Lucia and then also on Bonaire.

Stoffers (1956) also lists 35 plant species of exclusively West Indian origin that are known from the ABCs. As more research documents the less well-known animal groups found in the ABCs, including various arthropods such as dragonflies, moths, butterflies, pseudoscorpions, and spiders (Miller et al. 2003; Debrot 2006; Paulson et al. 2014), perhaps other fauna of West Indian origin will be discovered.

Although most plant and animal species of the ABCs are clearly of South American origin, many have been isolated long enough to have evolved distinct characteristics. Among the birds, Debrot (2006) lists 22 endemic subspecies of birds that occur in the ABCs but are of South American origin. Some of the more easily differentiated endemic subspecies (some may eventually be considered distinct species) include the three island-specific subspecies of Brown-throated Parakeet, Aruban Burrowing Owl, Curaçao Barn Owl, possibly the relatively recently discovered Bonaire Barn Owl, the ABC subspecies of Yellow Warbler, White-tailed Nightjar, Grasshopper Sparrow, American Kestrel, Eared Dove, Rufous-collared Sparrow, Aruba-Curaçao Bananaquit, and Bonaire Bananaquit (Voous 1983; Debrot 2006).

Examples among the mammals include the Curaçao White-tailed Deer, the Aruba-Curaçao Cottontail, the Curaçao Long-nosed Bat, and the Vesper Mouse (Debrot 2006).

Species or subspecies of reptiles (van Buurt 2005; Debrot 2006) that are of South American origin include the Aruban Rattlesnake (*Crotalus unicolor*), Aruba Cat-eyed Snake (*Leptodeira annulata bakeri*), Curaçao Three-scaled Ground Snake (*Liophis triscalis*), Aruban Whiptail (*Cnemidophorus arubensis*), the Curaçao Whiptail (*Cnemidophorus murinus murinus*), the Bonaire Whiptail (*Cnemidophorus murinus ruthveni*), Antilles Gecko (*Gonatodes antillensis*), Curaçao-Bonaire Leaf-toed Gecko (*Phyllodactylus martini*), Aruba Leaf-toed Gecko (*Phyllodactylus julieni*), and the Aruba-Curaçao Striped Anole (*Anolis lineatus*).

Debrot (2006) lists 40 species and 4 subspecies of insects and arachnids endemic to Aruba, Bonaire, and Curaçao. Plants, too, show some endemic species, with a total of at least nine found only in the ABCs, including three species of *Agave* and three of the *Melocactus*—one of the authors' favorite group of plants (Proosdij 2012).

Many species of birds on Aruba, Bonaire, and Curaçao move back and forth between the islands and the mainland of South America. For these species, the populations on the islands can likely be considered part of larger, regional metapopulations, spatially isolated groups that at least occasionally exchange breeding individuals among them. American Flamingos are known to regularly move back and forth between the islands and the mainland, although Bonaire is perhaps the sole primary nesting place for this regional metapopulation. Nesting terns can easily make long-distance movements and some individuals likely shift nesting locations, at least occasionally, from the ABCs to Venezuelan islands from year to year. Herons, egrets, and other wetland-dependent birds have been seen at sea among the ABCs, and they are known to appear in larger numbers on the islands when precipitation increases the volume of wetland habitat. These birds are clearly moving back and forth between the islands and the mainland. There are even observations of landbirds like Bare-eyed Pigeons and Ruby-topaz Hummingbirds flying above the sea at locations between the islands and mainland Venezuela, suggesting that for some species such movements may be much more common than we once thought.

Climate

Centuries past, even the first European colonizers of the islands noted their extremely hot and dry climate. Indeed, many landscapes on the ABCs often characterized as desert or desertlike are home to a variety of cacti and other plant species adapted to dry climates. This arid climate is shared with parts of the northern edge of the neighboring South American continent, so it is no wonder that a number of animals—including the Yellow Oriole, Ruby-topaz Hummingbird, and Bare-eyed Pigeon—and plants occur in both places.

Although the ABCs occupy a fairly small land area—and even though the climate on each of the islands is often rather uniformly arid—there is a suprisingly high number of plant community associations found here. For example, Stoffers (1956) described 18 plant community types on the ABCs, and de Freitas et al. (2005) classified and mapped 18 vegetation types on Bonaire. Although most bird species may not be sensitive to the full variety of vegetation types, there are, predictably, many species that show preference for certain structural aspects of the vegetation and landforms. Places with low scrubby thickets interspersed with open areas are important for Crested Bobwhite, Burrowing Owl (Aruba), Grasshopper Sparrow (Curaçao and Bonaire), and White-tailed Nightjar. Areas with flowering shrubs are critical for Blue-tailed Emeralds and Ruby-topaz Hummingbirds. Mangroves and thicker shrub or dry-forest vegetation often harbor nesting Brown-crested Flycatchers, Northern Scrub-Flycatchers, Caribbean Elaenia, and Black-whiskered Vireo (though the latter two species are widely distributed on Curaçao and Bonaire, on Aruba they have only been found in mangroves in recent years). Wintering or stop-over migrants that breed in North America tend to occur in greatest diversity in mangroves and shrubby areas near water, except during a major "fallout," when large numbers of migrating birds are blown off course by storms and seek food, water, and shelter on the islands, wherever they may find it.

For virtually all birds on the islands, whether breeder, migrant visitor, or winterer, survival depends on finding sufficient daily food and water and avoiding predators. In the arid climate of the islands, finding water sources is critical to survival. Historically, natural water seeps and springs would have been incredibly important during extended dry seasons; today, birds can also seek water at water treatment facilities, pools in backyards and hotels, and at other water sources created by humans.

Endemic Aruban whiptail lizards. These colorful reptiles are a common sight throughout the islands. They are unusual among whiptail lizards in that they are largely vegetarian.

But away from "civilization," massive numbers of birds visit water holes in Washington-Slagbaai National Park on Bonaire and in Christoffel National Park on Curaçao. Wetland birds appear on the ponds that develop behind dams after heavy rains, as at Malpais (Curaçao) and Onima (Bonaire). Food sources for birds are typically more abundant near wetlands or in vegetation near wetter areas, whether around the edges of a pond or waterhole or in arroyos (small valleys or ravines) where small streams flow after rains. Food includes seeds, insects, lizards, mice, and, in the case of bird-eating hawks like Merlins and Peregrine Falcons, smaller birds.

The bodies of water that harbor wetland birds contain small fish, freshwater shrimp, and other aquatic invertebrates as well as aquatic vegetation and seeds. All of these food sources reach their peak abundance after rains, and birds time their reproductive efforts to try to ensure that their young are hatched when there will be abundant food.

Although many birds on the islands breed throughout the year, nesting for many landbird species generally peaks around the time of the rainy season; nesting is sometimes delayed or possibly even suspended during protracted droughts. Wetland-dependent birds like the Caribbean Coot, Common Gallinule, Least Grebe, and Pied-billed Grebe will occupy freshwater ponds, sometimes within weeks after precipitation forms them, and then begin nesting. They will continue to nest as long as water remains; when water does begin to dissipate at the end of the rainy season, a poignant scene develops as adult birds and their young attempt to survive on an ever shrinking and crowded pond.

Rains stimulate the flowering of cacti and many other plants. In addition to providing nectar for the two resident hummingbird species, Bananaquits, and some migrant North American birds, these flowers also attract insects and bats. These animals all play a vitally important role as pollinators, assisting native plants in reproducing. In fact, some species of bats on the islands have seemingly co-adapted with certain cactus species in which flowers open only a single night; the bats actively search them out and move pollen among them as they feed on the nectar in the flowers (Petit 1995, 2001, 2011; Bekker 1996; Petit and Pors 1996; Petit et al. 2006).

The Surrounding Ocean

The birds and other animals whose natural history is interwoven with the waters that surround the islands live in synch with cycles entirely different from those of the animals that live exclusively on land. Herons, egrets, Brown Pelicans, and other resident bird species that feed along the shoreline and in shallow salinas and bays that are wholly or partly interconnected with marine waters may nest throughout the year. These same shallow intertidal and coastal habitats often support migrant wintering or stop-over Arctic or sub-Arctic breeding shorebird species (sandpipers and plovers, for example), wintering Laughing Gulls, Royal Terns from North American breeding locations, and sometimes other wintering terns such as Sandwich "Cayenne" Terns.

Ruddy Turnstones (*Arenaria interpres*) spend the winter months along the islands' coasts, far from their Arctic breeding grounds.

Another group of birds use these near-shore marine habitats generally only during the April–August period, when they come to the islands to nest. Snowy and Wilson's plover nest (at least on Bonaire and Curaçao) on beaches and dried out salinas, as do Least Terns, which also nest on reef terraces along the shore, wherever there is some sand or gravel. Sandwich "Cayenne" Terns, Common Terns, Roseate Terns, and Royal Terns arrive in the summer to nest on small islets in bays and in Bonaire's Pekelmeer, and on Aruba's San Nicolas reef islands, where they are joined by Bridled and Sooty terns as well as Brown and Black noddies.

The marine waters surrounding the islands, so well-known for their coral reefs and the snorkeling and scuba diving they afford, also play host to pelagic bird species that, for the most part, rarely come to land. Two pelagic bird species that do roost on land—and that regularly forage close enough to shore so that they are easily seen by birders—are the Magnificent Frigatebird and the Brown Booby. Frigatebirds roost in small mangrove islands in a few places, including just off Aruba's Oranjestad Harbor, where cruise ship passengers often see them coming and going to their roost. Brown Boobies roost on cliffs in a few favored spots on all three islands—and an attentive observer can sometimes get excellent views of the birds on the cliffs.

Most other pelagic bird species, however, are only occasionally seen from land or require hours of scanning out to sea with a telescope in order to have a chance of spotting them. Recently, off the coast of Aruba at least, observers on commercial sport fishing boats have expended considerable effort looking for whales, porpoises, dolphins, and pelagic birds (Luksenburg 2011, 2013, 2014; Luksenburg and Parsons 2013, Luksenburg and Sangster 2013). Not only have these efforts helped document the occurrence of at least 16 species of cetacean in the waters around Aruba (including Killer Whales, False Killer Whales, Spinner Dolphins, Atlantic Spotted Dolphins, and Pantropical Spotted Dolphins), they have also documented a number of pelagic bird species offshore from the islands (Luksenberg 2013, Luksenburg and Sangster 2013). Species photographically documented from these efforts include Black-capped Petrel, Cory's Shearwater, Red-billed Tropicbird, Masked Booby, South Polar Skua, and Pomarine Jaeger (Luksenburg and Sangster 2013). Some yet-to-be published data from offshore surveys conducted in Fall of 2014 also found Audubon's Shearwater, Masked Booby, and Red-footed Booby.

Land Use

Human society and industry have dramatically affected the ecology and animal communities on Aruba, Bonaire, and Curaçao, especially in recent history. The last 50 years have seen a rapid rise in the human population on the islands and increased numbers of tourists.

The introduction—often intentional—of non-native animal species has had a great impact on the ecology of the ABCs. The introduction of such herbivores as cattle, goats, donkey, sheep, and horses has likely had the most damaging effects—over the longest time period—on the vegetation communities, altering plant species composition, age, and structural components.

More recently, the introduction of the Boa Constrictor to Aruba appears to have diminished the size of bird and lizard communities, with perhaps the Aruban Brown-throated Parakeet and Rufous-collared Sparrow the most seriously affected. Many birds that are now commonly seen around homes and hotels on the islands are introduced (or likely introduced) species. These include House Sparrows, Carib Grackles, Shiny Cowbirds, and Saffron Finches.

The building of dams, especially larger dams for industrial use and agriculture, has greatly changed the ecology of the islands, as has the discharge from sewage treatment facilities. In a natural state of affairs, the bodies of water formed by rain come and go, thus affording habitat for birds only at certain times of the year. But dams and other artificial bodies of water maintain relatively stable water levels, which means that species that previously could not survive—or at least persist—on the islands now can. Neotropical Cormorants and Caribbean Coots, for example, now breed at permanent water bodies on Aruba, at both Bubali and the Divi Links golf course.

The loss of natural habitat through real estate and other development projects has of course dramatically changed the plant and animal communities on the islands. Less habitat means fewer birds, logically, although the lack of long-term monitoring data on bird populations has made this difficult to measure precisely.

Some bird species have taken advantage of new "built" habitats on the islands. In addition to the wetland-dependent birds that have taken advantage of human-made, permanent bodies of water, many species, some introduced and some native, now regularly inhabit landscaped and watered grounds of hotels and modern homes. Non-native Carib Grackles are seen foraging on the lawns of hotel grounds and stealing food from open-air dining rooms. They are often joined by native species like Bananaquits, Tropical Mockingbirds, and Eared Doves, sometimes even by Bare-eyed Pigeons and Troupials. In places where dining areas are near the shore, wintering Ruddy Turnstones will take food from dining areas, including from tabletops, if given the chance. It is always a surprise to see this bird, which breeds in the Arctic, begging for French fries at a dockside bar in the Caribbean! A few bird species that likely arrived on their own have found a foothold in new artificial habitats on the islands. Black-bellied Whistling Ducks, for example, now nest at the sewage treatment plant at Klein Hofje (Curaçao) and at Bubali (Aruba). Southern Lapwings and Cattle Tyrants have found that the large expanses of short grass on the Tierra del Sol golf course on Aruba provide just the kind of habitat they prefer.

Avifauna

Aruba, Bonaire, and Curaçao is an exciting place to go birding, in part because its species are an eclectic mix of residents, migrants, and marine birds. North American or European birders will undoubtedly be intrigued by the breeding tropical landbirds. For some of these landbirds, the ABCs are the northernmost outpost of their South American range. For other species, which breed in many regions of the Caribbean, the ABCs are their southernmost breeding site.

The Pearly-eyed Thrasher (Bonaire only), Black-whiskered Vireo, Caribbean Elaenia, Gray Kingbird, Scaly-naped Pigeon (Bonaire and Curaçao), and Black-faced Grassquit are essentially Caribbean species. Yet other birds are more closely identified with Central and South America, including the Yellow-shouldered Parrot (Bonaire only), Brown-throated Parakeet, Bare-eyed Pigeon, Eared Dove, White-tailed Nightjar, Northern Scrub-Flycatcher, Ruby-topaz Hummingbird, Blue-tailed Emerald, Yellow Oriole, and Troupial.

Some South American birds that have recently begun to breed on the ABCs include the Carib Grackle and, at least on Aruba, Southern Lapwing and Cattle Tyrant. The islands are also home to island-specific subspecies of American Kestrel, Burrowing Owl (Aruba), Barn Owl (Curaçao and Bonaire), Brown-throated Parakeet, Bananaquit, Yellow "Golden" Warbler, Rufous-collared Sparrow (Aruba and Curaçao), and Grasshopper Sparrow (Bonaire and Curaçao).

Resident birds that breed on freshwater ponds, salinas, and shallow bays are another attraction. Perhaps the most striking species is the American Flamingo, which displays a range of showy pinks; Bonaire has a breeding colony, while those that occur on Aruba and Curaçao probably are visiting mostly from

Bonaire. The White-cheeked Pintail, one of the most sought-after breeding species on the islands, regularly appears on its bodies of water. Many species of egret and heron—Snowy Egret, Great Egret, Cattle Egret, Little Blue Heron, Tricolored Heron, Black-crowned Night-Heron, Yellow-crowned Night-Heron, and Green Heron—are breeders, and most of them commonly occur in wetland habitats on all three islands. On the ABCs, freshwater habitats were once rare and ephemeral but, as the outflow from water treatment plants has created permanent bodies of water, some fairly large, they now support breeding species such as the Black-bellied Whistling Duck, Neotropical Cormorant, Least Grebe, Pied-billed Grebe, Black-necked Stilt, Caribbean Coot, and Common Gallinule.

A number of seabirds and shorebirds nest on the islands, mainly from April to August. Some of these, like the Laughing Gull, Brown Pelican, American Oystercatcher, and Sandwich "Cayenne" Tern, also occur regularly during winter, along the coasts of the islands. Other breeding species like the Snowy Plover, Wilson's Plover, Least Tern, Common Tern, Roseate Tern, Bridled Tern, Sooty Tern, Black Noddy, and Brown Noddy are present primarily only from March to August, when they nest.

A large proportion of the species that occur on Aruba, Bonaire, and Curaçao are migrants that do not breed there. Most of these nest in North America and migrate south to the Caribbean or Central and South America for the winter. The most commonly observed northern migrant wintering birds on the islands are probably the Northern Waterthrush, Blue-winged Teal, Lesser Yellowlegs, Sanderling, and Ruddy Turnstone. Other species, less abundant but regularly seen, include the Black-belled Plover, Whimbrel, Osprey, Merlin, Peregrine Falcon, and Belted Kingfisher. Yet other migrants are common on the islands but only for relatively short periods; these include the Barn Swallow, Bank Swallow, Blackpoll Warbler, and, occasionally, Yellow-billed Cuckoo.

An enjoyable aspect of birding on the islands is looking for the regular and not-so-regular northern migrants. You can find some of these birds, including the Black-and-white Warbler, American Redstart, Ovenbird, Hooded Warbler, and Prothonotary Warbler, with persistent pishing, especially in mangroves or shrubby areas near water. Over the last 20 years, we have noted a somewhat interesting phenomenon: there are a number of warblers found with increasing frequency in the ABCs that normally winter farther north and that are often quite rare on the South American continent.   These include the Northern Parula, Common Yellowthroat, Yellow-rumped Warbler, Palm Warbler, Prairie Warbler, Black-throated Blue Warbler, Cape May Warbler, Chestnut-sided Warbler, and Worm-eating Warbler. A few species that breed in the Greater Antilles and southern Florida have been recorded in the ABCs as migrants headed to and from South American wintering areas; some of these migrants—including the Black-whiskered Vireo and Gray Kingbird—also have resident breeding populations on the ABCs. Such species as the Mangrove Cuckoo and Caribbean Martin, however, only occur as migrants.

Although a variety of northern landbird migrants that winter in South America have been recorded in the ABCs, a number of them with some regularity (at least historically), including the Bobolink, Dickcissel, and Rose-breasted Grosbeak, most have only been recorded a handful of times. This includes species like the Swallow-tailed Kite, Eastern Kingbird, Olive-sided Flycatcher, Eastern Wood-Pewee, Veery, Gray-cheeked Thrush, Swainson's Thrush, Cerulean Warbler, Canada Warbler, Tennessee Warbler, and Summer Tanager.

Many shorebirds that breed in northern or even Arctic regions pass through the ABCs, with small numbers staying throughout the winter. From October through March or April, it is generally quite easy to find Lesser Yellowlegs, Greater Yellowlegs, Spotted Sandpiper, Solitary Sandpiper, Least Sandpiper, Black-belled Plover, Semipalmated Plover, Ruddy Turnstone, Sanderling, Whimbrel, Short-billed Dowitcher, and Wilson's Snipe in appropriate wetland or marine habitats on the islands. Shorebird species that are regular visitors but less common or more localized (easiest to find on Bonaire) include the Stilt Sandpiper, Semipalmated Sandpiper, Western Sandpiper, Pectoral Sandpiper, and Willet. Occasionally at the Pekelmeer, on Bonaire, relatively large numbers of Red Knots appear, but it is un-clear whether they are wintering there or just stopping over during migration. At least 10 other shorebird species have been recorded in the ABCs. Of the migratory waterfowl, only the Blue-winged Teal, a duck that breeds in North America, is a common migrant and winterer on the islands; all other species generally are much rarer, except for the Lesser Scaup, Ring-necked Duck, American Wigeon, and Northern Shoveler, which are now seen with some regularity at certain locations in the ABCs.

A smaller number of the migrant species recorded in the ABCs originate from South American breeding locations. Some are most likely birds that have made short-distance dispersal movements from the mainland, including the Masked Duck, Comb Duck, Fulvous Whistling-Duck, Whistling Heron, Boat-billed Heron, Roseate Spoonbill, Scarlet Ibis, Glossy Ibis, Black Skimmer, Limpkin, Wattled Jacana, Collared Plover, Yellow-headed Caracara, Amazon Kingfisher, Ringed Kingfisher, Ruddy Ground-Dove, Red-legged Honeycreeper, Yellow-headed Blackbird, and Red-breasted Blackbird. Other species are probably austral mi-grants, species that breed in southern South America and migrate to northern South America during the southern winter (June, July, and August). While a more complete understanding of austral migration—and the species that undertake it—requires further research, austral migrants recorded in the ABCs include the Gray-capped Cuckoo, Lesser Elaenia, Streaked Flycatcher, Vermilion Flycatcher, Fork-tailed Flycatcher, White-winged Swallow, Chilean Swallow, and Brown-chested Martin. Birders should keep watch for austral migrants, particularly June through September.

An in-depth understanding of the offshore marine bird life of the ABCs re-quires more research. While some dedicated seabird surveys were carried out in the 1970s, there has been little else except for occasional opportunistic records from passing cruise ships. Nonetheless, in 2010 and 2011, Luksenburg and Sangster (2013) went aboard commercial sports fishing boats that were taking visitors out for a day of fishing in waters around Aruba. They tallied a remarkable 415 boat-based surveys. The two saw and photographed many species unusual for the area, including the Black-capped Petrel, Cory's Shearwater, Red-billed Tropicbird, Masked Booby, and South Polar Skua (Luksenburg and Sangster 2013). Other offshore surveys carried out in 2013 and 2014 recorded sightings of tropicbirds, Audubon's Shearwaters, and Red-footed and Masked boobies. Birders may find it fruitful to go on commercial sport fishing trips to see seabirds. Nevertheless, newer, better optics make it possible to see more seabirds from strategically placed shore watches as well.

Like many birders, we have always been drawn to birding on islands. Islands are famous for drawing in exhausted migrants and concentrating them in a relatively small area. And in years of little rain, those migrants are further concentrated, as there are fewer locations to find water, food, and safety from predators. Such places are often hopping with a diversity and abundance of birds, and when birds are concentrated, it is easier to get close to spot any unexpected species. The ABCs have been host to their share of surprising finds. There is the single record of the Double-striped Thicknee on Curaçao in 1934, the Oilbird that occurred on Aruba in 1976 (Voous 1983), the two records of Northern Wheater (Voous 1983), the Brown Thrasher found on Curaçao in 1957 (Voous 1983), the Little Egret from Aruba in 2003 and 2005 (Prins et al. 2009), and the Western Tanager that the authors photographed on Bonaire in July, 2001.

Whether you enjoy watching colorful tropical resident species, scoping shorebirds and waterfowl at your favorite salina, scanning the sparkling waters and azure sky for seabirds, either from shore or from the deck of a boat, or pishing in the migrant songbirds at a waterhole, the ABCs are a birder's paradise.

Jeff Wells (right), one of the authors, birding at Aruba's Bubali Bird Sanctuary with resident birders and photographers Michiel Oversteegen (left) and his son Sven (center), members of the thriving local birding community.

Site Guide

The goal of the site-guide section of this book is to highlight the best locations to find birds. Describing every locale—on each of the three islands—would be impossible, of course; there are too many nooks and crannies, pull-over viewing spots, and other areas to list, and doing so would result in an unwieldy book with a dizzying array of options. With few exceptions, the goal of most birders is to see as many birds as they can, as easily as possible. To that end, we have mainly focused on time-tested, favorite birding spots that are relatively easy to get to. This is especially important for birders disembarking from cruise ships, many of them with just a few hours to spend on any given island.

At virtually any time of year, most of the featured hotspots offer great opportunities for seeing resident species; during migration seasons, these same spots also allow you to see migratory species, some headed north, others south. In addition to providing information about where to go, the site guide also tells you how to get there, usually starting from well-known, easy-to-find landmarks depicted on most maps worthy of the name. We also cite approximate distances (in miles and kilometers) between key itinerary locations, and let you know which journeys might require a vehicle with four-wheel drive and/or high clearance. (Always check to make sure that your rental car has a reliable spare tire, as unpaved side roads sometimes contain sharp, volcanic rocks that can cause punctures.)

As you plan your excursions, please keep in mind that both the natural world and human-made infrastructure undergo constant change. Flowering times and changing water levels, for example, can result in changes at a given location in the number of species and in their abundance. And, new roads are sometimes built, old roads rerouted, and the condition of any road will vary over time; also, with changes in land ownership, the access to a given birding site may change. So, as you set out on your birding itinerary, be prepared to improvise—and to consult locals—should you encounter changes in routes or property access.

Aruba

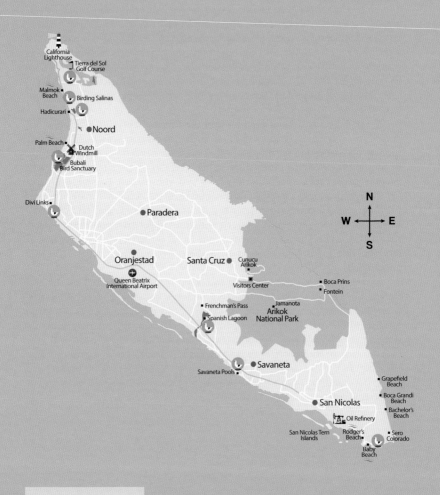

California Lighthouse
Tierra del Sol Golf Course
Malmok Beach
Birding Salinas
Hadicurari
Noord
Palm Beach
Dutch Windmill
Bubali Bird Sanctuary
Divi Links
Paradera
Oranjestad
Santa Cruz
Cunucu Arikok
Visitors Center
Boca Prins
Fontein
Queen Beatrix International Airport
Frenchman's Pass
Spanish Lagoon
Jamanota
Arikok National Park
Savaneta Pools
Savaneta
Grapefield Beach
Boca Grandi Beach
Bachelor's Beach
San Nicolas
Oil Refinery
Sero Colorado
San Nicolas Tern Islands
Rodger's Beach
Baby Beach

N
W   E
S

birding site

parking

birding site and parking

entrance

hotel and recreational spots

Aruba

Bubali Bird Sanctuary, Bubaliplas

Along the western coast of Aruba, just inland from its famous beaches, lies Bubali Bird Sanctuary, a large freshwater lake where a great number of bird species can be found. Originally a series of salinas (small, seasonal pools), the now-permanent wetlands of Bubali were created in 1973 from the freshwater outflow of the sewage treatment plant that serves the nearby hotels. The west side of Bubali is bordered by a thick band of mangroves. Elsewhere, particularly on the east side, there are thick reed beds.

Bubali not only offers a rich variety of bird species but also easy viewing—in 1998, a tower was erected on Bubali's north side, affording sweeping views of the birds that gather on the open water of the wetlands.

Birding Bubali

The best approach is to start in the thorn-scrub habitat on the west side of the lake, along the human-made canal that connects it with the ocean. This is a great area to see most of Aruba's scrub-loving species: Brown-throated Parakeet (listen for its harsh screeches), Tropical Mockingbird, Black-faced Grassquit, Common Ground-Dove, and, rarely, Burrowing Owl.

As you move from this scrub habitat into the abutting mangroves that border the lake, you'll probably notice a mysterious rustling sound that seems to repeat itself with every step you take. These are NOT the spirits of Bubali birds past! It's the sound of many sun-loving lizards as they scuttle away, one step ahead of you. Where scrub and mangrove meet, look for the tiny exquisite Blue-tailed Emerald, along with the Bananaquit, Northern Scrub-Flycatchers (less common), and, sometimes, Groove-billed Ani.

Entering the mangroves, you'll likely hear the low *whooo WHO-but-YOU, WHO-but-YOU* of Bare-eyed Pigeons and the sound of their noisy wing flaps as they move about. Common Gallinule, more commonly heard than seen, make clucking and trumpeting sounds as they shyly scurry away. You may flush a Green Heron, yellowlegs, or a snipe.

The mangroves host any number of overwintering warbler species. Northern Waterthrush is the most common; you'll hear their sharp *chink* notes. Ovenbirds are also fairly regular. Less common warblers include Cape May, Blackpoll, Black-and-white, Northern Parula, Prothonotary, and Hooded. Rare or one-time sightings have been made in Bubali of Black-throated Blue, Yellow-rumped, Magnolia, Palm, and Prairie warblers. It's nearly impossible to miss the many breeding Yellow "Golden" Warblers. Generally, the best way to find warblers here is to stand by a mangrove or shrub and pish for five to ten minutes. Their curiosity will bring them in.

The eastern side of Bubali is the ideal place to see groupings of Neotropic Cormorants, sunning themselves in the mangroves or diving for fish in the open

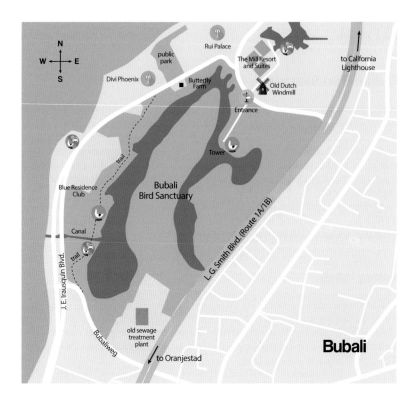

water. It's also the best place to find a wide variety of wading birds (including Tricolored and Little Blue herons), any one of which could be nesting in the mangroves. You'll have good looks at Caribbean Coot, Pied-billed Grebe, and Common Gallinule.

The viewing tower on the north side of Bubali provides an excellent vantage point. Here you can watch the birds as they go about their business, oblivious to your presence. The tower has a covered roof that provides much-welcome shade, making it an excellent place to set up your spotting scope. Look for Groove-billed Anis, herons, ducks, and other waterbirds. With luck, you may see a Sora during winter months or, more rarely, a Purple Gallinule striding along the reed edge. Check the water along the entrance road for herons, egrets, and Pied-billed Grebes. Pishing in the scrubby habitat near the tower entrance road can yield interesting birds. We have seen Common Yellowthroat and Bay-breasted Warbler in this area, which is usually full of Yellow "Golden" Warblers. The tower is ideal for children who have an interest in nature. Sometimes Bare-eyed Pigeons and Brown-throated Parakeets land at eye level in the tops of trees near the tower, giving stunning views. When water levels are high, parts of the road leading to the tower may be submerged, so use caution.

You'll also want to cross the street from the tower and look in the open water behind and beside the Mill Resort (where the big red Dutch windmill is located).

You'll likely have clear views of Blue-winged Teal, White-cheeked Pintail, Common Gallinule, Caribbean Coot, numerous herons and egrets, Black-necked Stilts, and smaller gatherings of other shorebirds as well (Killdeer, Greater and Lesser yellowlegs, Least and Semipalmated sandpipers, and occasionally other species). In recent years we have seen Lesser Scaup, Northern Shoveler, Black-bellied Whistling-Duck, and Scarlet Ibis here.

Directions

To reach Bubali from the bridge in downtown Oranjestad, follow L.G. Smith Boulevard (Route 1A) west, curving north for 2.8 mi (4.6 km) until you reach an intersection with a road named Bubaliweg. Turn left (west) toward the beach and follow this road (you'll pass the sewage treatment plant on your right) for about 0.3 mi (0.5 km) until it intersects with J. E. Irausquin Boulevard, which runs along the shore. Turn right on Irausquin Boulevard. About 200 yards (200 meters) before reaching a small bridge above the canal that connects Bubali and the sea, you'll come to a dirt road on your right. If you follow this it will take you into the thorn-scrub habitat bordering on the mangroves along the west side of Bubali. You can park here to explore the area on foot. Before venturing on, you might want to check some of the great vantage points along the coast to scan for terns and Brown Pelicans.

To reach the tower on the north side of Bubali, return to J. E. Irausquin Boulevard and turn right (north) and travel 0.75 mi (1 km) to a rather busy

The observation tower at Bubali provides an excellent vantage point for viewing the abundant water birds.

intersection. Turn right (east) heading toward the big, red windmill (visible on your left) and watch for the entrance to a dirt road on your right, within 200 yards (200 meters). This is the road to the observation tower, which itself should be visible a few hundred yards farther along.

Arikok National Park

Perhaps the most pristine place on the island is Arikok National Park, which lies in the hilly northeastern section of Aruba. The park consists of rolling hills covered with thorn-scrub vegetation and affords breathtaking views of the turquoise waters of the Caribbean; when the skies are clear, you can even see the Venezuelan mainland.

Officially incorporated in 2002, the park encompasses 7,900 acres (3,200 hectares), 17% of the land area of the island. When we first visited in 1993, the park was more of an idea than a reality. At the base of a hill was a dirt road and a small rustic sign, not too far from today's official park entrance. Today the park has a beautiful visitor's center located at the entrance. In addition to clean restrooms and a small gift shop, there is also a fascinating interpretive display that provides a bounty of information about the natural history of the park and the island in general. And, you will always find a ranger on duty, some of whom know their birds pretty well.

The main road through the park has specially engineered drainage dips that draw off water during tropical rains and also serve to keep visitors from driving too fast (thus providing some level of protection for the rare Aruban Rattlesnake). Today the areas along the main road are much more heavily vegetated than they were 15–20 years ago, likely because goats and donkeys have been excluded but perhaps also because the island now receives more rain.

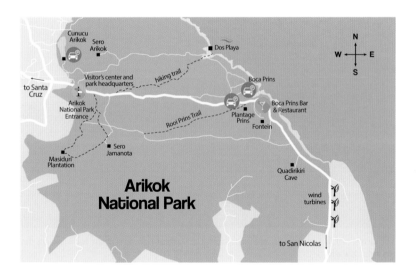

Birding Arikok National Park

Regardless of where you go in Arikok, it is best to begin early in the morning, when the park is pleasantly cool and the birdlife is most interesting—bird activity peaks just after dawn. From late morning on, the park gets hot, intensely hot, so remember to bring plenty of water. Also watch for snakes; although now very rare, the Aruban Rattlesnake does occur here as does the non-venomous Cat-eyed Snake. Boas have been introduced on the island and have devastated many bird populations, although they are essentially harmless to humans.

The park is large and has many hiking trails, so ideally you should plan on more than one day, especially when considering that the very best birding is restricted to early mornings. While the park does not officially open until 8:00 a.m., give a call to the Arikok National Park office ([+297] 585-1234) to see if you can arrange birding visits before then.

Black-faced Grassquits and Yellow "Golden" Warblers are ubiquitous here as they are across much of the island, and Bananaquits, although perhaps not as abundant as around some of the landscaped grounds of resorts and backyards, are still easy to find. Northern Scrub-Flycatchers and Brown-capped Flycatchers are widespread here, but can be difficult to see unless you can tune into their calls, as both species are amazingly adept at staying hidden in thick, thorny scrub vegetation. Troupials are often heard giving their loud piping calls and can be easily spotted as can Common Ground-Doves, Bare-eyed Pigeons, Eared Doves, and White-tipped Doves. Crested Caracaras may be seen gliding over the hills but don't be fooled into thinking you are seeing a White-tailed Hawk as they too have a

The beautiful visitors center at the entrance to Arikok National Park.

mostly white tail with a dark terminal band. Both Blue-tailed Emeralds and Ruby-topaz Hummingbirds seem to appear in numbers throughout the park when the shrubs come into full flower. Historically, the area that is now the park was home to at least occasional nesting pairs of White-tailed Hawks (the species nests on Curaçao and perhaps still on Bonaire), although it is unclear when they were last confirmed nesting on Aruba—certainly sometime before 1983. Two other species that are now gone from Aruba probably nested in the region that is now the park. The globally rare Yellow-shouldered Parrot disappeared from Aruba around 1947 (Voous 1983) but was documented historically from places like Fontein and Rooi Prins within what is now the park (Voous 1957). The Scaly-naped Pigeon apparently occurred as a resident breeder on Aruba until 1930, when it was last seen at Rooi Prins in what is now the park (there are two possible records of vagrant individuals on Aruba since that time).

Below we describe some of the sections of the park that we have always found to be productive birding spots, though we encourage you to check out less visited areas too.

Visitor's Center and Park Entrance

The big, shady trees that line the arroyo (usually dry) that runs in front of the visitor's center are a good place to begin. Approach quietly and try a little pishing under them to see what may come in. In these trees, we have seen uncommon to rare migrants in some years, including Yellow-billed Cuckoo, Hooded Warbler, Northern Parula, and Indigo Bunting. We have seen Blackpoll Warblers in this area even more frequently. This is one of the most reliable locations on the island to find Yellow Orioles. The area around the visitor's center and park entrance has been host to at least one pair of Burrowing Owls since at least the 1990s—inquire with the park rangers to get tips about where to see them. Please do your best to avoid frightening these rare birds.

In recent years, we have not seen two species that were once common in the area. The sweet song of the Rufous-collared Sparrow echoed in the surrounding hills until about 2003, but since then we have not found the species ourselves nor seen reports from other observers. Similarly, in the 1990s the hillsides here rocked with the squawks and screeches of the Aruban Brown-throated Parakeet, but we have not seen or heard them here in several years.

Cunucu Arikok

Heading north from the visitor's center toward Sero Arikok, on your left you will come to a gravel road (on which you can walk or drive) after a few hundred meters. It leads to the parking area at Cunucu Arikok, a charming example of the farming homesteads (cunucu) that were once common in the area. Like the original farms, Cunucu Arikok is enclosed by a stone wall (*tranchi*) and fencing to keep goats out. This location is protected from the constant trade winds and has always had thick vegetation as a result. There are easy walking paths throughout the area, an

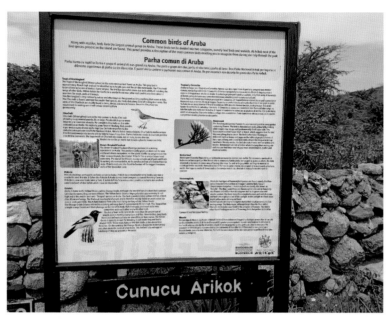

Arikok National Park has outstanding educational signs and other materials about the island's natural history. Cunucu Arikok is one of the best spots to enjoy birds in the park.

exhibition country house (*cas di torto*) built using historical materials and methods, and many benches ideally situated under shady trees and bushes. This has always been one of our favorite spots to leisurely enjoy close-up views of Aruba's birds in a natural setting. Virtually all of the characteristic birds of the island can easily be seen here, and we have rarely encountered more than a handful of other visitors, so it is typically very quiet. There are overhanging boulders on which the original indigenous inhabitants of the island made cave paintings of wildlife. A trail leads from the parking lot to the top of Sero Arikok, the second highest point on the island.

Cunucu Arikok is a pleasant 15-minute hike from the visitor's center or you can drive the steep gravel road to the parking area. Note that the road is often poorly maintained, and to avoid getting stuck—or ripping out an exhaust pipe—you should consider renting a four-wheel-drive jeep or other vehicle with high clearance.

Sero Jamanota and Masiduri Plantation

At 617 feet (188 meters), Sero Jamanota is the tallest peak on Aruba. Because of its elevation, there are a number of communication tower installations on its peak. A narrow service road that goes to the top is only open to foot traffic, but the view from there is outstanding and the surrounding habitat is all worth checking for birds. About a half mile (kilometer) beyond the visitor's center along the main road (Northern Loop), you can reach a parking area below the summit via a steep, mostly gravel, road (Southern Loop). Inquire at the visitor's center to confirm that this road is open and, if so, to check on its current condition. You can also reach the summit of Sero Jamanota by taking a hiking trail that can be accessed from the visitor's center. Expect the hike to take at least 30 minutes, depending on how many stops you make for birds. The Jamanota area was the last known spot on the island to have nesting White-tailed Hawks, and in this same area Tropical Kingbirds were reportedly found nesting in the 1980s.

From the parking lot at Sero Jamanota, you can walk a 1-mi (1.5-km) hiking trail to Masiduri Plantation, which lies southwest of Jamanota. This former experimental garden, which was started in the 1950s, nestles in a valley (*rooi*) in which several small dams were built to retain water. The area is generally wetter than surrounding areas. Very few birders have visited this spot so bird reports from here would be useful.

Plantage Prins (Plantation Prins), Rooi Prins, and Rooi Dwars

About 2.5 kilometers from the visitor's center, and just before you get to Boca Prins, is a parking area on the right with a trail that leads to the historic Plantage Prins (Plantation Prins) and two neighboring arroyos, Rooi Prins and Rooi Dwars. Through these arroyos flowed water once used to grow coconuts and fruit trees. Today, the coconuts are gone, as are all but a few of the fruit trees, but native vegetation grows here more thickly than on the surrounding hills. Although the thick trees and shrubs should attract both resident birds and migrants, this area, like many parts of the park, has received few visitors. Plan on at least 30–60 minutes to explore the area thoroughly.

Boca Prins

The term *boca* (spelled with a *k* on Curaçao and Bonaire, as in boka) refers to the mouth of river or the entrance to a harbor. Here on Aruba the term often refers to locations on the coast where a small stream occasionally (when there are heavy rains) reaches the ocean. These locations are often places where over thousands of years a streambed has eroded through the surrounding limestone terrace to make a deep canyon with a sandy beach along the water's edge. Boca Prins is one of these and sits at the point where the arroyo Rooi Prins terminates at the coast. In the geological past, water flowed through the arroyo and wore away the ancient, coral plateau. Today, Boca Prins is a beautiful sand beach that separates black, sharp-edged cliffs. From the parking lot (which is easy to spot), it is an easy walk down stairs to the beach. Beware that swimming is not allowed here due to the heavy surf and unpredictable currents that could sweep the unsuspecting out to sea or smash them against the razor-sharp rocks. We also recommend that you wear footwear, as the beach is unfortunately strewn with litter of all sorts.

Cactus- and shrub-covered hills are a striking feature of Arikok National Park and other places on the island.

As for the birds, Least Terns nest nearby in the summer. If the nesting colony is active, you'll see and hear the calls of the terns as they circle high above. Please maintain a safe distance from the nesting birds to avoid stepping on eggs or chicks or to frighten the adults (whose shade keeps eggs from baking in the heat) off of their nests. Search the rocks for the bright red bills of the American Oystercatcher—there is often a pair in residence here or nearby.

There are a restaurant and bar at Boca Prins (predictably called the Boca Prins Bar and Restaurant) that is a convenient place to grab a cold drink or eat a tasty meal, including local dishes such as *funchi* (a kind of cornbread), *keshi yena* (an empty Dutch cheese shell filled with spiced meat), and plaintain. In the garden next to the restaurant there are often temporary pools of water that host a shorebird or two.

Fontein

From Boca Prins—and with the little restaurant on your left—it's a short drive to Fontein. Fontein (in English, fountain) is a complex of caves whose name refers to a permanent natural spring that supplied water to the garden and house that formerly served as the official residence of the governor of the island.

The caves themselves are interesting—for their geological features, ancient indigenous drawings, and, sometimes, bats and other creatures. And for birders,

Common Ground-Doves sometimes nest on ledges just inside the caves. Rufous-collared Sparrows used to occur near the parking lot at the caves, but we have seen none of late, nor are we aware of any recent records. Just to the south of the parking lot are the freshwater pools and the gardens that some refer to, collectively, as Hofi Fontein; these sit beside the former governor's residence, now a privately owned house that is empty and for sale. These gardens, with their constant supply of freshwater, would seem likely candidates to host migrant and resident birds, especially in dry periods. However, we have mainly found Bananaquits, Yellow "Golden" Warblers, and Black-faced Grassquits, and not a lot more. On a visit in November 2013, we watched as one of the park rangers found and dispatched a 4–5 foot long boa constrictor—one of the most destructive animals introduced to the island—that was lying in wait for birds in the shrubs over our heads. Perhaps the presence of boa constrictors here has reduced bird diversity, though the area is still very much worth a visit for birders.

Directions

Although most maps bear the notation Arikok National Park, you may find it difficult to find the park entrance or to get a sense of its boundaries using these. While in recent years park staff have placed signage and created trails through the park, you've got to get there first! The best route is to start from Santa Cruz and, from there, follow signs for Arikok National Park (possibly indicated as Miralamar Pass on old maps). If you go by a yellow (formerly pink) building on the left called the Urataka Center, a bar-restaurant-ice cream stand (and a great place to stop for refreshments), then you're on the correct road. Eventually you'll see the gates, then the beautiful visitor center completed in 2010, and you'll know you're in Arikok National Park.

Once you've paid your entrance fee ($10 in 2015) in the visitor's center (a contribution that helps maintain the park and protect the birds and other animals), and passed by the gate house, you can go straight to Boca Prins, get to hiking trails for Jamanota, or detour to Cunucu Arikok. The main road in the park has now been paved.

When you are ready to leave the park, you can either retrace your route back on the main paved road past the visitor center or you can continue onward to exit the park via the city of San Nicolas. You will pass the Guadirikiri Caves and a set of tall wind turbines before you leave the park.

Savaneta Pool

In the town of Savaneta, just west of San Nicolas, there is an old salt pan that is virtually always covered with standing water. Savaneta Pool is located along a road that loops off Route 1A and that leads between the pool and the nearby Caribbean Sea. This large pool is a magnet for shorebirds and waders that feed in the water and roost in the clumps of mangroves growing in the pool and on bordering lands. They are seemingly heedless of the cars and trucks that pass by with frequency.

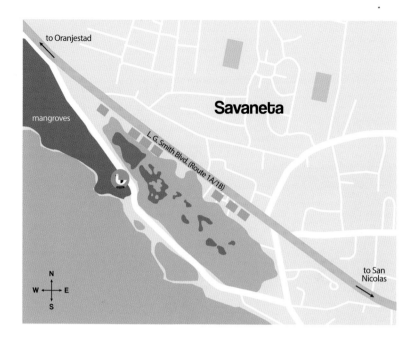

Birding Savaneta

As you approach the pool, park your vehicle off the road in one of the wide turn-outs; there are several along the left (north) side of the road. Find a good vantage point and set up a spotting scope. Once you've exhausted a given area, drive to the next point or, better yet, if you have the energy, walk the south perimeter of the pool, where you may flush some feeding shorebirds or get a better view of the waders. Be sure to look for birds wandering in and out of the mangrove clumps. In the winter, a variety of shorebirds feed here, though they can be difficult to pick out against the mangrove clumps, especially when they're roosting on the roots themselves, so scan carefully. Black-necked Stilts, which may breed as early as mid-February and as late as July, often use the mangrove roots for shelter for their downy young.

In winter, also watch for Greater and Lesser yellowlegs, Semipalmated and Black-bellied plovers, Killdeer, Short-billed Dowitchers, Whimbrels, Willets, Least Sandpipers, Stilt Sandpipers, and Spotted Sandpipers. Waders at Savaneta Pool are likely to include Tricolored, Great Blue, Little Blue, and Green herons, Snowy and Great egrets, and perhaps even a Reddish Egret. Of course, where there are shore-birds, watch for the Merlins and Peregrine Falcons that would like to make a meal of them. Both have occurred here.

If there is time, you might want to spend a few minutes birding the other side of the road (the ocean side). Bare-eyed Pigeons and Carib Grackles are common here, and you'll see Brown Pelicans, Royal Terns, and an occasional Osprey moving up and down the nearby coast.

The pools at Savaneta attract herons, egrets, and shorebirds.

Directions

From the bridge on Route 1A (also called L.G. Smith Boulevard) in downtown Oranjestad, head east toward San Nicolas, for about 8 mi (12.8 km). You'll see Queen Beatrix Airport en route; just before reaching the little settlement of Savaneta, on your right you'll see a turnoff to a road that runs along a large, metal pipeline. Turn right onto this road—and then left—and follow it as it bends around the salt ponds (salinas), which will be on your left, with mangroves and the ocean on your right. Park at the various turnoffs along the half mile (0.75 km) stretch of road and scan the ponds. Be careful to pull well off the road as locals drive fast despite the curves in the road. To investigate the ocean side, watch for one of the several concrete bridges that have been built over the metal pipeline to provide vehicle access to the shore. This beach area is used primarily by local residents for recreation and fishing boat access, so at times there may be a fair number of vehicles and pedestrians coming and going. To return to Route 1A, it's probably best to retrace your route as there are a number of confusing intersections in the residential section that you would need to pass through if you were to continue on the road around the salt ponds.

Spanish Lagoon

Spanish Lagoon is an expansive body of shallow, brackish water at the end of an inlet on the south side of Aruba, between the cities of Oranjestad and San Nicolas. It's almost completely surrounded by a dense tangle of mangroves, with gaps here and there affording limited views of the water.

Birding Spanish Lagoon

The real birding happens at the section of the lagoon that lies farthest inland, at the northern end. Birding here means walking along the dirt road that passes between the lagoon and the scrub-covered hillside to the west. You'll no sooner get out of your vehicle than you'll likely hear Tropical Mockingbirds—and perhaps a Troupial or two—singing loudly from the hillside. Watch for Crested Bobwhite scuttling across the road ahead of you. Bananaquits and Yellow "Golden" Warblers sing from the shelter of the mangroves. As you continue, look for breaks in the tangles. These offer welcome shade from the relentless sun and are also good places to pish for warblers that the mangroves sometimes harbor: Northern Waterthrush,

Prothonotary, American Redstart, Northern Parula, and Black-and-white are some of the species that will remind North Americans of home. You may see yellowlegs and Blue-winged Teal out on the water.

While poking around in these breaks, don't be alarmed by the occasional sudden splash—when startled, iguanas often drop from the branches into the lagoon's murky water. As you continue along the road, watch for Northern Scrub-Flycatchers and Grey Kingbirds perched atop the mangroves; pishing will attract both the Ruby-topaz Hummingbird and the Blue-tailed Emerald, which may zip up to within a foot of your face for close inspection. Also look and listen for Brown-throated Parakeets flying in from the hills to feed in the mangroves. Although in recent years we have not seen any, this species was once common at the lagoon. Black-whiskered Vireo has become more regular here in the last decade, and the area has long been a reliable spot to see Yellow Orioles. Since Spanish Lagoon provides both water and shelter, keep your eyes—and your mind—open to other possibilities as well. In April of 2000, we counted 5–6 Indigo Buntings near the end of our hike. Previous accounts showed no records of the species from Aruba before 1983. Other rare species we've encountered here include Worm-eating Warbler and Cape May Warbler.

Spanish Lagoon is a great place to watch for waterbirds, within the lagoon itself, and for landbirds in the mangroves and brushy hillside habitat.

Directions

From the bridge on Route 1A (L.G. Smith Boulevard) in downtown Oranjestad, head east for 5 mi (8 km). Continue beyond the intersection of Route 1A and Route 4B (to Santa Cruz) for 1.3 mi (2.1 km) and turn left for Frenchman's Pass (note that before you reach the turn you'll cross a bridge over an inlet that leads to Spanish Lagoon—we nicknamed this "Spanish Lagoon Bridge"). After the left turn, you'll pass an evangelical church (on your left) and a small community of residential houses before dropping slightly in elevation to the bottom of a broad flat valley. Although most of the time these flats are dry and sandy, rain flowing down the hillsides can sometimes fill them. At about 0.8 mi (1.3 km), turn left onto a dirt road with a sign for the Balashi Gold Mill Ruins. If you miss the turn and drive into a narrow canyon called Frenchman's Pass then you have gone too far. There are two options here. You can drive up the hill all the way to the parking lot for the Balashi Gold Mill Ruins and then walk the marked path down to Spanish Lagoon. We prefer the second option, which is to take a left about 100 meters in from the paved road onto another dirt road that leads you along the edge of the open lagoon. Depending on current park management plans, the road may be blocked here by large rocks, in which case you'll have to park and continue on foot. If the road is open, continue driving and park in front of a series of wooden post barricades where you'll see the sign for Arikok National Park, where you will park.

From the parking lot, walk on a road that wends along the edge of salt pans; in the winter rainy season, these fill with water and host Black-necked Stilts, Greater and Lesser yellowlegs, Semipalmated Plovers, Killdeer and various herons and egrets. Occasionally even a Roseate Spoonbill or American Flamingo. Off to the right there is a hill with the Balashi Gold Mill Ruins, where gold was once smelted; up ahead, you'll see the mangroves of Spanish Lagoon. Continue walking the road that runs beside the mangroves and through the shrubby thorn woodland to reach the lagoon.

Tierra del Sol

Located toward the western end of the island, not far from California Lighthouse, the resort of Tierra del Sol has a golf course, country club, and condominium resort. When we first starting visiting Aruba in the early 1990s, construction on the project had just begun. Today, the 600-acre complex includes a world-class golf course and hundreds of condominiums. The permanent bodies of water and the grass on the courses attract an abundance of birds, including many uncommon and rare visitors to Aruba. The natural habitat surrounding the golf course supports healthy numbers of many of the permanent resident bird species of the island. Encouragingly, the golf course managers seem to have worked hard to protect and support Burrowing Owls that nest along the edges of many of the sand traps. The nest sites are marked with short, wooden stakes. As many as three nesting locations are marked, with birds often visible near holes 1 and 12. The addition of green, grassy habitat to Aruba has also provided breeding locations for expanding populations of at least two bird species from South America, Southern Lapwing

and Cattle Tyrant. Like many parts of Aruba, unfortunately, the area is also home to several introduced species—notably Marine Toads and Boa Constrictors—both of which kill and eat many native species of birds and other animals. The golf course managers work hard to remove the boas, which are probably the species most destructive of bird populations in the area.

Birding Tierra del Sol

Birding at Tierra del Sol requires access to the golf course area, which is very popular, so it is important to ask permission at the golf pro shop in the club house. The golf course management have generally been very accommodating of birders. If you bird very early or late in the day—generally good times to bird anyway—you're more likely to stay out of the way of golfers. The course will sometimes provide birders with a golf cart, both for your convenience but also so they know where you are at all times (the carts have a GPS tracking device to help with managing the flow of golfers on the course throughout the day).

Near the club house and practice putting greens, keep your eyes open for Cattle Tyrants, as this has been a favorite area for them in the past. Perhaps the best birding area at Tierra del Sol are the freshwater pools (called "salinas" on the island) near holes 15 and 16, along the eastern edge of the course. Close to hole 15, there are several spots between the board walk and the restroom building where you can park a golf cart out of the way of golfers and scan the water for birds. At times the pools fill with hundreds of birds, including lots of Caribbean Coot, Common Gallinule, Blue-winged Teal, White-cheeked Pintail, Great Egret, and Snowy Egret, plus a variety of other birds in smaller numbers (Great Blue Heron, Black-crowned Night-Heron, Tricolored Heron, and Little Blue Heron).

44

Pied-billed Grebes nest here at the freshwater pools in some years, and in recent years these pools have been one of the best spots on the island to find uncommon or rare wintering ducks, including Lesser Scaup, Ring-necked Duck, Northern Shoveler, Northern Pintail, American Wigeon, and Green-winged Teal. The shallow waters of the freshwater pools, along with the shoreline and areas of marshy grass, usually also have a variety of shorebirds, including Black-necked Stilt, Greater and Lesser yellowlegs, Black-bellied Plover, Killdeer, Semipalmated Plover, Wilson's Snipe, Short-billed Dowitcher, Least Sandpiper, and Semipalmated Sandpiper. Other uncommon species that have occurred near the pools include American Golden-Plover, Solitary Sandpiper, Buff-breasted Sandpiper, Pectoral Sandpiper, and Stilt Sandpiper. Southern Lapwings have nested on or near the golf course, so be sure to look for them.

Incredibly, a few years ago Killdeer was confirmed as a breeder at the golf course, despite the fact that the nearest known nesting locations for the species are far to the north in the Greater Antilles. Other wetland-dependent species have occurred on the golf course, rarely and irregularly, including American Flamingo, Roseate Spoonbill, and Black Skimmer. Swallows, especially Barn Swallows, can often be seen coursing low over the freshwater pools and other parts of the golf course as they forage for aerial insects.

Anywhere on the golf course grounds, be on the lookout for the sometimes abundant Groove-billed Ani as well as resident Brown-capped Flycatcher, Eared Dove, Common Ground-Dove, Bare-eyed Pigeon, Troupial, American Kestrel,

Rare vagrants such as the Cattle Tyrant are among many unusual birds that have been found in and around the Tierra del Sol golf course.

and other species. Crested Bobwhite have been seen along the edges of the gold course and nearby, so listen for their "bob-white" whistles.

To see the resident nesting Burrowing Owls, look for the green, wooden nest-burrow markers at holes 1 and 12; scan for a pair of yellow eyes peaking over the edge of the green. You might see the owls at other holes as well.

As always, be on the lookout for oddities, as the combination of pools of water and grassy habitat could attract rare wandering vagrant species. Make sure you photograph or videotape anything unusual to provide documentation for other birders.

Directions

From the bridge on Route 1A (also called L.G. Smith Boulevard) in downtown Oranjestad, head west (and curve north), toward an area with high rise hotels. Continue on Route 1A for 7 mi (11 km), eventually passing a McDonalds and, farther, the Marriot Hotel, until you come to Malmokweg Street (Route 2 on some maps). Take the right here, heading away from the ocean, and continue for 0.5 mi (0.8 km) until you see the entrance to Tierra del Sol on the left. At the entrance booth, the attendant will wave you through. Stay on the main entrance road (Caya di Solo) for about 0.5 km to reach the club house and parking areas. Park, enter the club house, and go to the pro shop at the far end to inquire about birding.

Although not part of the golf course itself, there is a dirt road on the left about a mile past the entrance to Tierra del Sol that passes by a horse farm and then dead ends. Here birders have found Crested Bobwhite and, in the early evening, White-tailed Nightjar.

Divi Links

The Divi Links is a compact, nine-hole golf course within the Divi Hotel complex, about a mile from downtown Orenjestad. While the course itself is only open to golfers, birders can use an access road to reach a parking lot that serves the club-house restaurant and some condominiums, and from there visit and view a fresh-water pond and a portion of the greens nearby.

Birding Divi Links

From the circular drive beside the pond, you can view a nesting colony and roost of Neotropic Cormorants, located at a short distance in a large stand of mangrove and buttonbush. You might also find Great Egrets, Snowy Egrets, Tricolored Herons, Green Herons, Black-crowned Night-Herons, and other wading birds in and around the pond, while you will possibly see White-cheeked Pintails, Blue-winged Teal, Caribbean Coots, and Common Gallinules swimming in the water. Check the water's edge and nearby grassy areas for shorebirds, including Southern Lapwing, Killdeer, and Whimbrel. There are nesting Burrowing Owls on this golf course, but of course you cannot visit the nest sites unless you're on the course itself, which is only open to golfers.

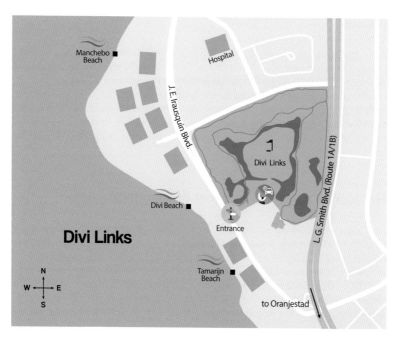

Manchebo
Beach

Hospital

J. E. Irausquin Blvd.

Divi Links

Divi Beach

L. G. Smith Blvd. (Route 1A/1B)

Divi Links

Entrance

N
W — E
S

Tamarijn
Beach

to Oranjestad

Aruban Burrowing Owls have adapted to nesting in unexpected places, including at the Divi Links golf course.

Directions

From the bridge on Route 1A (also called L.G. Smith Boulevard) in downtown Oranjestad, head west; go 1.7 mi (2.7 km), passing the cruise ship terminal, until you reach J.E. Irausquin Boulevard (near the Sun Plaza Business Center). Turn left (south) and you'll almost immediately note the first of the Divi hotel/condominium properties on your left. Drive about 0.3 mi (0.5 km) and turn right into the entrance to Divi Links, almost across from the entrance to the Divi Tamarijn Resort; keep driving until you see the pond in front of you and park in the parking lots to your left.

Hadicurari and the Salinas at the Skydiving Area

Hadicurari is a working marina that is sandwiched between two beach front hotels. When we first began visiting Aruba, the marina was a fairly quiet area, now it's surrounded by hotels and restaurants and is typically crawling with beachgoers. Many of the local fishermen bring their Caribbean bounty to market here, and the spoils are often tossed overboard, to the delight of hungry birds—and the birders who come to watch them. Hadicurari is an excellent place to see all plumages of the Magnificent Frigatebird, sometimes so close that you can even see salt

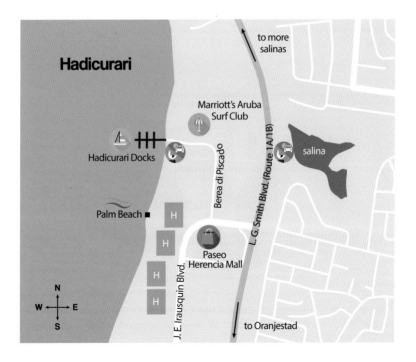

dripping from their bills. It is breathtaking to see these huge birds sweeping in for the fish. Brown Pelicans and Laughing Gulls also join the free-for-all. Terns and shorebirds frequently perch on a rock wall that shelters the marina 100 or so meters off shore. The Salinas are a short distance by road further west just across from and beyond the high rise hotels.

Birding Hadicurari and the Salinas

When we first started visiting Aruba, anyone could drive out to park along the beach and watch the boat traffic in and out of the marina. Today, in spite of all the surrounding development and resulting hustle and bustle, the local fishermen have fortunately retained access rights, and anyone can walk to the beach and marina from a parking lot a few hundred meters away. After the short walk from the parking lot, you'll see the beach and marina ahead. If fishermen are unloading their catch, you'll witness birds wheeling, hovering, and diving for the spoils. Climb onto the wharf and set up your spotting scope. In winter, you might see dozens of Sandwich Terns (mostly the yellow-billed "Cayenne" morph) and Royal Terns (immatures and adults). Look for Sanderlings and Ruddy Turnstones among the rocks and on the wharves and docked boats as well. Although gulls (other than the relatively abundant Laughing Gull) are rare anywhere in the southern Caribbean, Hadicurari is a likely location for one to

Although Hadicurari is no longer the quiet beach and marina it once was, local fishermen still retain access rights—and it remains a good place to look for overwintering Ruddy Turnstones, Sanderlings, terns, gulls, and other birds.

appear. During mid-winter, we've found single Ring-billed Gulls here on several occasions. When you're done birding, refresh yourself at the bar and grill next to the beach.

From here you can easily visit the two salinas that lie farther north and just inland from L.G. Smith Boulevard. They have afforded exceptional birding in recent years. The first salina you will reach is directly across from the Marriot Hotel; about 0.3 mi (0.5 km) farther north, to your right, lies the second salina (located behind a sky-diving business on our 2015 visit). There are a number of dirt turn-offs and roads to explore for different vantage points. American Flamingos and Reddish Egrets have been seen here quite regularly and we documented a Caspian Tern in 2012. There are almost always White-cheeked Pintails and Black-necked Stilts and a variety of herons, ducks, and shorebirds.

Directions

From the bridge on Route 1A (also called L.G. Smith Boulevard) in downtown Oranjestad, head west (Route 1A will soon curve north), passing through the city. After about 4.7 mi (7.5 km), you will come to J.E. Irausquin Boulevard, where you should turn left. Stay on it for just 200 meters, then take a right, onto Berea di Piscado, an unassuming road. Follow it (Berea di Piscado makes a sharp left hand turn at about 300 meters) for about 400 meters and park in the parking lot to your left. Make sure to lock your vehicle and do not leave any valuables in view.

Male Magnificent Frigatebird. Hadicurari is an excellent place to see all plumages of this species.

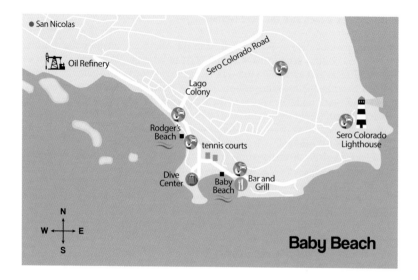

Baby Beach

To get to the salinas, drive back to Route 1A (L.G. Smith Boulevard) and turn left. After a few hundred meters, you'll see the first salina on your right, before you reach a roundabout. The turnout for the second salinas is about 500 yards (meters) beyond the roundabout; as of 2015, this second salina was located next to a sky-diving business.

Baby Beach, Rodger's Beach, and San Nicolas Tern Islands

Baby Beach, one of Aruba's renowned white sand beaches, is a popular destination for snorkeling enthusiasts. Within this sheltered, horseshoe-shaped cove, the Caribbean waters take on a turquoise color, and the surf is remarkably calm, making this a great place for families with young children. A gap between the eastern arm of the "horseshoe" and one of the reef islands leads to a dramatic coral reef, where you're likely to see dozens of snorkelers bobbing in the surf. So bring your bathing suit and snorkeling gear, but don't forget your binoculars! Particularly in the winter, Baby Beach is a good spot to look for shorebirds and Royal and Sandwich (including Cayenne-type) terns.

Even more exciting for birders is the opportunity to spot some of the tropical breeding terns, including Sooty and Bridled terns and Black and Brown noddies, which nest from March through July on the tiny reef islands in front of the oil refinery property in the city of San Nicolas, adjacent to Baby Beach. Generally known as the San Nicolas Tern Islands, one of the islands can be easily viewed from Rodger's Beach (note that it is spelled as "Roger's" on some maps), a beach only a few hundred meters to the west of Baby Beach.

Birding Baby Beach and the Reef Islands

From the parking area at Baby Beach, scan the island just offshore for roosting terns. If you've got a scope, this is a good place to set it up. We've had as many as 200 Royal Terns and up to 150 Sandwich Terns (usually the yellow-billed "Cayenne" morph) here. In the breeding season, look for Sooty and Bridled terns and for Brown and Black noddies. Out over the water near the islands, watch for Brown Pelicans, Magnificent Frigatebirds, and on occasion, Osprey. Check the beach, particularly the less-busy west end, for shorebirds, including Sanderling, Ruddy Turnstone, and Whimbrel.

If you're visiting Aruba anytime from late March through July, swing around the corner to Rodger's Beach. Just offshore is the easternmost of the San Nicolas Tern Islands, reef islands that are breeding grounds for Sooty and Bridled terns and Brown and Black noddies as well as Royal, Sandwich (including "Cayenne" morph), Common, and Roseate terns. From here it's possible to set up a telescope and get good views of the birds coming and going from the islands without any possibility of disturbing them. There's also a turn-out above the beach that provides elevated views of the islands.

The San Nicholas Tern Islands provide breeding habitat for an impressive diversity of species, including Sandwich "Cayenne" Terns, Black and Brown noddies, and Bridled and Sooty terns. Unfortunately, the islands remained unprotected as of 2016.

Directions

From the intersection of Route 1A and the road to Spanish Lagoon and Frenchman's Pass, follow the signs for San Nicolas. At 3.7 mi (6 km), you'll come to a roundabout. Take your first exit onto Bernhardstraat. After another 1.1 mi (1.9 km), the road forks. Take the left, staying on Bernhardstraat. In another kilometer or so, take a right onto Staringstraat. After another 0.2 mi (0.3 km), you'll come to an intersection with Route 1A. Turn right; this will lead you around a hairpin curve with a wall that blocks the unsightly view of the oil refinery on your right. Take your first right after the sharp curve onto Fortheuvel Street. Go one kilometer to a T-intersection. Turn right onto Sero Colorado road. Follow this for about one kilometer to another T-intersection. You will see a prison on your left. Turn right. You'll see a sign for Sero Colorado, leading through a walled gate. The landscape becomes barren—lots of cacti and thorn-scrub. As you drive along, keep your eyes ready for Crested Caracara. Stop here and there to look for Burrowing Owls standing on the parched earth in search of lizards. The birds are superbly camouflaged, so look slowly and methodically, and be sure to pull off to a safe place. This is the road to Baby Beach, remember, and is therefore quite busy. Once you've seen the owls, or exhausted your search for them, resume your drive, through a small housing development called the Lago Colony (once the homes for the executive managers of the oil refinery), until you come to another T-intersection. The left-hand turn will take you past old tennis courts and a baseball diamond—and over some serious speed bumps! In a minute or two, you'll come to the Baby Beach parking lot and the crystal-green sea reaching out in front of it.

To get to Rodger's Beach from back at the Lago Colony, take the right at the T-intersection instead of the left, which leads to the oil refinery. The road ends in the private entrance to the refinery, but there's a tarred parking lot on the side of the road that serves as a kind of overlook. From it, you'll see the island and the terns sailing around it. You can reach the beach by taking a right immediately before the baseball diamond on your way to Baby Beach, following the short road towards Rodger's Beach and the local marina.

Bonaire

Malmok

Boka Bartol

Washington-
Slagbaai
National Park

Mt. Brandaris

Boka Slagbaai

Salina
Slagbaai

Entrance

Boka Onima
Onima Reservoir

Fontein

Gotomeer

Rincón

Dos Pos

oil storage facility

N
W ← → **E**
S

Klein
Bonaire

Bonaire Sewage
Treatment Plant

Kralendijk

Flamingo
International Airport

mangroves

Lac Bay

Sorobon

Pekelmeer

Willemstoren
Lighthouse

- birding site
- parking
- birding site
  and parking
- entrance
- hotel and
  recreational spots

Bonaire

The Pekelmeer

The word *pekelmeer* (Dutch for "pickle lake") was used in the past to describe a single salt pond and the area surrounding it, but we use it here to refer to the entire group of salt pans and ponds (generally known as salinas on the islands) that span across much of southeastern Bonaire. It's the primary home for the bird for which Bonaire is most famous: the American Flamingo. The history of this part of the island is as fascinating as its bird life, and in fact the two themes are closely intertwined.

On the ocean side of E.E.G. Boulevard, the road that leads to the Pekelmeer, the beach is rocky and layered with dead coral, an indication of the superb scuba diving that awaits beneath the rolling blue-green surf. As you drive along E.E.G. Boulevard, your imagination may be piqued by the painted yellow rocks bearing the often whimsical names of the dive sites: Alice in Wonderland, Atlantis, and Vista Blue, for example. Along two sections of the boulevard, you will see small, stone huts—former slave quarters—that are relics from the early days of the salt industry, which for centuries has been integral to the economy of Bonaire. In recent times, this very same industry has had a major impact on the ecology of the Pekelmeer.

Salt produced at the Pekelmeer, on Bonaire, is shipped to the US and around the world. Although this is the primary home for the island's population of American Flamingos, its numbers have declined dramatically.

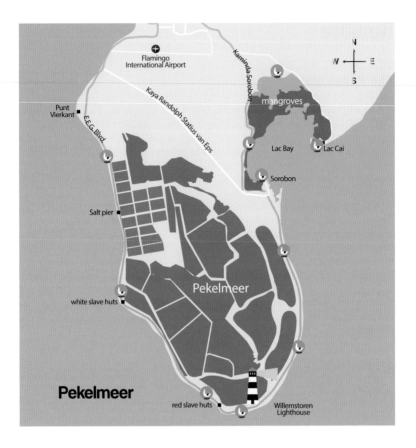

Pekelmeer

The Pekelmeer is the region that lies to the other side of E.E.G. Boulevard, away from the ocean. It is a broad, flat patchwork of salinas separated by human-made dikes. Where water is absent, the beds of salt (salt continues to be collected today) are a blinding white. In contrast, some of the standing pools are a rich red color resulting from brine shrimp, the flamingos' primary food source and the source of the birds' notorious pink-red plumage. Here among the pools, feeding in a manner that is often comical, are flamingos by the hundreds. A large number of shorebirds and herons also use this food-rich area, as do terns. Sandy shorelines of some of the pools also make ideal territory for the elusive Snowy Plover.

Bonaire's salt industry was developed in the early 1600s. Historically, the salinas of the Pekelmeer were flooded by the sea only when storms caused extremely high tides. As the water evaporated, the salt was removed manually by slaves. In the 1950s, nearly a century after the abolition of slavery, a privately owned salt plant, Antilles International Salt Company (AISCO), was built in the center of the Pekelmeer. With that came the creation of channels that let in the sea, regulated according to the needs of the plant. At the same time, fortunately, conservationists became concerned about the fate of the flamingos that breed and feed in the Pekelmeer. As a result of negotiations between bird conservationists, the Dutch government, and AISCO, the company created a flamingo breeding sanctuary. Although the birds accepted the sanctuary as a nesting area, the changes in the ecosystem caused by the salt plant resulted in less food being available to the birds. Some flamingos fly several hours, each way, to feed in Venezuela. Despite these changes, however, the Pekelmeer area is still a haven for flamingos and other wading birds, shorebirds, and terns.

Birding the Pekelmeer

A road traces the perimeter of the Pekelmeer, forming a loop and providing excellent viewing opportunities all along the route. When you see some birds, find an area to pull to the side of the road and get out for a closer look. A trip around the Pekelmeer starts by heading south from the city of Kralendijk, past the airport to the intersection of E.E.G. Boulevard and Kaya Randolph Statius van Eps. If you stay straight on E.E.G. Boulevard you will be following the western side of the loop road and will see signs for the Punt Vierkant lighthouse just beyond Bachelor's Beach. If you're up before the sun, be on the lookout for the red eye-shine of White-tailed Nightjars sitting in or beside the road, wherever it is bordered by vegetation. After you pass Punt Vierkant and see the salt mountains ahead, you'll begin to see shorebirds in spots with shallow water. Continuing south, you'll come to the salt pier in 2 mi (3.3 km); about another mile (1.4 km) beyond that point, you'll see a small bridge over the canal that was dug by the salt company. It's at this point that you'll start seeing more flamingos. Watch for the number of shorebirds and the variety of species to increase at the water's edge, near the yellow windmill. At about 1 mi (1.7 km) beyond the bridge, you'll come to the Witte (White) Pan slave huts. At any point between Punt Vierkant and Witte Pan, you might see a number of species feeding, including Brown Pelican, Tricolored Heron, Snowy Egret, Great Blue Heron, American Flamingo, Peregrine Falcon, American Oystercatcher, Black-bellied Plover, Collared Plover, Semipalmated Plover, Whimbrel, Ruddy Turnstone, Sanderling, Least Sandpiper, and Yellow "Golden" Warbler.

Another 3 mi (5 km) beyond Witte Pan sit the Red Slave huts, at the southern tip of the island. This area affords the closest viewing in the Pekelmeer of the American Flamingos as they feed in the pans. They are a spectacular site, forming a wash of pink over an array of background hues. This is a great area to watch small flocks of the flamingos as they fly off toward Venezuela, their pink-and-black wing pattern creating a dramatic impression. Watch for Magnificent Frigatebirds, too. Shorebirds such as Black-bellied Plovers, Greater Yellowlegs, Short-billed Dowitchers, Common and Royal terns, can also occur here, along with a few landbirds.

From the Red Slave huts, continue another mile (1.6 km), and you'll come to the Willemstoren Lighthouse and the remnants of a home. This is a great place to set up your scope and scan the pans, as the buildings provide shelter from the relentless wind blowing off the ocean. Here, among other shorebirds, you may see Red Knots. Don't forget to turn around and scan out over the sea for boobies and other pelagics—and to admire the evident courage of the local fishermen as they battle the wild sea in their small boats.

A bit more than 3 mi (5 km) north of the lighthouse is an especially reliable place to see Snowy Plovers scurrying about on the exposed dunes of the salt pans. (This is near the location marked Piedra Pretu on the ocean-side of some maps.) Be methodical in your scoping, as their pale, sandy colors provide good camouflage against the dry, salt-encrusted mudflats.

It's 4.3 mi (6.9 km) from the lighthouse (0.6 mi [1.9 km] from the Snowy Plover spot) to the Marculture Project. This is an excellent spot to get close-up, exquisite views of shorebirds and egrets—including various color morphs of Reddish Egrets—and in winter to note the fine distinctions between Western and Semipalmated sandpipers.

Continue north about 0.2 mi (0.3 km) to a three-way intersection. To close the loop and return to Kralendijk bear to the left onto a road with an impressively long name, Kaya Randolph Statius van Eps. If instead you would like to bird Lac Bay, you can bear right to Sorobon (you may see signs for the Sorobon Beach Resort here) and start your birding there.

Directions

Start in Kralendijk, from the TransWorld Radio (sign also says Radio Trans Mundial) roundabout (where Kaya Debrot and Kaya Amsterdam intersect); head south on Kaya Gobernandor Nicolaas Debrot (which later becomes Kaya Grandi) for 1.6 mi (2.5 km); you'll take a sharp right turn, where Kaya Grandi becomes a one-way street (if you don't take the right, you'll be driving into one-way traffic coming at you).

This right-hand turn will take you onto Kaya J.N.E. Craane. After the road veers right slightly, it becomes Kaya C.E.B. Hellmund. As soon as you pass Wilhemina Park, turn left onto Kaya Gilberto Croes. In another 0.2 mi (0.3 km) take a right onto Kaya International (you'll see a sports stadium on your left). After about 1.6 mi (2.5 km), you'll see the entrance to the airport; keep going straight and note that the road you are on is now named E.E.G. Boulevard. At 1.8 mi (2.8 km) beyond the entrance to the airport, you'll come to Punt Vierkant (note the salt mountains in the distance), where your birding day begins. From Punt Vierkant, stay on E.E.G. Boulevard to reach all of the birding locations described in the Birding section. As you move from site to site, essentially you will be circumnavigating the perimeter of the southeastern tip of the island.

Washington-Slagbaai National Park

Washington-Slagbaai National Park covers the very northern end of Bonaire. Established in 1969, the park is under management of STINAPA, Netherlands Antilles National Park Foundation. Its 8,300 acres (3,800 hectares) include a diverse array of landscapes—inland bays, lagoons, salinas, water holes, sandy beaches, rocky coasts, hills, and valleys are all part of the magnificent scenery. The

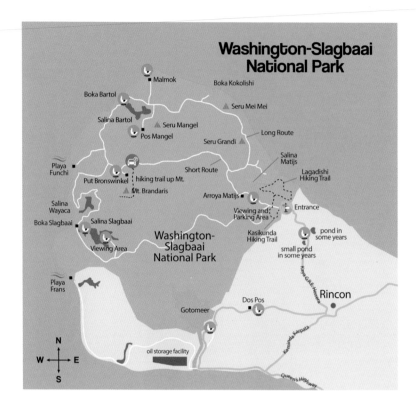

Washington-Slagbaai National Park

Malmok
Boka Kokolishi
Boka Bartol
Seru Mei Mei
Salina Bartol
Seru Mangel
Pos Mangel
Seru Grandi
Long Route
Salina Matijs
Playa Funchi
Short Route
Lagadishi Hiking Trail
Put Bronswinkel
hiking trail up Mt.
Mt. Brandaris
Arroya Matijs
Salina Wayaca
Boka Slagbaai
Salina Slagbaai
Viewing and Parking Area
Entrance
Washington-Slagbaai National Park
Kasikunda Hiking Trail
pond in some years
Viewing Area
small pond in some years
Playa Frans
Dos Pos
Rincon
Gotomeer
N
W — E
S
oil storage facility
Queen's Highway

park's highest point, Mount Brandaris, standing at 783 feet (241 meters), is in fact an extinct volcano. The vegetation in the park is in places very sparse, with a few cacti and low-growing shrubs, but there are also areas of scrub thickets three- to four-meters tall, and you will even find large, luxuriant trees (gum trees, brazil wood, acacia, for example) in the steep mountainous areas, especially on the western side of the park.

The 21 mi (34 km) of gravel roads through the park are winding and in some places quite steep. Although maintained as well as can be expected, they can become pockmarked during rainy periods, particularly where they pass through exposed mountainous areas. The park staff recommend a high clearance vehicle for touring the park, especially during rainy periods, but regardless of the vehicle you drive, go slowly and navigate any ruts or holes with caution. Most of the roads through the park are one-way and quite narrow. There are turn-outs throughout that afford spectacular views.

The park is open from 8:00 am to 5:00 pm every day. You'll need to stop at the gate to pay a $25 per person fee that helps maintain this ecological treasure. Be sure to pick up a park map and to inquire whether any of the areas are currently closed. You can lend further support to the park by purchasing your souvenir T-shirts and tote bags at the gift shop in the headquarters. It also offers a surprising variety of

Populations of Yellow-shouldered Parrots have been reduced to just a few locations, including Bonaire.

island-based natural history books as well as snacks and cold drinks. There is a small museum at the visitor's center.

Before you start your drive into the park, there are two hiking trails that start at the visitor's center, both of which provide wonderful opportunities for finding resident birds. One is a strenuous 50-minute round-trip hike; the other is an easier but longer 2-hour round-trip hike. Birders may prefer just to walk a shorter section of the trails and backtrack in order to have time to see the many other areas of the park. Or if you're staying on the island for an extended period, you can come back specifically just to walk the trails. Note that the shorter of the two trails leads to yet a third trail, a 45-minute to 1-hour trip to the top of Mount Brandaris, where there are stunning views of the island. This Mount Brandaris trailhead can also be reached by car from the short auto route. On clear days, it's possible to see Curaçao off to the west and sometimes the mountains of Venezuela to the south. Obviously, hiking is best done in the cooler morning hours, and always carry plenty of water, a hat, and sunscreen.

In rainy years, a small pond forms about a mile (1.4 km) before the entrance to the park, on the left-hand side of the road. Watch for it about 1.5 mi (2.4 km) after you leave Rincon on Kaya G.R.E. Herrara. There is also sometimes a pond accessible via a 400-meter dirt road on the right-hand side of the road across from where these roadside ponds form. Any freshwater ponds that you come upon are worth checking carefully as they could be hosting all kinds of oddities. In November 2010, for example, a Sungrebe was photographed in a pond on Bonaire.

Birding Washington-Slagbaai National Park

Once you've paid your fee, picked up your park map, and stocked up on water, you're ready to move on to the first of many locales that are promising for a number of birds. Many birders come to Washington-Slagbaai National Park to look for the rare Yellow-shouldered Parrot. As you travel through the park, watch for birds sitting on cacti or flying in small flocks. Remember that, in flight, one of the most obvious features to distinguish Yellow-shouldered Parrots from Brown-throated Parakeets is the short, blunt tail of the parrot compared to the proportionately much longer, wedge-shaped tail of the parakeet. Parrots and parakeets can both appear at the two water holes (Pos Mangel and Put Bronswinkel) within the park but are often hidden in the foliage surrounding the water holes, so listen for their occasional calls; the parrot makes throaty, resonant squawks, while the parakeet produces thinner, screechy calls. Note that the Yellow-shouldered Parrot can be especially difficult to see in the park under dry conditions, when a large proportion of the population sometimes travels to the suburbs of Kralendijk to feed in fruit trees in backyards.

Salina Matijs

From the park entrance, it's about a 0.6-mi (1-km) drive to Salina Matijs. You'll see a parking lot on your right overlooking this large salina, which can occasionally hold shorebirds and waterfowl when there's plenty of rain. As many as 40 American Wigeon occurred here on one occasion!

Arroyo Matijs

To our knowledge, this arroyo (stream valley) does not have a formal name, but since it is just a few tenths of a km up the road (on the left) from Salina Matijs, we have come to refer to it as Arroyo Matijs. Even if the streambed appears dry, the soil may be damp in places and thus a haven for insects, which attract birds. Pish the trees along the edges. You may gain the attention of Blue-tailed Emeralds, Caribbean Elaenias, Northern Scrub- and Brown-crested flycatchers, Black-whiskered Vireo, Yellow "Golden" Warblers, Bananaquits, Troupials, and Yellow Orioles. You may also intrigue any unusual migrants in the area—we've had Chestnut-sided Warbler here, for example. Watch for Bare-eyed Pigeons and Brown-throated Parakeets, and look overhead for Crested Caracara.

Continuing a short distance—a football field or two—beyond Arroyo Matijs, you'll come to a fork in the road. To your right is the long route through the park, which meanders along the coastline; it is about 21 mi (34 km) and takes 2.5 hours. To your left is the short route, which cuts across the interior of the park near the base of Mount Brandaris; it is about 15 mi (24 km) and takes 1.5 hours. One advantage of the long route is that it takes you to two of the more unique birding spots on the island, Pos Mangel and Malmok (described below), in addition to some other interesting spots. Note that the two routes meet up just beyond Put Bronswinkel.

The Long Route: Area Before Turnoff to Boka Kokolishi

After turning right at the fork in the road, drive for about 2 mi (3.5 km) and look off to the right for a barren open area with scattered, low-growing brush and cacti (if you come to the sign for Boka Kokolishi, you've gone too far). Recent rains may mean a hatch of insects that will attract a short list of birds. One year, we had a flock of six Barn Swallows, eight Gray Kingbirds, two Northern Scrub-Flycatchers, and a few Yellow "Golden" Warblers, all foraging low, even perched on the ground.

Pos Mangel

On the long route, the next productive birding spot is Pos Mangel, about 1.6 mi (2.5 km) beyond the turnoff to Boka Kokolishi. When you come to the sign for Pos Mangel, you'll turn left off the long route, and continue down this side road for 0.6 mi (1 km) until you reach Pos Mangel. The site itself is a bird magnet, despite its diminutive size. It is a shaded water hole, about six meters in diameter, that leans into an open muddy-sandy area on one side and is bordered by a dense tangle of trees around the remaining two-thirds or so. Because it is one of the few sources of fresh water within the park, you'll see a great diversity of species here. The ground around the water hole will likely be roiling with dozens of Common Ground-doves and Eared and White-tipped doves. In addition to the doves, in a single visit one year we saw Lesser Yellowlegs, Spotted Sandpiper, Scaly-naped and Bare-eyed pigeons, Brown-throated Parakeet, Yellow-shouldered Parrot, Blue-tailed Emerald, Caribbean Elaenia, Northern Scrub and Brown-crested flycatchers, Tropical Mockingbird, Pearly-eyed Thrasher, Black-whiskered Vireo, Yellow "Golden" Warbler, Northern Waterthrush, Bananaquit, Black-faced Grassquit, Troupial, and Yellow Oriole. A few trees held so many Common Ground-doves that the branches were bending under the weight of the birds—a dozen or more to a branch, perched like feathery brown Christmas bulbs.

The best way to make sure you don't miss anything is to sit where you've got a good panoramic view, then wait and watch. Many birds will wander in quite close, and your quiet presence just might peak the curiosity of Pearly-eyed Thrashers, which will possibly inch their way along a branch to investigate you. During one visit, a persistent Yellow "Golden" Warbler (the resident subspecies) not only fed from our open hands but found his way into the cab of our truck, from which he refused to leave, flying back in repeatedly. A flock of more than a dozen Troupials gathered to within a few feet of us as we peeled our oranges, and two or three of the largest iguanas we've ever seen wandered over, hoping for a hand-out of bread and cheese.

On your way back out to the long-route road, pull into the parking area near the edge of Salina Bartol. Look for Least Sandpiper, Brown-throated Parakeet, Tropical Mockingbird, Caribbean Elaenia, Northern Scrub-Flycatcher, Yellow "Golden" Warbler, Northern Waterthrush, Bananaquit, Troupial, and Yellow Oriole.

Malmok

Return to the long-route road and continue in the same direction as before. Almost immediately—after about 200 meters—you'll see a turn-off on your right for Malmok. Follow this road for a little less than a half mile (0.75 km) and stop as you approach the coastal cliffs. The cliff ledges here are a traditional Brown Booby roosting site. Approach the edge very cautiously as a slip off them would send you down into the water and onto sharp rocks. If you scan the cliffs here you may see immature and adult Brown Boobies and sometimes a few coursing by offshore as well. Although unlikely, you never know if a rare Red-footed Booby or Masked Booby could join them as well, so scan carefully!

Although Brown-throated Parakeets can be seen throughout Washington-Slagbaai National Park, the areas around the waterholes Pos Mangel and Put Bronswinkel are especially good places to see them and many other birds.

Boka Bartol

From Malmok, return to the long-route road. Continue in the same direction as before, for about another half mile (0.75 km). The road will lead you between the rocky coast on your right and a shallow lagoon at the foot of some small cliffs on your left. This is Boka Bartol. Pull off to the side of the road and glass for Pied-billed Grebe, Neotropic Cormorant, Brown Pelican, Green and Tricolored herons, Snowy Egret, Black-necked Stilt, Greater and Lesser yellowlegs, Least Sandpiper, and any number of other waders. The water is wide enough that you may need to set up a scope to ID distant shapes, to make sure you're not in the presence of some rare South American heron.

Put Bronswinkel

From Boka Bartel continue on the main road (toward Funchi Beach) for 1.6 mi (2.5 km), until you reach the intersection where the long route through the park reunites with the short route. Turn to your left onto the short route, as though you were heading back to the visitor's center. Continue for about 0.3 mile (0.5 km) until you reach the turn-off for Put Bronswinkel, on your right. (If you have a good sense of geography, you will have figured out that you can get from the visitor's center to Put Bronswinkel by either the long or the short route.)

Like Pos Mangel, Put Bronswinkel is another fantastic water hole. Depending on the condition of the road, you may need to park several hundred meters before the water and hike the short distance.

From a rise on one side of the hole, a trickle of water dribbles in; we've had as many as 50 Bananaquits bubbling here at one time, along with a few Black-faced Grassquits and other species. There is an old stone bench near the water that makes a nice spot from which you can watch for Scaly-naped Pigeon, Common Ground-dove, Eared and White-tipped doves, Brown-throated Parakeet, Blue-tailed Emerald, Caribbean Elaenia, Northern Scrub-Flycatcher, Gray Kingbird, Tropical Mockingbird, Pearly-eyed Thrasher, Black-whiskered Vireo, Yellow "Golden" Warbler, Northern Waterthrush, Hooded Warbler, Bananaquit, Black-faced Grassquit, Troupial, and Yellow Oriole. Watch for these not only around the water's edge but also in the patch of native greenery off to the left as you approach the hole, fenced off to keep out the goats. One year we watched a Green Heron forage by hopping into the air for insects. Another year, up by the dribbling stream, we watched as an apparently flightless juvenile Crested Caracara screeched and begged, and a parent obliged with food.

Where the Long Route and Short Route Rejoin

When you're convinced you've seen everything that's hanging out at Put Bronswinkel (or more likely, feel pressed for time), return back to the junction of the long and short route, and continue straight ahead, staying on this road through its various twists and turns.

Boka Slagbaai

After a few minutes—and before reaching Boka Slagbaai—you'll come to a large salina-bay called Wayaka. Although usually not extremely fruitful for birds, we did see three Osprey together during one of our visits. About 2 mi (3 km) from the intersection of the long route and short route, you'll round a corner, drive down a small hill, and lose your breath! You have arrived at Boka Slagbaai, where you might see hundreds of flamingos in all hues of pink feeding in the cove, some right along the shore, less than a dozen meters away. Once you've recovered from the amazing show of color, take some time to watch the fascinating feeding behaviors of the flamingo, each bird with a slightly different style of sweeping the bill back and forth through the water. Unlike the flamingos at the Pekelmeer, the Slagbaai birds are generally tame. You can get so close that you can hear their low grunts as they feed. Nonetheless, please keep in mind that approaching too close will disturb the feeding habits of these birds, so please stand at a sensible distance from them. Make sure to scan the cove thoroughly, because there will certainly be a rich array of wading birds among the American Flamingos. Look for shorebirds on the old coral piled along the shoreline.

On the side of the road opposite the cove, you can't help noticing the turquoise sea lapping under the overhanging cliffs and spilling onto the coral beach (you'll probably see snorkels bobbing along the glistening surface, too—the underwater life here is spectacular).

You will also notice a brightly colored set of historical buildings called Landhuis Slagbaai, between the road and the sea. During our first visit to Bonaire in 1998, a restaurant within the buildings provided us with cold drinks and a tasty lunch of satè and rice. Nothing fancy, just a place to get a little refreshment and a break from the sun. The experience is one of our most cherished memories, combining as it did a view of the flamingos, simple delicious food, and the friendly local woman who ran the place and spoke nary a word of English. Over the years it seems that the management of the establishment has changed periodically, so we can't vouch for food or service—or even if the place will be open—but hopefully it will be as memorable and enjoyable when you visit as it was for us.

The last 6.2 mi (10 km) from Slagbaai back to the park headquarters is not known for any notable birding attractions, although there is a beautiful turn-out that offers a view of the far end of Boka Slagbaai, about 1.25 mi (2 km) from Landhuis Slagbaai. By the time visitors reach this point of the journey, often, the afternoon heat is at its most intense, so the tendency is to drive through with the air conditioner on. But there are spots to see birds here as well, so keep your eyes open and stop and check any spots that seem promising.

Directions

Driving from Kralendijk to Washington-Slagbaai National Park is a short trip of about 30 minutes. In the first leg of the journey, you need to get to the town of Rincon, which is 9–10 mi (15–16 km) from Kralendijk. You have two options. You can go inland, following a road whose name officially seems to change every few kilometers. Whatever the name, just keep following it to Rincon. The other route goes along the coast following the Queens Highway before turning inland on Karminda Karpata, which will take you to Rincon. In Rincon, signs for the park will lead you onto Kaya G.R.E. Herrera, from which it is about 2.5 mi (4 km) to the park headquarters.

Dos Pos and Gotomeer

Dos Pos means "two wells" in Papiamento, although the name officially refers to a somewhat larger area that includes what was originally a small farm (the site is now managed by the parrot conservation organization ECHO). In this book, we use the name more narrowly to refer to the roadside spot where one of the original wells is located (accessible by the public). The area has the look and feel of a funky experimental-art exhibit. A bright yellow concrete structure, painted with pastel blue trim, covers the well itself. A tall, silvery windmill looms above it as though standing guard, pumping the water from the ground to a nearby tank whose enclosure continues with the yellow-and-blue motif. Off to the side, there's an open structure, like a bird bath, and near that, a series of spigots. Traditionally, Dos Pos was an important source of fresh water for the residents of Rincon. With the advent of modern technology, however, Dos Pos is no longer the primary watering hole for humans, although many people still arrive with their empty jugs to collect this precious resource.

Near Dos Pos is the gorgeous inland bay of Gotomeer, where American Flamingos once nested and still today feed. The bay also hosts shorebirds like Snowy Plover and Wilson's Plover and sometimes terns, including Sandwich "Cayenne" Terns.

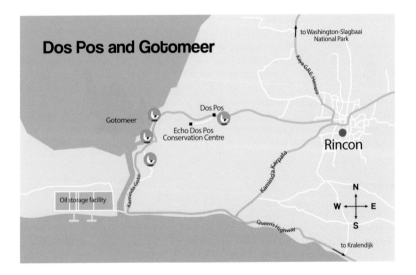

Dos Pos means "two wells" in Papiamento, although in this case the name describes the general area. The authors have seen Pearly-eyed Thrasher and Caribbean Elaenia at point-blank range here.

Birding Dos Pos and Gotomeer

Like most places on the ABC islands, where there is fresh water, there are birds, and at Dos Pos there are lots of them. When water drips or seeps out from the well, as it often does, you will almost always encounter birds coming in for a much-needed drink. We have seen Pearly-eyed Thrashers and Caribbean Elaenia at point-blank range here. Migrant and wintering songbirds are attracted to the water too: Hooded Warbler, Prothonotary Warbler, Palm Warbler, Blackburnian Warbler, Magnolia Warbler, Black-throated Blue Warbler, Scarlet Tanager, Summer Tanager, and other species have been recorded at the site. After you have spent some time to see what comes in, try pishing to see if you can induce a new species to show itself. Given that Yellow-shouldered Parrots nest, feed, and roost in the general area, keep your eyes open for them. This was a favorite spot of long-time local bird guide Jerry Ligon, who passed away in 2015.

The ECHO parrot conservation headquarters, located near the well, offers tours of hiking trails that they have developed; their staff are experts at knowing where to see Yellow-shouldered Parrots in the wild. See the ECHO website (http://www.echobonaire.org/) for contact information and other details.

The inland bay of Gotomeer (see directions below) offers amazing views. Often, you can see American Flamingos and various herons and egrets close-up, as well as Snowy Plover and Wilson's Plover along the shore. In some years, Sandwich "Cayenne" Terns have nested here.

Directions

In Kralendijk, start your trip from the Transworld Radio building at the roundabout (where Kaya Debrot and Kaya Amsterdam intersect) and head northwest on Kaya Gobernandor Nicolaas Debrot. After about 1.3 mi (2 km), the road takes a radical 90° right turn away from the shore and then in a mile and a half (2.7 km) a 90° left turn, becoming Queen's Highway. Continue on this highway, which skirts the shore, for about 4.8 mi (7.8 km) until you reach the mouth of Gotomeer; along the way you might want to stop at a wonderful overlook called 1000 Steps.

When you reach the mouth of the Gotomeer, you'll see in front of you a set of large oil storage tanks, where you should turn right onto Kaminda Goto. This road goes along the edge of the large inland bay and offers great birding along its length, for American Flamingo, Snowy Plover, Wilson's Plover, and sometimes Sandwich "Cayenne" Tern. There are a number of viewing areas and turn-outs along the way. After you have left sight of the Gotomeer, about 2.5 mi (4 km) from where you turned onto Kaminda Goto, watch for the windmill and the bright blue-and-yellow Dos Pos installation on your left. Park here and find a place to sit quietly and watch for birds. If time permits, you can continue on another few minutes to the town of Rincon and on to Washington-Slagbaai National Park.

Lac Bay Area

Lac Bay is situated on the southeastern edge of the island, just north of the Pekelmeer. Its calm, shallow waters are startlingly green, with beaches enjoyed by swimmers, windsurfers, and parasailors. The mangroves provide excellent birding from several access points. These provide habitat for Green and Tricolored herons and Reddish Egrets (known to have nested in the past) as well as Snowy and Great egrets and Great Blue Heron. Keep a watchful eye for the possibility of something rarer like Scarlet Ibis (recorded here in 2011), Little Egret (recorded on nearby Aruba), and other species. Magnificent Frigatebirds historically had a large night roost at Lac Bay, though it's unclear whether this roost continues. Regardless, during the day you have a good chance of seeing them cruising by as they search for food. The mangroves are also an excellent place to look for wintering migrants (Prothonotary Warblers, Northern Waterthrush) and breeding resident birds (Caribbean Elaenia, Gray Kingbird, Yellow Oriole, among others). From April through July, sandy open areas and salinas around the outskirts of the bay can host nesting Least Terns, Snowy Plover, and Wilson's Plover, and in winter various shorebird species may be found.

Birding Lac Bay Area

For the best birding coverage of Lac Bay, you'll want to stop at a number of strategic locales. Start in the beach area of Sorobon, on the southern side of the bay. Depending on whether or not the beach is being used for watersports (which often depends on the time of day you make your visit), there might shorebirds along the

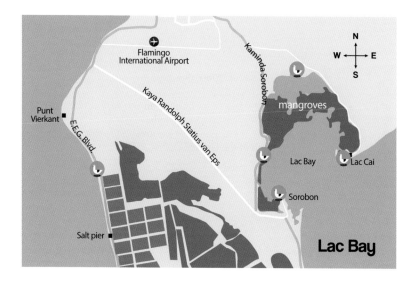

beach. We've seen common winterers like Semipalmated Plovers and Ruddy Turnstones on or near the docks of the marina, as well as uncommon species like Willet along the shore. Explore the mangroves for Bare-eyed Pigeons, Black-faced Grassquits, Bananaquits, Tropical Mockingbirds, and wintering migrants such as Northern Waterthrush. Reddish and Great egrets may be feeding in the waters that reach the area in front of the mangroves. Watch for Laughing Gulls and Royal Terns out over the bay.

From Sorobon, go back to the main road and turn right. In a few tenths of a kilometer, pull off the road to the right in front of a salt pan; this will be wet or dry, all depending on the season and the how rainy it's been. Behind this you'll see a dense stand of mangroves that abuts Lac Bay. Poke around in them for Tricolored Heron, Greater Yellowlegs, Black-necked Stilt, and landbirds such as Brown-throated Parakeet, Blue-tailed Emerald, and Gray Kingbird.

If time is pressing, skip the salt pan, and continue until you see a sign for Kaminda Sorobon, where you'll turn right. After about 1.25 mi (2.0 km), turn into a clearing next to which mangroves border Lac Bay. As you park, you'll likely see among the mangroves a rowboat here and there, half in, half out of the water. Scour this area, as it has proven very good for island breeders like Brown-throated Parakeet, Bare-eyed Pigeon, Common Ground-Dove, Blue-tailed Emerald, Northern Scrub-Flycatcher, Gray Kingbird, Brown-crested Flycatcher, Caribbean Elaenia, Tropical Mockingbird, Yellow Oriole, and Bananaquit (in abundant numbers). It's also good for migrants such as Northern Waterthrush and the occasional Prothonotary Warbler; both species have a preference for mangrove habitat in their wintering grounds, though the two rarely overlap in their breeding range. Farther along to the north of the parking area there are several openings in the mangroves that sometimes harbor a few shorebirds—we've seen Least Sandpiper, Short-billed Dowitcher, and Ruddy Turnstone here.

Lac Bay offers many places to bird. Scan for shorebirds, herons, and egrets along the water's edge; explore the mangroves for Black-faced Grassquits, Bananaquits, and wintering migrants such as Northern Waterthrush. Watch for terns and gulls out over the bay.

Continuing along this road you'll either see more shallow pools edged by mangroves or dry salt flats, depending on the season and other factors. In this area, sometimes there are flocks of American Flamingos feeding, as well as herons and egrets.

About 3 mi (4.8 km) from where you turned onto Kaminda Sorobon, you'll come to a right-hand turn with a sign for Lac Cai. Follow this dirt road (which can be rutted and bumpy) through low thorn-scrub habitat until it opens up, at approximately 2 mi (3.0 km), into yet more areas with mangroves and salt flats. Again, there can be American Flamingos, Tricolored Herons, and, in the twilight, Yellow-crowned Night-Herons feeding here. In summer (April–July) you will see and hear Least Terns, some feeding and a few nesting in the middle of the broad, barren expanse. A little farther along, the road passes closer to the mangroves, often with Wilson's Plovers and, occasionally, Snowy Plovers—watch for other shorebirds—watch for movement against the blazing sand. In about another 0.6 mi (1 km), on the left (north side), there are a series of old dammed salt pans as you near the rough coastline. In the winter there can be a diversity of shorebirds feeding and roosting. During the summer season, along with a surprising diversity of summering shorebird species, there are breeding Snowy and Wilson's plovers, Least Terns, and roosting Royal and Common terns. Even right along the road, the Snowy and Wilson's plovers sometimes have young and may try to lure your vehicle away from their chicks with a distraction display, which seems either incredibly brave or foolish considering the difference in size between a plover and a vehicle!

Your final birding stop along Lac Bay is Lac Cai, situated at the end of a peninsula, across the bay from Sorobon and about 0.3 mile (0.5 km) beyond the last of the salt pans. As you approach, you'll see small mountains of conch shells near the water, large enough that they've made it onto several island maps. Lac Cai, on the ocean side beach, is wild and windswept; the water is a swirl of current and chop—swimming is discouraged. Scan here for Magnificent Frigatebirds, Brown Pelicans, Royal Terns, and other seabirds.

Directions

In Kralendijk, begin at the Transworld Radio building at the roundabout (where Kaya Debrot and Kaya Amsterdam intersect) and head south on Kaya Gobernandor Nicolaas Debrot (which later becomes Kaya Grandi); after 1.5 mi (2.5 km), you'll take a sharp left just at the point where Kaya Grandi becomes a one-way street (if you don't take the left, you'll be driving the wrong way on a one-way street). This left-hand turn will take you onto Kaya L.D. Gerharts, although it may not be marked as such. You'll come to Plaza Reina Juliana, where you'll see a Catholic church the color of coral; turn right onto Kaya Nikiboko-Zuid (which about halfway through your journey changes to Kaminda Sorobon) and continue for 6 mi (9.7 km), until you come (half-way through, the road name changes) to a T-intersection, where you'll see a large salina. Go left onto Kaya Randolph Statuis Van Eps; after just 0.6 mile (0.9 km) you'll come to another intersection, where you take a left-hand turn onto a road that will lead you to Sorobon; en route, on your right you'll pass the Sorobon Beach Resort, a clothing optional beach that presents another chance to use your binoculars, should you be distracted from birding. The road ends at the tip of a small peninsula with a sandy beach on the north side and a marina on the inland portion. This is Sorobon, your first birding stop. From here, follow the directions given in the previous section.

Boka Onima

This is one of many places on Bonaire known for their ancient Indian inscriptions painted on the cliffs. It is near what used to be a reservoir, a somewhat dilapidated structure that happens to be a good place to bird. Boka Onima itself is situated several hundred yards from the coastline, in a dry streambed (arroyo) that runs from the old reservoir ("upstream") and ends in a narrow inlet in the rough wild surf of the northeastern side of the island. The reservoir itself is surrounded by dense buttonbush; beyond the circumference, vegetation is sparse, the ground parched, all lorded over by high cliffs that bear the scars of the violent surf that has long since receded.

In its heyday, the reservoir hosted breeding Pied-billed and Least grebes, Caribbean Coot, Common Gallinule, White-cheeked Pintail, and other waterfowl. In recent years, the concrete dam has fallen into disrepair, with a deep break preventing water from reaching levels of more than a couple of feet. But this is enough to attract more than a few species, especially during migration.

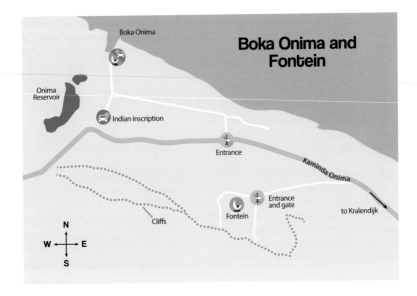

The beach at the end of the inlet is a reliable place to look for American Oystercatchers and Ruddy Turnstones. In the breeding season, a Least Tern colony settles in along the coast between the Indian inscriptions and the inlet.

Boka Onima makes a nice trip when combined with a visit to the fascinating Indian inscriptions, and the drama of the landscape alone, like the backdrop to an eerie science fiction movie, is worth the trip.

Birding Boka Onima

At the inlet, pull off to the side of the road at a place that affords a clear view of the beach. Stay in your car (oystercatchers are easily startled) and look for oystercatchers and other shorebirds that might be feeding along the surf's edge. Also look for them up on the rocky plateau on the other side of the inlet. If oystercatchers are present, make a point to move the car away from the inlet before you get out so that you won't scare them—they may have nests or young nearby.

From the inlet, walk up the arroyo until you come to the reservoir; you'll see that the dam wall bears a crack. Because of the dense vegetation skirting the circumference of the reservoir, good views into it are impossible without navigating through the tangle near the crack in the dam and hoisting yourself up onto the dam. If you suspect that there's enough water in the reservoir to attract waterfowl, move slowly so as not to startle the birds into the vegetation and climb through or over the crack in the dam. If the reservoir is empty, hop down into it (as long as you're sure you can climb back out) and systematically pish among the buttonbush for Northern Scrub and Brown-crested flycatchers, Bananaquits, Yellow "Golden" Warblers, Northern Waterthrush, and other migrants.

To view Least Terns, park near the sharp 45° bend in the road (see below) and approach slowly on foot. Make your stop brief, and be sure to stay far enough away to not flush adults off nests, exposing the eggs to the deadly sun.

Directions

From the Transworld Radio building at the roundabout in Kralendijk, drive east along Kaya Amsterdam. After a little less than a mile (1.5 km), you'll come to a T-intersection, where you turn right and then—almost immediately—turn left onto Kaminda Gurubu (the name of this road changes several times, eventually becoming Kaminda Onima). After 5.8 mi (9.3 km) on the road of many names, just a hundred yards (meters) past a sign for Fontein on your left, turn right onto a dirt road that will lead you to Boka Onima. In 0.1 mi (0.2 km) you'll come to a 45° turn in the road, bending left. If you stop here, it's a short walk from the road to the Least Tern colony, down near the ocean; above the colony, terns will likely be zipping around. Staying left from here, it's another 0.8 mi (1.3 km) to a T-intersection. Turn right to get to the inlet and the old reservoir, but since you're practically already at the Indian inscriptions, just to your left, why not take a left and scoot over a few hundred meters to the parking area for a look? (The inscriptions, in red, are on the cliff faces, on the right side of the parking area.) From the inlet, the old reservoir is a few hundred meter's hike up the arroyo.

At Boka Onima, you can enjoy watching birds such as American Oystercatcher and Least Terns while also taking in some of the island's indigenous history, as represented by these ancient Indian inscriptions.

Fontein

Once a fruit plantation, today Fontein functions, very informally, as a privately owned park. Marked by a hand-painted sign on an old board nailed to a small dinghy, the site is known for a natural spring that runs from a narrow, shallow cave in the cliffs that overlook what is left of the plantation. For birders, the main draw is the Yellow-shouldered Parrots that can sometimes be seen here. The old fruit trees still bear fruit, so many other species also arrive to feed. In fact, according to Voous (1983), Fontein was the first—and for many years the only—place on Bonaire where the Pearly-eyed Thrasher was found. The antiquated irrigation system that starts in the cave leads to shaded pool of water that is enticing to birds, particularly during dry years. Paths lead through recently planted garden plots. A friendly array of chickens, ducks, and geese will emphatically greet you upon your arrival.

Fontein is situated at the end of a dirt road, behind a locked gate that interrupts an otherwise impenetrable barrier of cacti. A second, locked gate guards walk-in access to the running water. There are no set visitation hours. At the time of this writing, the owner was living on Aruba, spending only a few days each week in a rented house on Bonaire. His phone number is posted at the gate. We called him, drove to his house in Kralendijk, picked up the keys (he graciously lent us his flashlight so we could peak into the cave), and drove to the park. There is a nominal fee, which he trusts you to leave in a marked box near the entrance. He is in the process of building a small residence on the premises and plans to make other improvements as time and resources allow.

A visit to Fontein often yields a "fountain" of bird sightings that might include the striking Scaly-naped Pigeon (shown here) and many other species.

Birding Fontein (See map, p. 72)

The best time to visit is either early morning, when the parrots are leaving their roosts, or early evening, when they are returning for the night. Unless you get lucky and show up when the owner is present, you'll need to make arrangements to pick up the key ahead of time.

Once through the gate, stop in the open area and scan the cliffs. Anticipate the squawk of parrots and their green flash as they fly against the dramatic backdrop of the ridge. Watch for where they land and set up scope for a better view. Be on the lookout for Brown-throated Parakeets, maybe even a Merlin, against the cliffs as well. You'll see Gray Kingbirds, Yellow Orioles, Black-faced Grassquits, and other resident birds in the nearby trees.

Move on to the spring behind the walk-through gate, in the left-hand corner; you'll hear the running water as you approach. As you move among the chickens and geese, listen (over the clucking and honking) for Scaly-naped and Bare-eyed pigeons; Eared, Common Ground, and White-tipped doves; parakeets and parrots, of course; Northern Scrub and Brown-crested flycatchers; Pearly-eyed Thrashers; Black-whiskered Vireos, Groove-billed Anis, and other residents. Large flocks of doves in particular cluster around the water.

As you concentrate on the sounds and movements emanating from the trees, don't be alarmed by fruit falling on or near you—this may be the work of parrots silently feeding in the canopy above!

In migration and winter periods, hope for some migrant thrushes, cuckoos, and other birds that might be familiar to North Americans. Historically, Fontein was among the most reliable locales to find migratory warblers that are generally rare or uncommon on the island.

Directions

From the Transworld Radio building at the rotary in Kralendijk, drive east along Kaya Amsterdam for 0.9 mile (1.5 km) to a T-intersection. Go right. Almost immediately, turn left onto Kaminda Gurubu—the name of this road evolves several times through its length, eventually becoming Kaminda Onima. After 5.8 mi (9.3 km), look for a handmade sign reading Fontein, pointing down a dirt road on the left. Take this road; it winds through thorn-scrub habitat for 0.6 mi (1.0 km) and ends at the gate.

Bonaire Sewage Treatment Plant and Area

Just on the outskirts of Kralendijk is a relatively new sewage treatment plant where treated water is released to form freshwater pools. This attracts a variety of wetland birds, including White-cheeked Pintail, Black-bellied Whistling Duck, Blue-winged Teal, Least Grebe, Pied-billed Grebe, Sora, and Black-necked Stilt, as well as variety of herons and egrets. Other species may also be attracted to the area; in fact, the first record for the ABCs of Lined Seedeater (2013) and Pied Water-Tyrant

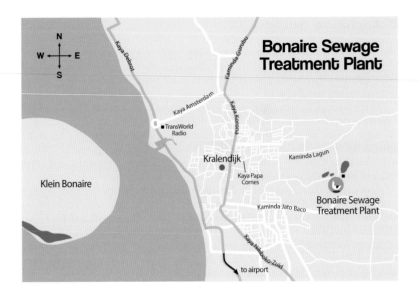

(2015) came from this spot. A rare Yellow-hooded Blackbird was photographed here in October 2015 as well. After birding the freshwater pools, it's worth continuing farther down the road to check the nearby fields for the elusive and rarely seen subspecies of Grasshopper Sparrow.

Birding the Bonaire Sewage Treatment Plant

Upon arrival, it's probably best to scan the pools from your car to ensure that birds close to the road won't fly away before you have a chance to identify them. Park off the road so that trucks and cars can get by without difficulty. It is best to stay on or near the main road unless you've gained explicit permission from personnel at the treatment plant.

Directions

From the Transworld Radio building at the roundabout in Kralendijk, drive east along Kaya Amsterdam for 0.9 (1.5 km) to a T-intersection. Go right. Almost immediately, turn right onto Kaya Karona and follow for 0.9 mile (1.4 km). Take a left onto Kaya Papa Cornes, which will become Kaminda Lagun. Follow this road for 1 mile (1.7 km); turn right at the entrance road to the treatment plant, and make a short drive to the end, where you'll see a complex of buildings, including those of the Bonaire Department of Agriculture, Livestock and Fisheries. Scan from the car at first to avoid flushing nearby birds, then park somewhere out of the way. Call the plant at 717-8836 to check on current access conditions.

Black-bellied Whistling-Duck (*Dendrocygna autumnalis*).

Curaçao

Watamula

Westpunt

Christoffel
National Park
Visitor's Center

Mt. Christoffel

Sint Willebrordus

Hato International
Airport

Salina
Sint Marie

Bullenbaai

Malpais

traffic circle near
Mahuma

Salina
Sint Michiel

Klein Hofje

Muizenberg

Blue Bay Resort
Golf Course

Schottegat

Piscadera
Bay

Sint Joris
Bay

Willemstad

Jan Thiel
Lagun

Spaanse
Water

Santa Barbara

Oostpunt

N
W ← → E
S

- birding site
- parking
- birding site
  and parking
- entrance
- hotel and
  recreational spots

Curaçao

Christoffel National Park

The 4,500 acres (1,800 hectares) that compose this natural area were designated a national park in 1978. Christoffel encompasses the highest peaks of the island, including the 1,230 ft (375 m) Mt. Christoffel (often just referred to as Christoffelberg on the island), and contains some of its thickest and most luxuriant habitat. The park staff have worked hard to control goat populations; as a result the vegetation tends to be diverse, thick, and tall. The many kilometers of trails and roads provide opportunities to visitors to explore without bumping into lots of other people. The varied views within the park are beautiful, composed as they are of many hues of green vegetation, reddish hills, and—visible between the peaks—the sparkling Caribbean waters stretching to the horizon. The vegetation is thicker and lusher in the stream valleys (arroyos), with epiphytes and orchids that are surprisingly large given that the streambed is usually dry or carries only trickles of water.

The Christoffel Mountain Trail, which follows one of these arroyos, provides excellent opportunities to appreciate this striking habitat. From the top of Mt. Christoffel you can sometimes see Bonaire and the Venezuelan mainland. Along this trail and at other places in the park, the black basaltic base geology is readily visible. Altogether, there are eight official hiking trails and more than 15.5 mi (25 km) of road. The main road to Westpunt (Weg Naar Westpunt) bisects the park. The northern section has the park headquarters, gift shop, restrooms, refreshment stand, and plantation house museum; the larger, southern section houses Mt. Christoffelberg and many smaller mountains.

While many of the bird species that occur in the park can also be found at other locations on the island, few of these locations match the beauty of Christoffel. So, if you only have time to go birding at a single spot in Curaçao, Christoffel is exactly what you are looking for.

There are many habitat types in the park, and the more areas you explore, the more likely you are to see not only all the common birds of the island, but also to discover something unusual. Areas in the arroyos that have lush vegetation, for example, harbor breeding Black-whiskered Vireo and Caribbean Elaenia and may be more likely to hold a migrant or wintering warbler. Dry habitat, however, is a place to see Northern Scrub-Flycatcher and Brown-crested Flycatcher. Various flowering trees and shrubs attract both Blue-tailed Emerald and Ruby-topaz Hummingbird as well as large gatherings of Bananaquits. When conditions are generally dry, water holes can attract hundreds of birds; these are also ideal places to see an abundance of young birds after the breeding season.

Unless you have made special arrangements, the park does not open until 7:30 a.m., except on Sunday, when it opens at 6:00 a.m. This means that you'll probably only have time for a walk or visit to one area during the cool morning hours, when birds are more active. If you want to brave the heat, you can bird throughout the day, of course; most people find the heat a bit much to bear and prefer to bird in the morning and late afternoon or evening.

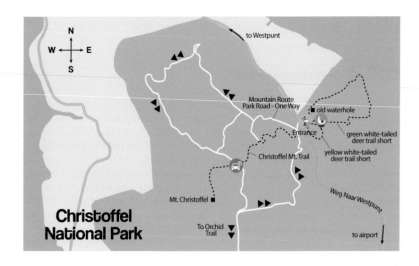

Christoffel
National Park

Birding the North Side of Christoffel National Park

The park headquarters and several of the shorter trails that start from near there are good places to bird. In fact, close to the refreshment stand there are bird feeders that are kept stocked throughout the year, so you will almost always see Troupials, Bare-eyed Pigeons, Rufous-collared Sparrows, and other common birds there. If you walk down the road about 100 meters, to a spot just beyond the museum building, there's a large sign showing the various park trails and roads in the northern section of the park. The yellow "white-tailed deer trail short" and the green "white-tailed deer trail long" are both excellent birding trails, each of which can be covered in an hour or two (if you don't stop, you can walk them in only 30–40 minutes). Both trails offer Bare-eyed Pigeon, Common Ground-Dove, and White-tipped Dove, and as the trail names suggest, possibilities for sightings of the rare Curaçao White-tailed Deer. Families of Brown-throated Parakeets can be heard screeching as they zoom by or observed as they feed or preen in the treetops. Blue-tailed Emeralds should be abundant, and if you watch closely you may encounter the less common Ruby-topaz Hummingbird. During migration periods, liberal pishing will likely bring in migrant warblers, especially species like Blackpoll Warbler, which are common in the fall, and Northern Waterthrush, which regularly winters, but perhaps some uncommon or rare species and possibly some other landbird migrants. You will undoubtedly draw in loads of resident Yellow "Golden" Warblers and Bananaquits with your pishing, and possibly a Black-whiskered Vireo or two.

Another must-see spot for birders is the small Dos Pos picnic area with a water hole and well nearby. This area is also sometimes referred to as Pos di Pia (a water hole that livestock can walk into). There are actually two water holes here (hence Dos Pos). One is now encased by a well and windmill pump, the second is a dilapidated open water hole behind an outdoor auditorium.

Curaçao's Christoffel National Park offers superb birding opportunities and easy access to incredible views of the natural landscape of the island and the sea.

The Dos Pos picnic area may not be well marked clearly on maps and has not been recognized for its birding potential. It has been used mainly as a staging point for field trips by classrooms of local grade school children, virtually all of whom are given at least one guided tour and educational experience during their tenure in Curaçao public school. To get here from the visitor's center, walk the road (closed to motor vehicles) that branches off from the "Plantation North Coast Route" at the first bend, just beyond a large concrete cistern on the left. This section is the combined "Plantation/Boka Grandi Trail" and the "White-tailed Deer/ Boka Grandi Trail." The walk down to the classroom area and water holes is only about 0.5 km. You'll know you're there when you see the hiking trail branch off to the right at a clearing where the open air auditorium is located.

The open water hole (the better spot for birds) will require a little bushwhacking, as it is in the woodland behind the outdoor auditorium and the water hole itself is quite overgrown. Watch for a concrete or stone wall that was apparently once part of the water collection or irrigation system; follow it to the right. This is another area where liberal pishing should yield numbers of migrants. In November 2003, 17 Blackpoll Warblers responded to our pishing, along with a Caribbean Elaenia, a couple of Black-whiskered Vireos, and other common species of the island. One particularly wet year we even found a Least Grebe skulking around the water hole.

Birding the South Side of Christoffel National Park

On the south side of Weg Naar Westpunt road is the largest section of the park, and the more mountainous. Some places are so steep that it might seem that your vehicle is on the verge of tipping end over end. Appearances aside, there is no need to worry, though best to use caution when it rains as the roads might become slippery.

One of our favorite strategies in this section of the park is to stop at the crossings over the streambeds (arroyos), which are usually dry, and to pish and see what's in the area. Some arroyos might even have a trickle of water, making them even more attractive to birds.

One of the best hikes is the Christoffel Mountain Trail, which extends from the park headquarters to the top of Mt. Christoffel. You can either hike the entire length of the trail or drive to a parking lot near the base of the mountain, from which you can access the final third of the trail. There are Rufous-collared Sparrows near the parking area at the base of the mountain that are incredibly tame and always watching for food handouts—they may even hop into your car through an open window to look for crumbs. The last third of the trail follows a streambed up the mountain that passes through surprisingly lush vegetation, including large epiphytes; pay attention to water trickles or damp areas in the streambed, as birds will come in to drink there. Ruby-topaz Hummingbirds, Blue-tailed Emeralds, Caribbean Elaenias, Black-whiskered Vireos, Yellow Orioles, and many other species seem to be in higher abundance in this habitat than in habitats in many other parts of the island. Once the trail starts up the mountain, there are also some outstanding views, ideal spots to watch for soaring resident White-tailed Hawks as well as migrating Ospreys, Peregrine Falcons, and Merlins.

There are many other places within this section of the park where you can stop and bird, though few in which you are likely to see a dramatically different set of bird species. The Orchid Trail is another location where the vegetation is thicker and more luxuriant; it's likely to hold many of the same species as the Christoffel Mountain Trail.

Directions

From the roundabout near the airport (and near the community of Mahuma), travel north on Weg Naar Westpunt Road (sometimes referred to as the West Route). After 17 mi (27 km)—or a drive of 30–40 minutes—you will come to the entrance to Christoffel Park, on your right. The entrance and park headquarters are within the grounds of the former Savonet Plantation, a collection of yellow adobe houses and barns. There is a small check-in station ($12 entrance fee as of 2015), gift shop, museum, and refreshment stand that serves light lunches and cold drinks. After paying your entrance fee, you can walk any of the several kilometers of trails on both sides of the road or drive over 20 mi (32 km) of road that were paved in 2003 through a multi-million dollar construction project. The park is open Monday–Saturday, 7:30 a.m. to 4:00 p.m, Sunday, 6:00 a.m. to 3:00 p.m.

Malpais

For birders, the main draw is a large freshwater pond (Lago Disparse, according to the sign) that forms behind an earthen dam when it rains. At such times, the pond can be filled with grebes, gallinules, coot, waterfowl, and other species.

Away from the water, the land itself—a conservation area recognized by the government—is excellent habitat for terrestrial birds. Below the dam, there is a stand of trees, some of which are quite tall for the island and were likely planted long ago on the site of what used to be the Hofi Plantation. Within the trees, which compose a small forest really, many Brown-throated Parakeets roost every evening. And at certain times of the year, there is also a large roost of doves and pigeons. At first light, the stream of parakeets and doves flying from the trees is an impressive sight. Note that the forest has at least one water hole, named Pos di Pia, though years with heavy rains may create additional water holes.

The hills surrounding Malpais have thorn scrub and cactus, and in the early morning, with the pink-tinged sky, it is a beautiful spot. Although not in view, the island's dumpsite is just to the north of the area and, depending on the direction of the winds, you might smell the refuse.

Birding Malpais

Walking in from the gate (directions below), listen and watch for birds in the surrounding thorn-scrub habitat, including Black-whiskered Vireo (they breed here),

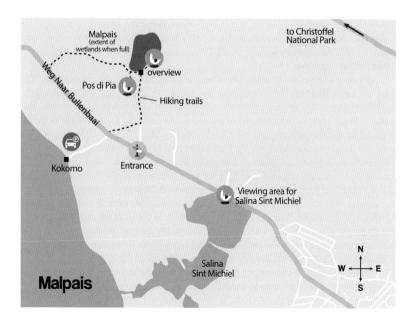

In wet years, the dam at Malpais creates a wetland often filled with Pied-billed and Least grebes, White-cheeked Pintails, and other water birds.

Brown-throated Parakeet, Scaly-naped Pigeon (generally rare but they roost here and sometimes feed at the farm nearby), Bare-eyed Pigeon, White-tipped Dove, Eared Dove, Caribbean Elaenia, Scrub-Flycatcher, Brown-crested Flycatcher, Gray Kingbird, Blue-tailed Emerald, Ruby-topaz Hummingbird, Groove-billed Ani, Rufous-collared Sparrow, Black-faced Grassquit, Troupial, Yellow Oriole, and other common species.

As you get closer to the dam, you'll note a metal structure on your left and the sign for Lago Disparse. Walk up onto the slight rise; this is the top of the dam, stretching about 100 meters across the southern end of the depression, where the water collects. Start by scanning from the top of the dam toward the north (a scope is useful here, especially if there are waterbirds present). If there is indeed water behind the dam, you should check for Pied-billed and Least grebes, Blue-winged Teal, White-cheeked Pintail, Common Gallinule, Caribbean Coot, Black-necked Stilt, and herons and egrets. Check carefully for rare waterfowl and other species as the site is a potential magnet for uncommon species when it contains water. Even when the area is dry, scan for White-tailed Hawk, Crested Caracara, Peregrine Falcon, Merlin, and American Kestrel. During migration, scan the skies for swallows—we've noted flocks of Barn Swallow and Caribbean Martin passing overhead, and several historical records of vagrant swallow species have come from this site. If you are scanning from the dam at dawn or dusk, look carefully at the departing or arriving pigeons for the scarce Scaly-naped Pigeon. Don't be fooled by any Rock Pigeons that might fly by here on their way to the pig farm just a little farther down the road.

To get to the Pos di Pia Waterhole, go to the western end of the dam and look for a sign that marks the trail. The trail is not well-maintained and may be a bit hard to follow, but by picking through the brush and generally heading downslope you'll find the Pos di Pia, which in rainy years will be bordered by other water holes. By staying on this trail and heading in an easterly direction, you will circle back to the dirt road you came in on and will also get to the Hofi Malpais picnic area.

Throughout your walks, try pishing regularly as this is virtually the only way to see northern migrant warblers and vireos (and sometimes Yellow-billed Cuckoos). In the fall you'll likely find at least a few Blackpoll Warblers (sometimes many), but from fall through spring you might come up empty-handed. We've had Philadelphia Vireo, Chestnut-sided Warbler, Northern Parula, and Common Yellowthroat—species either considered rare on Curaçao or without prior records. Pishing will also bring in for closer inspection most of the common species.

Directions

From the roundabout near the airport (and also near Mahuma), travel south on the main road (Weg Naar Westpunt). Continue for about 1.4 mi (2.3 km) until you get to the second traffic light; take a right onto Weg Naar Bullenbaai Road. Follow it for 2.8 mi (4.5 km) until you come to a sign for Hofi Malpais, where you'll see a dirt road on the right blocked by large stones. Park your car on the side of the road and continue by foot the half kilometer or so to the dam on your left. On the way, you'll pass a turn-off and sign for Hofi Malpais on the left, where there is a small picnic area under the trees, and signs for several other trails that you might want to explore. You can also park across the street at Kokomo and walk in from there.

Klein Hofje

Klein Hofje is a sewage treatment plant, which means it smells bad. But the good news is that sewage treatment plants are often ideal places to look for birds, and Klein Hofje is no exception. It consists of a series of small, stinking ponds interconnected by low, flat concrete borders. The area is enclosed by a chain link fence through which you can easily view birds. Inside the plant facility, near the main building, there are a few relatively large palm trees and other non-native tree species that are sometimes used as roosting sites for pigeons and doves, and may harbor other species as well. The surrounding thorn-scrub habitat is sometimes busy with both local breeders and migrants. You'll probably see a few treatment plant employees, who might look at you strangely—what kind of wierdo comes to Curaçao to visit a sewage treatment plant? Just the same, they are not likely to deny you permission to poke around for birds.

Birding Klein Hofje

Pull off to the side of the road at an angle that affords a view of the ponds. The fence is virtually next to the road, and the ponds are just on the other side of the fence. First observe from your vehicle, focusing on the near side of the ponds, then

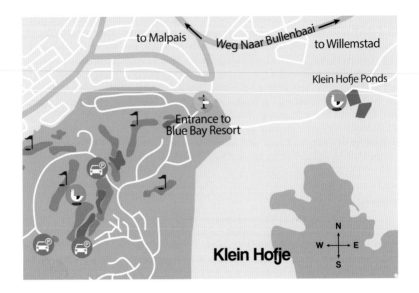

Klein Hofje

hop out and scan the ponds for Least Grebe, Neotropic Cormorant, Black-bellied Whistling-Duck, White-cheeked Pintail (we've seen them here with downy young), Blue-winged Teal (in winter), and possibly other ducks. Scour the concrete borders for Green and Tricolored herons, Caribbean Coot, Spotted Sandpiper, Black-necked Stilt, and other waders and shorebirds.

Search among the nearby trees for residents such as Bare-eyed Pigeon, Common Ground-Dove, White-tipped Dove, and Tropical Mockingbird. The scrubby areas will likely yield Rufous-collared Sparrow, Black-faced Grassquit, Yellow "Golden" Warbler, Blackpoll Warbler (especially in the fall), and Saffron Finches. You'll probably see and hear Yellow Orioles and Troupials, too. And expect the unexpected— we had a nearly full-breeding-plumaged Chestnut-sided Warbler here one year, the first for the island. Check the power lines for Gray and Tropical kingbirds—we've had Tropical Kingbirds here several times and have rarely found them anywhere else on the island. Get permission from a plant employee, if possible, to walk inside and peruse the trees in case they are harboring birds.

Directions

From the roundabout near the airport (and near the community of Mahuma), travel south on the main road (Weg Naar Westpunt). Drive for about 2 mi (3.2 km) until you reach another roundabout. Take the first right onto an unmarked road. Follow this road for 0.5 mile (0.8 km) until you see the Klein Hofje sewage treatment plant on your left. Park off the road somewhere along here. (If you continue on this road for another few tenths of a kilometer, you'll come to some other scrubby areas that you should check for migrant warblers.)

Muizenberg

A potential magnificent wetland area, Muizenberg has been poorly maintained of late. Nevertheless, in years of relatively abundant rainfall, it becomes what is probably the largest shallow freshwater wetland on the island of Curaçao. It is sometimes filled with waterbirds, shorebirds, and foraging American Flamingos. Unlike other freshwater wetlands on the island, however, Muizenberg retains tiny pools and moist areas (a small pool usually persists at the dam in the southeastern corner) even in very dry years; these can harbor Wilson's Snipe and yellowlegs. Despite the great birding at Muizenberg, its proximity to the built-up area of Willemstad can make some areas aesthetically unappealing. The easily accessible portions along the roadsides have become an informal (and likely illegal) dumping ground for household trash and discarded appliances. Still, the area has great potential for finding odd and interesting species of birds.

Birding Muizenberg

The birding at Muizenberg is partly dependent on the amount of rainfall that the island has received in the weeks or months preceding your visit. When the area is filled with water, scanning from openings along the roadside that afford unrestricted views will provide great views of Caribbean Coots, Blue-winged Teal, White-cheeked Pintail, various herons and egrets, and possibly American

Bare-eyed Pigeons are just one of the northern South American specialties that can be seen in the ABCs, including at places like Muizenberg and Klein Hofje on Curaçao.

Muizenberg

Muizenberg

farms along this section

soccer field

extent of water when full

Kaya San Juan

Kaya Muizenberg

baseball field

Kaya Belize

Van Krimpenlaan

Kaya Muizenberg

Suffisantweg

dam

N
W ← → E
S

Flamingo. The area is quite extensive, with scrubby vegetation scattered throughout, so birds can remain hidden, and the numbers of birds and species can shift frequently as birds move from place to place and come in and out of view.

The scrubby vegetation along the roadsides can harbor a variety of landbirds, especially migrants and winterers. Your best chances of finding these is to pish for 5–10 minutes, especially while standing under the cover of a large bush. Watch carefully as the birds come in for a closer look; some species are often rather skittish and will fly in for a look and then quickly depart. Along with all of the resident landbirds, we've found Blackpoll Warbler, Northern Waterthrush, American Redstart, and Prothonotary Warbler. Throughout the area, watch for various shorebird species in whatever shallow pools may be present. When the water level is high, there will likely be flocks of Black-necked Stilt and both species of yellowlegs. A variety of other shorebird species are possible, including Pectoral Sandpiper, Least Sandpiper, and who knows what!

When the water level is low, you could encounter shorebirds or ducks in ditches, small grassy pools, or any other scattered bodies of water that remain. To get to these, you might have to walk out a bit into the general area. It can be muddy, so use caution and prepare to get wet. If you get near enough to these little pools and muddy wetlands, you may flush a Wilson's Snipe, Least Sandpiper, or a yellowlegs. Snowy Egrets sometimes feed in these spots but check them carefully as Little Egret is a distinct possibility and has already been recorded on Aruba.

At the southeastern end of the wetland, where the dam is located, there's almost always at least a small pond (and in rainy years, a large one). This will have flocks of ducks, largely White-cheeked Pintail, Black-bellied Whistling Duck, and, in season, Blue-winged Teal. Check through them carefully for the rare (for here) species like Northern Shoveler, Lesser Scaup, American Wigeon, or something else! The pool near the dam also usually hosts Caribbean Coot, various shorebirds along the shallow edges, and feeding herons and egrets. Along with the commonly occurring species and the aforementioned Little Egret, always keep an eye out for species like Western Reef Heron (an African wintering species that has occurred on Barbados) as well as a number of South American herons like Boat-billed Heron and Whistling Heron. Merlins and Peregrine Falcons frequent the area, attracted by the concentrated food supplies. In November 2003 we watched a Peregrine Falcon eating a pigeon on a perch in the middle of the pond while ducks and shorebirds continued feeding almost right below it. The area south of the dam is a wide ditch into which people have thrown much garbage and even old cars. It abuts the backyards of what seem to be relatively nice houses, though you might see scrawny dogs running around in here. Like birding in much of the Muizenberg area, this is not an aesthetically appealing location. As always, use caution when birding along the roadside anywhere in Muizenberg as many of the roads are very busy and at least a few people will beep and yell as they pass you. Park well off the road and stay off it when you are birding. As you would anywhere in urban locations, it is probably wise to stay relatively close to your car at all times; lock it, and do not leave any valuables in view.

Directions

From the roundabout near the community of Mahuma, take the road (Weg Naar Hato) to the airport. Continue past the airport (at which point it becomes Franklin D. Rooseveltweg) for 3 mi (4.9 km), until you come to Cabo Vereweg, where you turn left. Continue on Cabo Vereweg for about 1 mi (1.8 km) until you come to Seru Fortunaweg, where you will turn right. Shortly after the turn, after about 0.4 mile (0.7 km), Seru Fortunaweg makes a sharp right turn, at which point you will turn left, onto Kaya San Juan. Continue for 0.3 mile (0.5 km) until you come to a T-intersection and stop sign. Turn left onto Kaya Muizenberg and follow for 0.5 mile (0.8 km) to another T-intersection (just before coming to the T-intersection, you will see a Little League recreational field and a soccer field on your right).

As you come to a stop at this T-intersection, you will see the Muizenberg Wetlands directly in front of you. Turn left here to follow the road that encircles the wetland, always taking right turns in order to ensure that you remain on the road closest to the wetland area. Park at intervals along the road, wherever you see good habitat and wide areas for safe parking. Just a few hundred meters after the last T-intersection, near the baseball and soccer fields, there is a sharp left, which is a good place to park and look for birds. After the first 0.6 mi (1 km) or so, you'll be driving around the north side of Muizenberg, which is a farming area with no views or public access to the wetlands. Continue (taking right-hand turns to stay on the road closest to the wetlands) for about 1.7 mi (2.7 km) and take a right onto a road that runs atop an asphalt dam. Follow this until you have views of the water on your right. To return to the T-intersection, you can retrace your route.

Salina Sint Michiel

Between two reddish hills dotted with green thorn scrub, lies this wide salina, an expanse of water that connects to the sea. Early in the morning, while it's still cool, birdsong often floats up from the surrounding thorn scrub, and shorebirds actively feed. In the heat of the day, the character of the salina changes dramatically, becoming hot and rather inhospitable; then, shorebirds and ducks are more likely to be huddling in whatever shade they can find. This salina, like many on the islands, has been influenced by humans for centuries. Various low, rocky walls have been built across different parts of it and at the mouth that opens to the sea in order to lower levels so that the water evaporates and produces salt; it is also possible that the walls have been used to capture fish that swim in from the sea. Although Sint Michiel is largely devoid of vegetation, on the eastern side there are thick bushes along the shore, among which shorebirds and ducks often seek shelter.

Sint Michiel sits at the base of a dramatic series of hills. Weg Naar Bullenbaai Road runs right next to the salina, so you can get close-up views of the flamingos, without leaving your car! The cars that zoom by just meters from the dazzling pink hues to these birds go virtually unnoticed by them. Naturally, the flat expanse of this wetland is also is a good place to scope for other waders and shorebirds.

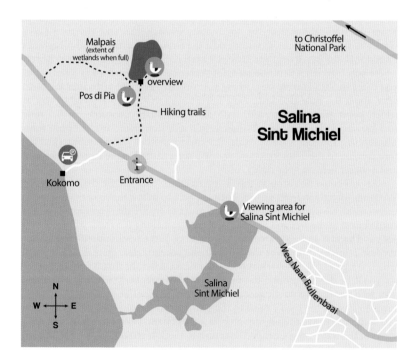

Close-up views of flocks of American Flamingo are one of the highlights of a visit to Salina Sint Michiel on Curaçao.

Birding Salina Sint Michiel

Pull a safe distance off the road, park, and scan methodically from one end of this hotspot to the other. You can get so close to the flamingos that you'll hear their goose-like grunts. Avoid getting too close, however, as they are here to feed. Along with the flamingos, you may find Tricolored Heron, Reddish Egret, Snowy Egret, Great Egret, White-cheeked Pintail, Blue-winged Teal, Black-necked Stilt, Lesser and Greater yellowlegs, Least Sandpiper, and perhaps other shorebird species. The surrounding thorn-scrub habitat harbors resident species like Crested Bobwhite, Brown-throated Parakeet, Northern Scrub-Flycatcher, Gray Kingbird, and other common resident species. This is also a great place to scan for raptors; we've seen White-tailed Hawk, Crested Caracara, American Kestrel, and Peregrine Falcon here.

Directions

From the roundabout near the airport (and near Mahuma), travel south on the main road (Weg Naar Westpunt). Drive for about 1.4 mi (2.3 km) until reaching the second traffic light; take a right at this light onto Weg Naar Bullenbaai Road. Stay on this road for 2 mi (3.2 km), until you see the salina on your left. Park at any one of the several wide places along the roadside where you can move your car completely and safely off the road. Observe the salina from several spots so that you can see from various angles and notice any birds that may be hidden from view.

Watamula

Watamula is located on the northwestern tip of the island. This flat limestone terrace is part of an ancient coral reef; as the terrace nears the coast, cliffs plunge down 30–60 feet (10–20 m) to the surging ocean. The terrace is barren except for a few scattered bushes. Like many of the ancient coral reefs on the island, the ground is characterized by small pockmarks and jagged edges. A strong wind blows in from the ocean, and near the edges of the lower cliffs you will often become covered by the sea spray rising up from the pounding waves. Besides being a great place to scan the ocean for seabirds, the cliffs are a traditional roosting site for numbers of Brown Booby and, occasionally, other seabird species. Less common species to watch for at the roost (or nearby) include Red-footed Booby, Brown Noddy, and Black Noddy. Although the largest numbers would be expected to occur at dawn and dusk, a few Brown Boobies may occur throughout the day, diving for fish within 100 meters off shore.

Birding Watamula

When you reach the somewhat inauspicious dirt parking area, first scan the seas to find out if there are any interesting birds just offshore. The most likely species is Brown Booby, but, in addition to the birds listed above, also look for Bridled Tern, Sooty Tern, and even Pomarine Jaeger, Parasitic Jaeger, and Audubon's Shearwater. Looking for seabirds is often a challenge. Sometimes there's not much to see; the birds you do see can be distant and blurry in the heat; and it can be notoriously difficult to identify them. But who knows what you might see? Before you settle down into your lawn chair for a morning of

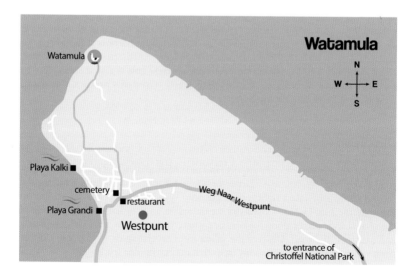

The cliffs of Watamula are great spot to look for Brown Booby. Large numbers some-times roost here for the night and a few can also be seen flying by during the day.

scanning the ocean, walk over to the shore to the southeast of the parking lot and scan the cliff face visible across a small cove. Brown Boobies roost along these cliffs; the immatures can be surprisingly difficult to pick out against the dark, brownish rock. Of course if you see blotches of white guano, look above them, as the location is clearly a popular roosting spot. If you can be here at dawn or dusk, you might find larger numbers of birds, either at their cliff roost or coming and going.

Directions

From the roundabout near the airport (and near the community of Mahuma), head north on the main road (known both as Weg Naar Westpunt Road and the West Route), toward the town of Westpunt. After about 16.8 mi (27 km), you'll come to the entrance to Christoffel National Park. Continue on for another 3.7 mi (6 km); immediately after a restaurant, you'll see a sign for Playa Kalki. Turn right (there's a cemetery at the corner); continue several hundred yards, ignore the turn-off for Playa Kalki, until you come to a cactus fence. Turn left immediately after the cactus fence. Stay on this dirt road for 1.6 mi (2.6 km), remembering to keep right when you come at any intersections. You should emerge on the coast at a parking area with a green-and-white sign indicating Watamula and describing its history. The drive from the airport should take about an hour.

Blue Bay Resort Golf Course

Just about anywhere, bodies of water, even those made by humans, attract birds. But on a naturally dry island like Curaçao, they're bird magnets. The Blue Bay Resort, a private resort complex, is located in a little valley, with condominiums along its upper edges.

The primary attraction for birding is the chain of four human-made freshwater ponds that run through the center of the golf course. The resort is located along the southern side of the island, just a few kilometers west of Willemstad.

Birding Blue Bay

Because this is a private resort, you need to first go the guardhouse at the front gate and ask the security guards for permission to enter. They will likely say yes, but might request that you to stay near the road and avoid the golf course. Please follow their directions, and don't abuse the privilege, to help ensure that they will continue to allow birders on the property.

Although you're likely to see many of the island's common resident birds at any location in the resort that has habitat for them, the birding highlights are generally found in and around the ponds. During migration and winter, the ponds will almost certainly attract dozens, if not hundreds, of Blue-winged Teal. Look through them carefully, not only for White-cheeked Pintail and Black-bellied Whistling Duck, both of which breed on the island, but also for the rare and uncommon waterfowl species for the island. Lesser Scaup and Northern Shoveler were on the ponds in November 2003—the former an uncommon visitor, while the latter was a new island record. For the North American birder, such species may not be terribly exciting, but documenting species new to the island is important for historical records. There's also the possibility of several South American waterfowl species that have occurred in the ABCs in the past, including Comb

Duck and White-faced Whistling Duck. Other waterbirds known to use the ponds include Pied-billed Grebe, Neotropic Cormorant, and Common Gallinule; Least Grebe and Caribbean Coot are also likely. Shorebirds use the shallow sections of the ponds and edges; depending on the season, just about any of the North American breeding species that winter in South America are possible. Resident Black-necked Stilts should be present, sometimes in relatively high numbers depending on the season. Because the ponds are surrounded by the short-clipped grass of the golf course, they provide seemingly attractive habitat for any of the grassland-specialist shorebirds that migrate through on their way to their southern South American wintering grounds. Species like American Golden-Plover and Buff-breasted Sandpiper, which have been infrequently recorded from the ABCs, are worth scoping for here. A single American Golden-Plover was feeding along the edge of the first pond in November 2003. The Southern Lapwing, a grassland shorebird specialist resident in Venezuela, and now established on the islands, should occur here as well.

Although you wouldn't normally associate landbirds with such a place, it's worth checking the birds hopping around the surrounding vegetation and passing overhead. Migrating swallows may descend to look for food over the water or to snatch a passing drink. Migrant or wintering warblers could be in the area as well. With the abundant birdlife in the area there's a great likelihood of spotting one of the bird-hunting specialist falcons—Peregrine Falcon or Merlin—darting by. Finally, keep your ears and eyes open for some of the resident species that occur here like Crested Bobwhite, Brown-throated Parakeet, Bare-eyed Pigeon, Groove-billed Ani, and Troupial.

Directions

From the roundabout near the airport (and near the community of Mahuma), travel south on the main road (Weg Naar Westpunt). Drive for about 1.4 mi (2.3 km) until reaching the second traffic light; take a right at this light onto Weg Naar Bullenbaai Road. Continue 1 mi (1.6 km) on this road (you'll pass the Centrum supermarket) until you come to a sign for Bullebaii; turn left and follow this road for 0.2 mi (0.3 km) until you come to the gatehouse for Blue Bay Resort and Golf Course. Ask the gatehouse staff for permission to look for birds at the ponds. Find locations to view the ponds from the roadside, out of the way of cars, golfers, and golf carts. The second pond is best viewed from the top of the hill to the east of the pond. Do not walk around the ponds, or on the golf course or walkways, without specific permission to do so.

Anatomical Features

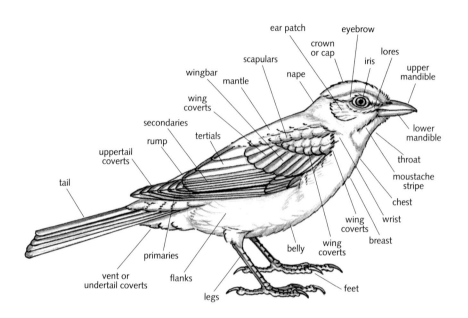

upperwing of duck

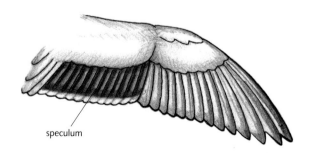

Anatomical Features

upperwing

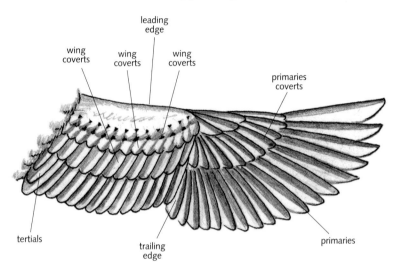

leading edge

wing coverts

wing coverts

wing coverts

primaries coverts

tertials

trailing edge

primaries

underwing

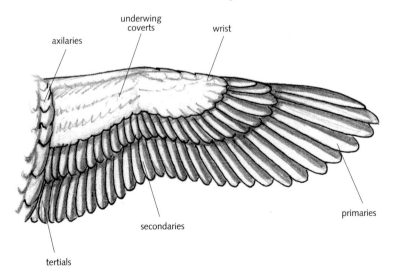

underwing coverts

wrist

axilaries

tertials

secondaries

primaries

Species Accounts

Family Anatidae: Ducks

White-faced Whistling-Duck (*Dendrocygna viduata*)   Plate 1

Papiamento: *Pato pigigi cara blanco, Patu pidjidji kara blanku*
Dutch: *Witwangfluiteend*

Description: Has typical long-legged, erect whistling-duck posture. Adult with white face and patch on throat (sometimes dirty); black on back of head and nape extends around lower neck as thin band; also has rufous neck and upper breast, dark brown back, barred flanks, and black belly and undertail. Black bill and legs. In flight, underwing dark, upperwing dark, with rufous leading edge. Immature birds similar but drabber, with brown instead of black on head and nape. **Similar species:** Adult Black-bellied Whistling-Duck lacks white face and has bright pink (not black) bill and pink (not black) legs. Immature Black-bellied Whistling-Duck shares dark bill and legs but does not show contrast between white face and dark neck and has white wingstripe (wings dark on White-faced). Fulvous Whistling-Duck lacks face contrast; has cinnamon-colored face and underside, white flank streaks, and white rump. **Voice:** High-pitched, three-noted whistle. **Status:** Four records; one at Bubali wetlands, Aruba, Aug 1982; two at Groot Santa Martabaai, Curaçao, May, 1957; 28 reported from Malpais, Curaçao, Apr 1972; reportedly occurred at Klein Hoje sewage treatment ponds several times in 1990s (all records from Prins et al. 2009). **Range:** Tropical Africa and Central and South America. In Americas, from Costa Rica south to Argentina. Also Trinidad.

Black-bellied Whistling-Duck (*Dendrocygna autumnalis*)   Plate 1

Papiamento: *Pato pigigi barica preto, Patu pidjidji barika pretu*
Dutch: *Zwartbuikfluiteend*

Description: Has typical long-legged, erect whistling-duck posture. Adult with bright pink bill and legs, black belly, gray face, rusty back. In flight, shows striking white wingstripe outlined with black trailing edge of wing. Immature with dark bill and legs. **Similar species:** See White-faced Whistling-Duck and Fulvous Whistling-Duck. **Voice:** High-pitched, four-noted whistle. **Status:** Established as breeding resident on Curaçao at Klein Hofje sewage treatment plant since approximately 1990 (B. de Boer pers. comm.), with numbers there reported to have reached as high as about 90 birds (B. de Boer, pers. comm.). Highest number we have counted there was 43, on March 29, 2000. Birds occasionally seen at other locations on Curaçao (55 at Malpais, Jan 4 2016; 26 at Blue Bay Golf Club, Sept 23, 2009 fide eBird) and have bred at the LVV location near Seru Lora. Prior to 2009, rare on Aruba but now perhaps a breeding resident. Early records: two near Arashi, Sept 1977; one at San Nicolas reef, July 1981; five at Bubali wetlands, Aug 2002, one near Palm Beach, Nov 3, 2007

(S. Mlodinow); 17 at Bubali, June–July 2009 (M. Barrett), 12 at Ciribana near Santa Cruz, June 2013 (J. Pippen), several in wetlands across the street from Bubali, from Nov 2013 through at least March 2014 (AJW and others). Maximum from Aruba of 16 on Aug 8, 2015, behind the Mill Resort (R. Wauben fide eBird). Have now bred on Bonaire; specific records include: 13 at Sabana, Jan 1981; one near Lagun, May 2004 (Voous 1983, Prins et al. 2009); pair with 12 downy chicks photographed, Jan 13, 2014, wetland at corner of where roads Hanachi Amboina and Kaya Nikiboko Noord come together (Jerry Ligon, Bonaire Reporter, July 21–Aug 4, 2014); 12 near Lagoen Hill, Jan–Feb 2014 (B. and N. de Kruijff). **Range:** From southern US, south through Central America to Argentina.

Fulvous Whistling-Duck (*Dendrocygna bicolor*)   Plate 1

Papiamento: *Pato pigigi kane, Patu pidjidji kane*
Dutch: *Rosse Fluiteend*

Description: Has typical long-legged, erect whistling-duck posture. Cinnamon-colored underparts and face. Thin, dark stripe down back of neck. White flank streaks; back feathers dark at center, with cinnamon edges, white rump stripe. Dark legs and bill. **Similar species:** Lacks black belly of White-faced and Black-bellied whistling-ducks, white wingstripe of Black-bellied Whistling-Duck. **Voice:** High-pitched, two-noted whistle. **Status:** Two records; four birds (one shot and mounted) Aruba, Feb–May (and perhaps as late as Sept 1965); four birds at Onima and Playa Grandi, Bonaire, from March–May 1975 (Voous 1983). **Range:** From southern US, south through Greater Antilles and Central America, to Argentina.

Greater White-fronted Goose (*Anser albifrons*)   Plate 1

Papiamento: *Ganso cara blanco, Gans, kara blanku*
Dutch: *Kolgans*

Description: Large goose unlike any regularly occurring bird in the islands. Note black belly markings, pink bill, white face, bright orange legs. **Similar species:** Similar Greylag Goose, which is a common domestic farm animal, has been reported occasionally on the islands as an escaped domestic species. It lacks the black belly markings, is larger and bigger bellied. **Status:** A very unlikely species to occur on any of the islands. There is a single record of an adult photographed at Bubali, Aruba, in early June 1980, that was said to have stayed only a few days (Voous 1983). Along with being far out of range, the date would make the record more likely to pertain to an escaped captive, since in June the natural population is in the Arctic breeding range. Voous (1983) reports examining the photo but it was never published and apparently never archived for access by other researchers. **Range:** Breeds across far north of Alaska and Western Canada, Greenland, and Asia. North American birds winter along Pacific Coast of US and in US Gulf States south into Mexico.

Comb Duck (*Sarkidiornis melanotos*) Plate 1

Papiamento: *Pato bolobonchi, Patu bolobonchi*
Dutch: *Knobbeleend*

Description: Large, heavy-bodied duck. Back black and wings with glossy greenish sheen; male has black flanks, female flanks grayish. Male with dark crown and freckled foreneck and face. Blue-gray legs, feet, and bill. Bill of male has large exaggerated crest; female bill lacks crest. **Similar species:** Most likely to be confused with domesticated Muscovy Ducks or other domesticated ducks with a lot of white on underside. These birds usually have white in wings, which is lacking in Comb Duck, and usually also have dark heads. **Status:** Rare. Three records from Aruba: four males, Bubali, Aug 1982; four males, Bubali, Aug 1992; three (one male photographed) near Bubali, Aug 28, 2007 (G. Peterson fide eBird). Three records from Bonaire: eight at Washikemba, Apr 1975; one adult male, Mona Pasashi, Apr 1981; six at pond near entrance of Washington-Slagbaai NP, March 2000. One record from Curaçao, nine at Malpais, March 1971 (Voous 1983, Prins et al. 2009). **Range:** Tropical Asia and Africa; in Americas, from Panama south to Argentina.

American Wigeon (*Anas americana*) Plate 1

Papiamento: *Pato americana, Patu merikano*
Dutch: *Amerikaanse Smient*

Description: Medium-sized dabbling duck with distinctive rounded head and short, blue-gray bill, chestnut breast and flanks, white wing linings, and dark green speculum on trailing edge of wing. Male with white crown, green patch behind eye, white patch on upperwing. Female with plain face, gray patch on upperwing. **Similar species:** Male's white crown, chestnut breast and flanks, and white wing patch are distinctive. Female is superficially similar to other female dabbling ducks but note distinctive rounded head shape, chestnut breast and flanks. **Status:** Uncommon to rare winter visitor in small numbers to freshwater pools on all three islands. Largest number recorded was an exceptional 40 at Salina Mattijs, Bonaire, March 1955. Recent Bonaire records include one, Washikemba, Bonaire, Nov 2001; several at pond near entrance of Washington-Slagbaai NP, Bonaire, Dec 2004 (Prins et al. 2009); four, near Lagoen Hill, Bonaire, Feb 2014 (B. and N. de Kruijff); 26 at Lac Bay, Jan 20, 2014 (C. Sheeter fide eBird); one, Bonaire Sewage Treatment Plant, Nov 6, 2015 (R. Laubach fide eBird); one, Bonaire Sewage Treatment Plant, Feb 16, 2016 (J. Gerbracht fide eBird. On Aruba, records include: one male, Tierra del Sol salina, March 2006 (M. Eising); one male, one female, Tierra del Sol salina, Nov 24, 2011 (AJW); 12 at Tierra del Sol salina, Feb 22, 2012 (J. Wierda); five, salina behind Windmill Resort, Nov 24, 2012 (AJW); five, Tierra del Sol salina, Nov 26, 2012, perhaps the same birds as seen behind Windmill Resort on Nov 24 (AJW); five, Tierra del Sol salina, Nov 27, 2013 (AJW); one, salina behind Windmill Resort, Nov 29, 2013 (AJW); maximum 11, Tierra del Sol salina, Feb 2014 (S. Mlodinow, J. Wierda); 13, Bubali, Dec 5, 2015 (G. Peterson fide eBird). Recent Curaçao records include: one, Muizenberg, Curaçao, Dec 4, 2014 (unknown observer fide eBird); one, Salina

Sint Michiel, Curaçao, Dec 6, 2014 (M. Timpf fide eBird). **Range:** Breeds from northern Canada and Alaska south to Colorado. Winters from US, south through Caribbean and Central America to Colombia and Venezuela.

Mallard (*Anas platyrhynchos*)   Plate 1

Papiamento: *Pato rabo di krul, Patu rabu di krul*
Dutch: *Wilde Eend Ar*

Description: Breeding male very distinctive, with green head, yellow bill, chestnut chest, and white collar contrasting with pale gray back, sides, and belly. Females, immatures, and males in eclipse plumage are mottled brown (like many dabbling ducks), with orange legs and blue speculum bordered by two white bars. Female bill has dark center with orange tip and base. **Similar species:** Breeding male malard unmistakable, not readily confused with any other bird. Female and immature superficially similar to a number of species. Larger than Blue-winged and Green-winged Teal, both of which have all-dark bills and different wing patterns. Female Northern Shoveler smaller, with massive bill and different wing pattern. Female Northern Pintail has all-dark, longer bill, long-necked profile, and dark speculum. Female Gadwall most similar but not recorded from the islands; differs from female Mallard in having white rather than blue speculum and steeper forehead, giving head more domed appearance. **Status:** Very rare. Only four records and apparently none documented with photos or video, all from Bonaire and all females; one, Mona Pasashi, Sept 1983; one at Lac, Apr 1985; one at Mona Pasashi, July 1985: one at Playa Grandi, Dec 1989 (Prins et al. 2009). **Range:** Occurs in Europe, Asia, and North America. In North America, breeds in Canada and US. Winters in US, south to northern Mexico, with occasional records in Caribbean.

Blue-Winged Teal (*Anas discors*)   Plate 2

Papiamento: *Pato moreke, Patu moreke*
Dutch: *Blauwvleugeltaling*

Description: A small duck. Male in breeding plumage distinctive, with white crescent on head. Females, immature, and eclipse plumage males mottled pale brown, with dark bill. In all plumages, note large, pale, blue wing patch and dark green speculum. **Similar species:** The most similar species are all rare on the islands. Female American Wigeon has rounded head, short, more silvery-gray bill, and lacks blue wing patch. Female Green-winged Teal very similar but note smaller, less flattened bill and lack of blue patch in wing. Female Cinnamon Teal virtually identical but with subtle differences in head pattern and slightly longer, more shovel-like bill. **Voice:** Reedy quacks; males in courtship give high, ascending whistle. **Status:** Common winterer and migrant, the most abundant and widespread duck on the islands except in summer. Common to see groups of 5–20 in freshwater ponds; Voous (1983) reports that all three islands have hosted flocks numbering as many as 200–300 birds. Ligon (2005) reports a flock of 65 at Washikemba, Bonaire, on Nov 16, 2001. Highest numbers we have recorded were 250–300 at Tierra del Sol salina, Aruba, on Nov 24, 2011; 120 at Bubali, Aruba, on Jan 15, 2003; 81 at Blue Bay Golf Course ponds, Curaçao, on Nov 22, 2003. **Range:** Breeds from Canada south through US. Winters in southern US, Central America and Caribbean, south to Chile and Argentina.

Cinnamon Teal (*Anas cyanoptera*)   Plate 2

Papiamento: *Pato kolo kane, Patu kolo kane*
Dutch: *Kaneeltaling*

Description: Male in full breeding plumage is unmistakable; note reddish-cinnamon color with no barring or spotting on breast or belly, red eye, dark bill. Female and immature almost identical in appearance to female and immature Blue-winged Teal. **Similar species:** Breeding male unmistakable but beware molting male Blue-winged Teal, which can show very reddish breast and belly but always has some dark spotting or barring. Male in nonbreeding plumage very similar to female Blue-winged Teal but has more reddish tone overall, red eye; nonbreeding male and female Cinnamon Teal show less contrasting facial pattern and slightly larger, more spatulate, bill. **Status:** Very rare. Single record is of a male at Tierra del Sol ponds, Aruba, on Oct 30, 2007 (Prins et al. 2009). **Range:** Breeds in western North America from southern parts of British Columbia, Alberta, and Saskatchewan south through Great Plains to Mexico; winters from California and Texas south, rarely as far as Costa Rica and Panama. South American resident race occurs from Colombia south through Andes Mountains to Chile and Argentina.

Northern Shoveler (*Anas clypeata*)   Plate 2

Papiamento: *Pato boca hancho, Patu boka hanchu*
Dutch: *Slobeend*

Description: Long, spoon-shaped bill very distinctive. Breeding male with dark green head, white breast, rusty flanks and belly, large, light blue wing patch and dark green speculum on trailing edge of wing, and dark bill. Nondescript female and immature with mottled brown head and body; also note white underwing, light blue wing patch, and green speculum. On female and immature, bill with dark center and orange edges. Adult male in nonbreeding plumage is similar to female but has cinnamon flanks, dark head and neck, and, as molt progresses, is increasingly patchy until full breeding plumage is acquired. **Similar species:** With good look, bill is virtually unmistakable in both sexes as is male in breeding plumage. Female, immature, and nonbreeding males can look quite similar to many female and nonbreeding plumaged dabbling ducks but most other expected species will show a black, not orangey, bill. **Status:** Uncommon and irregular winter visitor with approximately 20 records since 1970s, spanning from Sept to Apr (Prins et al. 2009); perhaps becoming more regular, at least on Aruba. Eight published records from Aruba as of 2009 (Prins et al. 2009), from Bubali and Tierra del Sol wetlands, with maximum of 15–20 birds in

March 1988. One additional pre-2009 record of Northern Shoveler from Aruba (un-specified location), Oct and Nov, 2003 (M. Victoria). Recent records include: two female-plumaged birds at Tierra del Sol salina, Nov 26, 2012 (AJW); five at Tierra del Sol salina, Feb 12, 2013 (J. Wierda); one female-plumaged bird at salina behind Windmill Resort, Nov 29, 2013 (AJW); one at Tierra del Sol salina, Feb and March, 2014 (J. Wierda); one near Palm Beach, Aruba, Nov 13, 2014 (A. Silveira); one male, Bubali, Aruba, May 21, 2016 (G. Peterson fide eBird). At least nine records from Bonaire from at least six locations with maximum numbers of five birds at Washikemba, Sept 1985 (Prins et al. 2009), and seven at Lac Bay mangroves, Oct 12, 2010 (J. Ligon fide eBird). Four published records from Curaçao (Prins et al. 2009), one to two birds from each of three locations (de Savaan, Malpais, Klein Hofje), plus sighting of one female-plumaged bird photographed by J. Wells and G. Phillips on Nov 22, 2003, at Blue Bay Golf Course. **Range:** Breeds in Europe, Asia, and North America. Winters in Europe south to Africa and in Western Hemisphere south in small numbers to Caribbean coasts of Colombia and Venezuela.

White-cheeked Pintail (*Anas bahamensis*)   Plate 2

Papiamento: *Pato di ana, Patu di ana*
Dutch: *Bahamapijlstaart*

Description: White cheek and throat show striking contrast to otherwise brown-and-black mottled plumage of both males and females; green speculum on trailing edge of wing with buffy borders; bill bluish-gray with bright red sides. In imma-ture, bill apparently all dark. **Similar species:** Unmistakable. **Status:** Resident breeding species at wetlands on all islands, usually in relatively small numbers, which can fluctuate depending on rainfall. Largest numbers we have recorded at a single location are: 65 adults with 3 downy young on Apr 14, 2000, on the fenced-in sewage treatment plant ponds at Bubali, Aruba; 35 at Bubali wetlands, Aruba, Jan 12, 1998; 25 at Salina Sint Michiel, Curaçao, Nov 22, 2003; 20, including two adults with broods of seven and five young, at Klein Hofje sewage treatment plant ponds, Curaçao, March 29, 2000. J. Ligon (2005) reports a concentration of 25 at Gotomeer, Bonaire, in Feb 2003. **Range:** One population breeds in Caribbean basin south to northern coast of South America. Other spatially separate populations occur in southern South America and along Pacific Coast of South America.

Northern Pintail (*Anas acuta*)   Plate 2

Papiamento: *Pato rabo largo, Patu rabu largu*
Dutch: *Pijlstaart*

Description: A long-necked, slender dabbling duck. Breeding male has white neck stripe contrasting with brown head and a pale gray back; also note black undertail coverts and white on flank and on long, thin, pointed tail. Female, immature, and nonbreeding male show mottled brown with bluish-gray bill, very plain face, green speculum on trailing edge of wing, and dark legs. **Similar species:** Male unmistakable. Female, immature, and nonbreeding male generally similar to females of other dabbling duck species but note long slender neck, bluish-gray bill, green speculum with no blue in forewing, and dark legs. All three plumages larger, longer necked, lighter colored, and more plain-faced than female Green-winged Teal. **Status:** Rare. At least four records on Aruba: one male, Tierra del Sol salina, March 1997; one male, Bubali wetlands, July 2002; one male, Tierra del Sol salina, March 28, 2003 (all Prins et al. 2009); one female-plumaged bird at Tierra del Sol salina, Nov 27, 2013 (AJW). Two records on Bonaire: one male and one female, Playa Grandi, Feb 1980; one near Cai, March 2001 (Prins et al. 2009). No records for Curaçao. **Range:** Breeds across northern Asia, Europe, and North America and winters south to Africa and southeast Asia. In North America, winters from southern US south through Mexico and Central America, regularly to Panama and rarely to northern South America.

Green-winged Teal (*Anas crecca*)   Plate 2

Papiamento: *Pato ala berde, Patu ala berde*
Dutch: *Wintertaling*

Description: A very small, dark dabbling duck. Breeding male has rusty-colored head with large, green patch behind eye, gray back and sides, white vertical bar on side, and green speculum on trailing edge of wing. Female, immature, and nonbreeding male dark, mottled brown, with dark bill and legs, green speculum, buffy streak along side of tail. **Similar species:** Breeding male distinctive. Female very similar to Blue-winged Teal, but latter shows pale blue wing patch in flight, lacks buffy streak near tail, has more white in face, and a larger bill. **Status:** Rare, with only four records from Aruba and two from Curaçao. On Aruba: two males and three females at Tierra del Sol salina, March 2003 (S. Mlodinow); one male at Tierra del Sol salina, March 27, 2004 (S. Mlodinow); two female-plumaged birds at Tierra del Sol salina, Nov 24, 2011 (AJW); one female at Tierra del Sol salina, Feb 2, 2014 (S. Mlodinow). On Curaçao: one male at Muizenbuirg, March 1996 (Prins et al. 2009); one male photographed by Carel de Haseth at the LVV site in Willemstad near Seru Loraweg, from March 21 to Apr 1, 2016 (pers. comm.). All records either of the North American form or assumed to be so. **Range:** North American form, sometimes regarded as a separate species, breeds in Canada and northern US; winters in southern US and, regularly, south to southern Mexico; rarely winters farther south (one record for Colombia).

Ring-necked Duck (*Aythya collaris*)   Plate 3

Papiamento: *Pato boca mancha, Patu boka mancha*
Dutch: *Ringsnaveleend*

Description: Male in breeding plumage has black head, upper breast, back, and rear that contrast with gray sides; white stripe descends from near shoulder to water line in sitting bird. Bill gray with white ring and black tip. In all plumages note the erect posture and peaked head. Nonbreeding male with brownish sides; white mark on each side reduced or lacking. Female and immature tan-brown overall; gray head has darker crown; white eye-ring and white on face, at base of bill; white ring on bill reduced or lacking. **Similar species:** Most likely to be confused with the somewhat more regular Lesser Scaup. Male Lesser Scaup in breeding plumage has white (not gray) sides and gray (not black) back; lacks large, black tip and white ring on bill. Female and immature Lesser Scaup are darker brown overall than female and immature Ring-necked Duck; their crown does not contrast with their face; the white at base of bill is often more pronounced; lack white eye-ring of Ring-necked Duck. Nonbreeding male Lesser Scaup lacks bill pattern of Ring-necked. **Status:** Five records up to 2009 (Prins et al. 2009) but in recent years seen annually on Aruba, especially at Tierra del Sol salina. Records up to 2009 are: one adult male at Washington-Slagbaai NP, Bonaire, Jan 1971; eight adults at Muizenberg, Curaçao, Nov 1999; seven at Muizenberg, Curaçao, Jan 2000; one at Muizenberg, Curaçao, Nov 2000; two females at Tierra del Sol wetlands, Aruba, March 2005 (Prins et al. 2009). Recent records from Aruba: one female-plumaged bird at Tierra del Sol salina, Nov 24, 2011 (AJW); four female-plumaged birds at Tierra del Sol salina, Nov 28, 2011 (AJW); nine at Tierra del Sol salina, Feb 23, 2012 (J. Wierda); two female-plumaged birds at Tierra del Sol salina, Nov 26, 2012 (AJW); two at Tierra del Sol salina, Feb 3, 2013 (J. Wierda); one female-plumaged bird at Tierra del Sol salina, Feb 2, 2014 (S. Mlodinow); two, near Bubali, Dec 12, 2015 (R. Wauben fide eBird). **Range:** Breeds in northern North America; winters from US south to Panama, with several records from Colombia and Venezuela

Lesser Scaup (*Aythya affinis*)   Plate 3

Papiamento: *Pato toper chikito, Patu toper chiki*
Dutch: *Kleine Topper*

Description: Breeding male with black head (sometimes with purplish or greenish gloss); upper breast and rear black; sides white, back gray; bluish bill with small black tip. Male in nonbreeding plumage with brownish sides and overall dingier appearance. Female and immature chocolate-brown, with bold white spot at base of bill. **Similar species:** Male Ring-necked Duck has black (not gray) back and gray (not white) sides; also note white mark coming down from shoulder; bill has white ring against large black tip. Female and immature Ring-necked Duck show white eye-ring, lighter plumage, and head with dark crown contrasting with lighter face. **Status:** Rare but regular winter visitor. Voous (1983) reports sightings between Oct and March in numbers normally ranging from single birds to flocks of six. We have recorded the species four times on Aruba and are aware of at least five other recent records: one female-plumaged bird at Bubali wetlands, Jan 6, 1999 (AJW); one female-plumaged bird at Bubali, March 24–20, 2003 (S. Mlodinow); one at Tierra del Sol, March 2005 (S. Mlodinow); three female-plumaged birds at Tierra del Sol salina, Nov 24, 2011 (AJW); one female-plumaged bird at Tierra del Sol salina, Nov. 26, 2012 (AJW); three at Tierra del Sol salina, Feb 3, 2013 (J. Wierda); one female-plumaged bird at salina behind Windmill Resort, Nov 28, 2013 (AJW); one female-plumage bird photographed at a wetland said to be near Palm Beach, Nov 17, 2014 (A. Silveira); one, Divi Links, March 29, 2016 (R. Wauben fide eBird). On Curaçao we found one female-plumaged bird at Blue Bay Golf Course ponds, Nov 22, 2003; one was seen at Muizenberg, Jan 2006 (J. van der Woude). Ligon (2005) reports one female-plumaged bird on Bonaire at a freshwater pond near the airport, Nov 10, 2004. Largest number recorded was 18 on Bonaire in Jan 1971 (Voous 1983). **Range:** Breeds in Alaska, western Canada, and south to northwestern US; winters from US south to northern Colombia.

Bufflehead (*Bucephala albeola*)   **Plate 3**

Papiamento: *Pato kabez di bagon, Patu kabes di bagon*
Dutch: *Buffelkopeend*

Description: Tiny diving duck. Breeding male has black back and white body; head black, with white wedge on back; short bill; puffy head. Female and immature sooty brown overall with white patch behind eye. Nonbreeding male similar to female but with larger white area on head. **Similar species:** With good views, unlikely to be mistaken for any regularly occurring species. See North American field guides to compare to Common Goldeneye and Hooded Merganser. **Status:** Single record of what was reported to be a male in nonbreeding plumage at Malpais, Curaçao, Nov 30 to Dec 7, 1998 (Prins et al. 2009). This is an exceptional record, as the species is rare south of central Mexico and northern Florida. **Range:** Breeds across North American Boreal forest, from Alaska east to Ontario and western Quebec. Winters from coastal Alaska on Pacific Coast and Maritime Provinces on Atlantic Coast south through much of US to northern Florida, across Gulf Coast to central Mexico.

Masked Dusk (*Nomonyx dominicus*)   **Plate 3**

Papiamento: *Pato mascard, Patu maskara*
Dutch: *Maskerstekelstaart*

Description: A dark, shy, low-riding duck. Breeding male reddish, with brown mottled body, black face, and light blue bill. Female, immature, and nonbreeding male rather nondescript, with mottled brown bodies, two dark lines on face, dark crown, and grayish or blue bill. **Similar species:** Male in breeding plumage should be unmistakable. Females perhaps could be confused with female Blue-winged Teal or other dabblers but note two distinct lines across face. **Status:** Four records for the islands: two from Curaçao, two from Bonaire. One immature at Salina Martinus, Bonaire, Feb 1981; one at Klein Hofje sewage treatment plant ponds, Curaçao, Oct–Nov 1994; male and female at Muizenberg, Curaçao, Apr–May 1997 (Voous 1983, Prins et al. 2009); two on a freshwater pond at Dos Pos wetlands, Bonaire, May 11, 2012 (J. Ligon). **Range:** Permanent resident from Greater Antilles and Mexico south to Brazil.

Family Odontophoridae: New World Quail

Crested Bobwhite (*Colinus cristatus*)   Plate 15

Papiamento: *Patrishi, Cucui, Patrushi, Sloke*
Dutch: *Kuifbobwhite*

Description: Small, short-tailed, short-legged quail with crested head; most often seen walking or running. Brown, black, and white mottled plumage allows the bird to blend in well with its surroundings. Male has prominent black eye stripe and moustachial stripe against white or tan face; female has plainer, unmarked, brownish face. **Similar species:** No other quail species occurs on the islands. Doves walking on the ground might be mistaken for quail at a distance but lack crest and do not have white spotted underparts and mottled plumage. **Voice:** Male's song a loud, two-parted whistle with second note shorter and higher, often rendered as *bob-white* or *coo-kwee* (Voous 1983). Often much easier to hear than see. Birds in flocks give a variety of subdued chirping calls. **Status:** A resident breeding bird of Aruba and Curaçao; does not occur on Bonaire. Voous (1983) expressed concern that the species was in decline but there has been no systematic monitoring to document the status. Since the introduction of Boa Constrictor to Aruba many local birders feel that the species has had a major decline in numbers. Generally seen in small flocks in the countryside. On Aruba, has been recorded in Arikok NP, near Spanish Lagoon, and near Tierra del Sol wetlands. On Curaçao more widespread, including from Malpais, Koraal Tabak, Jan Thiel, and Salina Sint Michiel. Although one set of authors has suggested that the species may have been originally introduced to Aruba and Curaçao (Madge and McGowan 2002), there is no direct evidence for this from any published source (Voous 1983; Prins et al. 2009). **Range:** Guatemala south through Central America to Ecuador and northern Brazil. As many as 14 different subspecies have been recognized, based mainly on differences in head coloration.

Family Podicipedidae: Grebes

Pied-billed Grebe (*Podilymbus podiceps*)   Plate 3

Papiamento: *Zambuyadó pico diki, Sambuyadó pìk diki*
Dutch: *Dikbekfuut*

Description: A small, stocky waterbird familiar to many North American birders. Overall nondescript, brownish but note characteristic, fluffy white "cottonball" rear end. In breeding season, combination of pale bill with black ring and white eye-ring makes it distinctive; throat black (white in nonbreeding season). Often swims with only head and neck above water; may slowly sink below surface if alarmed. Juveniles are "zebra-striped." **Similar species:** Least Grebe shares white rear end but is smaller and darker, with small, pointed bill, black cap, and golden eye. A cursory glance at distant birds could lead to confusion between Pied-billed Grebe and Blue-winged Teal. **Voice:** Though vocalization is not often heard on the islands, most common territorial male call is either a rapid-fire *COO-COO-COO-COO-COO-COO* or a loud, maniacally yodeled *CAoo-CAoo-CAoo-CAoo*, dropping in pitch on second syllable. **Status:** First recorded in ABCs in 1950s, appearing in salinas formed after heavy rains and remaining until salinas dry up. Has become regular breeder year-round in small numbers (typically 3–10) in permanent Bubali wetlands and irregular breeder at Tierra de Sol wetlands, on Aruba; irregular breeder on bodies of water on Bonaire and Curaçao. **Range:** Throughout Western Hemisphere from southern Canada through Central and South America to Argentina, and including Caribbean.

Least Grebe (*Tachybaptus dominicus*) Plate 3

Papiamento: *Zambuyadó chikito, Sambuyadó chikí*
Dutch: *Amerikaanse Dodaars*

Description: Small, drab, brownish, secretive grebe with dark cap and, in breeding season, dark throat and back. Cheeks, neck, and sides pale brown. Also note golden eye and thin bill (thinner and longer than bill of Pied-billed Grebe). Shy, often stays close to vegetation, quietly disappearing under water when disturbed. **Similar species:** See Pied-billed Grebe. **Voice:** Makes a high-pitched, nasal, churring sound, perhaps more reminsecent of a call produced by an amphibian—and unlikely to be confused with other bird species; also makes a nasal, tinhorn, single-note call superficially similar to notes of Caribbean Coot and Common Gallinule. **Status:** Historically irregular, appearing in salinas formed after heavy rains. Now, probably regular, however secretive behavior makes status difficult to document. Of 14 surveys, for example, conducted in early Jan in Bubali wetland, Aruba, from 1993 to 1999, Least Grebes were noted in only 3 years, on 7 occasions (AJW). Breeding, which can occur throughout the year, confirmed on Curaçao at Malpais, on Bonaire at former Onima reservoir, and, more recently, in March 2000 (an exceptionally wet year) at temporary ponds at Onima (JCL) and in June 2000 at Dos Pos (JCL). Maximum numbers: 50 on July 5, 1971, Malpais, Curaçao (Voous 1983); 46 on Sept 19, 1971, Malpais, Curaçao (Voous 1983); 20 on Jan 8 1993, Bubali, Aruba (AJW); 20 (mostly immatures) on Nov 1, 2000, at pond near entrance to Washington-Slagbaai NP, Bonaire (JCL). On Curaçao, regularly occurs at reservoir at Malpais; in wet years, probably occurs irregularly at temporary ponds such as those at Muizenberg. On March 31, 2000 (an exceptionally wet year), two birds observed at small pond at Dos Pos, Christoffel NP, Curaçao (AJW). **Range:** From southern Texas south through Central and South America to northern Argentina, and including Greater Antilles and Bahamas.

Family Phoenicopteridae: Flamingos

American Flamingo (*Phoenicopterus ruber*) Plate 11

Papiamento: *Flamingo Cu, Flamingo, Chogogo*
Dutch: *Rode Flamingo*

Description: Unmistakable. Huge, pink bird with long legs, curved neck, and large, black-tipped bill. Juveniles gray rather than pink; immatures, which take three to five years to reach breeding age, generally duller pink than adults. **Similar species:** Only other tall, pink bird that has occurred on the islands is Roseate Spoonbill, which has a distinctive spoon-shaped bill and naked, unfeathered head. **Voice:** Loud, ringing honking, remarkably like the sound of geese. **Status:** Successful breeding in ABCs confirmed only on Bonaire, where numbers range from an estimated high of 10,000, in the 1980s, to 1,500–3,000 in more recent years. Largest numbers found at the Pekelmeer, where they regularly breed. Birds frequently seen, sometimes in the hundreds, at Lac Bay and Gotomeer, as well as other salinas, especially within Washington-Slagbaai NP. On Curaçao, where birds also sometimes number in the hundreds, they occur at Muizenberg, Salina Sint Michiel, and Jan Kok, where the American Flamingo has attempted, unsuccessfully, to breed (Prins et al. 2009). Formerly quite rare on Aruba, with only three records in 1986 and one in 2004. By 2011 (Luksenburg and Sangster 2013), small groups of 1–11 began to be seen on Aruba, particularly at Tierra del Sol salina and the salinas behind Palm Beach (near Rooi Santo, on the way toward Malmok, close to the sky diving hut), and were reported multiple times between 2011 and 2015. **Range:** Spotty distribution around the relatively few known breeding colonies, from Caribbean and coastal Mexico south to coast of northern South America, and from eastern Colombia to Brazil.

Family Procellariidae: Petrels, Shearwaters

Black-capped Petrel (*Pterodroma hasitata*)   Plate 4

Papiamento: *Parha di tormenta pèchi preto, Para di tormenta pèchi pretu*
Dutch: *Zwartkapstormvogel*

Description: A relatively large pelagic seabird unlikely to be seen close to land. Distinctive black cap separated from lighter brown back by white nape and from black, bulky bill by broad white forehead. The "white-faced" form has white extending around eye while the "dark-faced" form has dark area extending down from cap to enclose the eye (Howell 2012). White rump contrasts with brown back and dark brown tail. White on underside of body. Dark gray on upperwing with faint dark bar, white on underside with distinctive black bar on front (leading) edge and black outline along back (trailing) edge of underwing. Flight pattern and shape distinctive, with wings held closer to front of body and angled back at wrist. In flight under strong winds (as likely in waters off the ABCs), birds arc up steeply from ocean surface, then turn and glide down again. **Similar species:** Most similar is rare Great Shearwater, which is larger with less white on rump and nape, lacks white forehead; some Black-capped Petrel have less white on nape and rump, making them more similar in appearance to Great Shearwater. Underwing of Great lack distinctive black bar along leading edge but do show indistinct brown markings along front and back edges. Great usually has dirty-looking brownish patch on belly, unlike sparkling white belly of Black-capped Petrel. Experienced observers at fairly close range may note that bill of Great is longer and slimmer. At a distance, most obvious difference is flight pattern and shape, with Great's wings appearing straight along front edge, lacking bend at wrist while gliding. While all shearwaters and petrels have characteristic arcing flight pattern, Great's upward arc is usually not as steep and high as Black-capped Petrel, and larger size and proportionately longer wings make flapping appear more languid and smooth. **Voice:** Silent at sea. **Status:** Classified as globally endangered by the International Union for the Conservation of Nature, with only approximately 1,000 breeding pairs of this species left in existence, most breeding in mountains of Haiti and the Dominican Republic. Records from the ABCs from the 1970s suggest it is an uncommon visitor Apr–May (Voous 1983); in 1970, seabird researcher R. van Halewijn recorded it 13 times (including one sighting in Dec, off Aruba) during offshore cruises. The Apr–May period corresponds to the period of peak abundance off North Carolina, when birds apparently disperse from breeding grounds after Nov–May nesting period. While conducting intensive surveys from sportfishing boats to document cetaceans off Aruba, three dark-faced Black-capped Petrels were photographed near killer whales on Apr 14, 2011, about 1.9 mi (3 km) from the northern tip of Aruba (Luksenburg and Sangster 2013). A Black-capped Petrel carrying a satellite tag (affixed to the bird at the breeding grounds in the Dominican Republic, Apr 8, 2014) was tracked to within about 6.2 mi (10 km) of the northern tip of Aruba on June 11, 2014 (www.atlanticseabirds.org). Seabird watching from sportfishing boats off Aruba may provide opportunities to see this and other pelagic seabird species. **Range:** Once bred on at least six Caribbean islands but now known only from Haiti and Dominican Republic, and possibly Dominica. In nonbreeding season (June–Oct), occurs offshore, from Virginia south through Caribbean to northern South America waters, eastern Colombia to northeast Brazil.

Bulwer's Petrel (*Bulweria bulwerii*)   Plate 4

Papiamento: *Parha di tormenta Bulwer, Para di tormenta Bulwer*
Dutch: *Bulwers Stormvogel*

Description: Relatively small, dark brown petrel, with buffy bar on upperwing. Wingspan long; tail long, often appearing pointed. As on other tubenoses, bill black, short, and stubby. Flight characteristic, with bird remaining close to ocean surface (within 2 meters) and veering erratically from side to side, like Leach's Storm-Petrel or nighthawks. **Similar species:** If close enough to see buffy bars on wings, only likely confusion is with Leach's or Wilson's storm-petrels, which are entirely brown and, in worn plumage, sometimes show buffy bar on upperwing. Most individuals of both species show white rump patch—generally quite prominent unless very distant (very rarely, white rump is so pale it is virtually absent, O'Brien et al. 1999). Both species much smaller than Bulwer's Petrel, with shorter wings and tail; tail is squared or slightly-forked (but not pointed). Generally, compare to storm-petrels, noddies, and young sooty and bridled terns. **Voice:** Silent at sea. **Status:** One record within territorial waters of ABCs, from observations at sea, May 13, 1970 (four nautical miles northeast of Klein Curaçao), but several other observations from other regions of the Caribbean (Voous 1983) and from well-surveyed waters off North Carolina, indicating small numbers reach beyond expected wintering range, including offshore waters of northeastern South America and northwestern Africa. The breeding season extends from Apr through Sept, while the record from ABCs was in May and record from North Carolina was in Aug, suggesting these birds are nonbreeders or early dispersing failed breeders. Other records by R. van Halewijn (in Voous 1983) of birds well north of Curaçao and Bonaire. **Range:** Occurs in the tropical and subtropical waters of the eastern Atlantic and from the central Pacific across to the Indian Ocean. In addition to breeding sites in Pacific and Indian oceans, breeds on small islands off northwestern Africa, including islands near Azores, Madeira, Canary, and Cape Verde.

Cory's Shearwater (*Calonectris diomedea*)   Plate 4

Papiamento: *Capiadó Cory, Kapiadó Cory*
Dutch: *Cory's Pijlstormvogel*

Description: Relatively large, long-winged seabird of the open ocean. Pale sandy-brown on back and wings extends onto head and cheeks. Wingtips darker than rest of wing. Narrow white rump patch. White underwing outlined in dark; white belly and undertail white. Bill rather heavy; pale yellow or pinkish-yellow. **Similar species:** Black-capped Petrel is stockier, with slimmer wings, much darker on back, shows obvious white collar and large white rump patch; and has thick, black bill. Great Shearwater has obvious cap, dark belly patch, dark undertail coverts, and a thin, dark bill. **Voice:** Silent at sea. **Status:** Single record, a bird photographed 4.5 mi (8.8 km) off the southwestern coast of Aruba, Feb 18, 2011 (Luksenburg and Sangster 2011). Note that two very similar taxa could conceivably also occur in these waters. These are the Cape Verde Shearwater (*Calonectris edwardsii*) and the so-called Scopoli's Shearwater (Howell 2012), yet to be granted species status; consult Howell (2012) for more information. **Range:** Breeds on islands off northwest Africa, ranges widely at sea in North Atlantic and South Atlantic, but generally thought to be quite rare within the Caribbean basin west of the Lesser Antilles and Trinidad and Tobago.

Great Shearwater (*Ardenna gravis*)   Plate 4

Papiamento: *Capiadó grandi, Kapiadó grandi*
Dutch: *Grote Pijlstormvogel*

Description: Relatively large, long-winged seabird of the open ocean. Brown cap and scaly brown back and wings contrast with white nape, dark bill; white rump patch contrasts with dark brown tail and wings. Underside white but note dark smudge on belly. Underwing white, with dark along trailing edge extending to wingtip, narrower dark outline along leading edge, and with some dark mottling closer to body. Dark undertail coverts. **Similar species:** Most similar species known to have occurred in waters around ABCs is Black-capped Petrel, which has a darker cap and lacks scaly appearance of back and wings; also much whiter on underside, lacking belly smudge, and has white undertail coverts. On underwing, definition between dark and white less stark in Great Shearwater than in Black-capped Petrel; on underwing of Great Shearwater, the narrower dark front edge and dark mottling near body are diagnostic. Cory's Shearwater is slightly larger than Great and lacks capped appearance and belly smudge. Cory's has much reduced white rump and, at close range, a yellowish-pinkish (not black) bill. **Voice:** Silent at sea. **Status:** Single record of dried carcass found on southeast coast of Bonaire, Aug 1976 (Voous 1983). Birds apparently regularly pass northward off Trinidad in June (ffrench 1991) and a few birds may occasionally move into Caribbean. **Range:** Breeds in the southern hemisphere, on islands of South Atlantic including Tristan da Cunha, Gough, and Falkland Islands. Migrates north to spend southern hemisphere winter in the North Atlantic, from about May through Sept. Recent work in Guadaloupe has shown thousands of Great Shearwaters annually move north in June and July in waters along eastern edge of Lesser Antilles (Levesque and Yesou 2005).

Audubon's Shearwater *(Puffinus lherminieri)* Plate 4

Papiamento: *Chokwèkwè, Paloma pia di patu*
Dutch: *Audubons Pijlstormvogel*

Description: Small shearwater with dark, blackish upperparts; white below except for dark undertail coverts, wingtips, and border of underwing. White extends up side of face to behind eye, creating what at a distance might look like a partial neck collar. At very close range, note white in front of eye. Like all shearwaters, flies low to the water on stiffly held wings, regularly hooking up into the air and gliding back down again, interspersed with rapid flaps. Compared to other shearwaters, Audubon's has a proportionately longer tail and shorter wings, such that a birder from northern climes might see a resemblance to an alcid in flight. **Similar species:** Although not yet recorded for the ABCs, the Manx Shearwater is most similar in appearance; it should not be ruled out as it has been found to occur regularly along the eastern edge of the Lesser Antilles (Levesque and Yesou 2005). Manx is larger, with proportionately shorter tail and longer wings; it thus looks less alcidlike and more closely resembles a large shearwater profile in flight. At closer range, note Manx's white undertail coverts, more reduced white on face, and lack of white in front of eye. Flight also different, with slower wingbeats and more and higher gliding than in Audubon's. Apparently some young Great Shearwaters can have surprisingly dark and unmottled upperparts that could fool an observer at a distance if characteristic shape, size, and flight difference are not noticed. **Voice:** Silent at sea but at breeding sites, which are frequented at night, gives eerie bleating calls of four syllables (Howell 2012). **Status:** Regular but uncommon resident in offshore waters surrounding the islands; may breed (though unconfirmed). Reports from all months of the year, but calls of breeding birds heard Sept–Oct 1977 and Feb and May 1960. Note that three subspecies have been described, including one *(loyemilleri)* that is thought to frequent the waters of the ABCs (Howell 2012). Recorded from all times of year in waters between and near Bonaire and Curaçao but not yet definitely from Aruba. Stranded birds and carcasses have been found on both Curaçao and Bonaire. On Bonaire, birds have been heard at night in hills within Washington-Slagbaai NP and limestone escarpment along windward side of island. **Range:** Often treated as a single species that is widespread in tropical oceans worldwide but likely a number of different species with taxonomy still in question (Howells 2012). Known to breed on islands throughout Caribbean, including Venezuelan islands of Los Roques, Los Hermanos, and La Orchila.

Family Hydrobatidae: Storm-Petrels

Wilson's Storm-Petrel (*Oceanites oceanicus*)   Plate 5

Papiamento: *Parha di tormenta Wilson, Para di tormenta Wilson*
Dutch: *Wilsons Stormvogeltje*

Description: A small, dark bird of the open sea. At a distance, can appear swallow-like, though it is decidedly larger than most swallows. Bold, white rump contrasts with dark brown body, wings, and tail, but it may not be obvious at distance. Underwing pale. Flies low over the water, often skimming surface and stopping to dip legs into the water. Flight very butterflylike, with shallow wingbeats. Wings comparatively short and rounded, with little "bend." Tail square (but can appear shallowly forked) and relatively short. Feet project behind tail in flight. At close range, pale patches on upperwing are obvious. **Similar species:** At sea, migrant Purple Martin or Caribbean Martin (especially males) could superficially resemble a storm-petrel, though shape and flight manner are markedly different. Leach's Storm-Petrel shares same plumage pattern (white rump contrasting with brown body, wings, and tail) but has longer, more angled, wings, nighthawklike flight pattern, dark underwing, and longer, more forked, tail. Band-rumped Storm-Petrel, while not yet recorded for the ABC islands, is entirely possible and has characteristics intermediate between Wilson's and Leach's, so take it into account with all storm-petrel sightings. **Voice:** Silent at sea. **Status:** Five definite records from ABCs, at sea, several involving multiple birds (up to nine); despite limited sightings, probably regular during some portion of species' May–Sept nonbreeding season, as evidenced by regular sightings of unidentified storm-petrels by local fishermen (Voous 1983). Generally rare in Caribbean, but uncommon in Bahamas (Raffaele et al. 2003), one record for Trinidad (ffrench 1991). Aruba: Nine birds followed ferry from Curaçao to just outside Oranjestad harbor, June 11, 1970, and three seen following the ferry departing Punta Basora, Aruba, June 24, 1970. Single bird seen halfway between Aruba and Curaçao on Sept 7, 1971. Bonaire: Single bird sighted 6 mi (9.7 km) northwest of Bonaire on July 14, 1970. Curaçao: In addition to above, single bird seen June 17, 1970, near Punta Kanon, Curaçao (all Voous 1983). **Range:** Breeds on sub-Antarctic islands and Antarctic coast Nov–May and migrates north to spend nonbreeding season (May–Sept) in northern parts of Atlantic and Indian oceans and midddle latitudes of Pacific Ocean. Recent work in Guadaloupe notes that thousands of Wilson's Storm-Petrels annually move north in the March to May period in waters along the eastern edge of the Lesser Antilles (Levesque and Yesou 2005).

Leach's Storm-Petrel (*Oceanodroma leucorhoa*)

Papiamento: *Parha di tormenta Leach, Para di tormenta Leach*
Dutch: *Vaal Stormvogeltje*

Description: Like the closely related Wilson's Storm-Petrel, a small, dark bird of the open sea; at a distance, can appear swallowlike, though it is decidedly larger than most swallows. Bold, white rump (not always obvious at a distance) contrasts with dark brown body, wings, and tail. Underwing dark. Flies low over the water, often skimming the surface and stopping to dip legs in the surface. Flight very direct, with erratic sudden turns and relatively deep wingbeats, often likened to flight of nighthawks. Wings, comparatively long and pointed, with characteristic backward "bend." Tail deeply notched and relatively long. Feet do not project behind tail in flight. At close range, pale patches on inner part of upperwing (carpal bar) are obvious. **Similar species:** See Wilson's Storm-Petrel. **Voice:** Generally silent at sea. **Status:** Known to occur regularly, though rarely, in waters of the South Caribbean. Expected to occur Sept–May, when these birds vacate northern Atlantic breeding grounds. Four reports from ABCs, including one specimen. <u>Aruba</u>: No records. <u>Bonaire</u>: Single injured or exhausted bird captured Nov 17, 1979, at Playo Abao (Voous 1983); injured bird captured off Klein Bonaire July 2, 1996, was said to be this species (Ligon 2006) but date makes Wilson's Storm-Petrel more likely—no photos or specimen available for review. <u>Curaçao</u>: Single bird observed Jan 26, 1952, off Klein Curaçao (Voous 1983); one injured bird captured Jan 17, 1967 at Sint Anna Bay, Willemstad Harbor, was preserved as a specimen (Voous 1983). **Range:** Breeds on islands in both North Atlantic and Pacific oceans. North Atlantic population thought to largely winter in waters off west coast of Africa. Uncommon but regular in the eastern Caribbean beginning in Jan, when birds start northward migration from wintering areas off west coast of Africa, and continue to early June. Recent work in Guadaloupe has shown hundreds of Leach's Storm Petrel annually move north, mainly March–May, in waters along the eastern edge of the Lesser Antilles (Levesque and Yesou 2005).

Family Phaethontidae: Tropicbirds

Red-billed Tropicbird (*Phaethon aethereus*)

Plate 5

Papiamento: *Bubi rabo largo shouru, Bubi rabu largu shouru*
Dutch: *Roodsnavelkeerkringvogel*

Description: Superficially ternlike in appearance, roughly same size as large tern; tropicbirds are very white birds with (in adults) elongated tail streamers. Adults show mottled back (appears whitish at a distance) and black outer primaries that extend onto "elbow" (primary coverts). Black line extends from base of upper mandible through eye and across nap. Bill reddish in adult. Juvenile similar but with yellow bill and without tail streamers. Tropicbirds fly with snappy, shallow wingbeats. **Similar species:** At a distance, Royal Tern and Sandwich Tern are easily mistaken for a tropicbird, as they appear very white under the bright tropical sun, but tropicbirds have less bouncy, more direct flight, with snappier, shallower wingbeats, and with wings seemingly set farther back on body, giving heavier look in front compared to terns. Adult White-tailed Tropicbird has bold black marks on inner part of upperwing forming distinctive W pattern and lacks barring on back. Juveniles more difficult to distinguish as both have dark barring on back, but black on wingtips extends onto "elbows" on Red-billed, lacking on White-tailed. In both adults and juveniles, black eye line extends across nape in Red-billed but not in White-tailed. Species differ in size, shape, and flight style; Red-billed, with a larger bill and body, has comparatively slower wingbeats than White-tailed. **Voice:** Generally silent at sea. **Status:** Rare pelagic bird; 13 records, six of them far from land. <u>Aruba:</u> Four records of single birds photographed from sport fishing boats, 0.6–13.7 mi (1–22 km) off the island, June 2010 and Apr and June 2011 (Luksenburg and Sangster 2013). <u>Bonaire:</u> One record 20 nautical mi northwest of Malmok, May 25, 1970; one record 12 nautical mi southwest of southern coast of Bonaire, May 12, 1977 (Voous 1983); one dead adult found near Sorobon, May 14, 1986; one juvenile collected and photographed on southeast coast, July 20, 1991 (both Prins et al. 2009). <u>Curaçao:</u> One landed on a ship near Curaçao on Feb 7, 1939; one caught on board a ship in harbor on Feb 4, 1966; one oiled bird found on north coast, Oct 30, 1968; one seen flying past Watamula, Westpunt, Dec 1978; one at sea near Klein Curaçao, Jan 29, 1992 (Voous 1983; Prins et al. 2009). **Range:** Pacific Ocean, from Baja California south to Peru; in Caribbean, breeds in Lesser Antilles, Tobago, and Venezuela's Los Roques islands (only about 93 mi [150 km] from Bonaire). Also in tropical and subtropical Atlantic off Brazil and northwestern Africa, and in northwest Indian Ocean. Highly pelagic, so sometimes wanders far offshore and far from breeding areas.

White-tailed Tropicbird (*Phaethon lepturus*)   Plate 5

Papiamento: *Bubi rabo largo blanco, Bubi rabu largu blanku*
Dutch: *Witstaartkeerkringvogel*

Description: Superficially ternlike in appearance, roughly the size of a large tern; tropicbirds are very white birds with (in adults) elongated tail streamers. Adult shows a white back, black outer primaries that do not extend onto "elbow" (primary coverts), and black line on inner part of upperwing that extends from "elbow" back toward tail. This gives general impression of a black W on the upperwing. Black line extends from base of upper mandible through eye but not to nape. Bill yellowish. Juvenile has black mottling on back as in Red-billed Tropicbird and reduced version of dark W marking on upperwing; yellow bill; lacks tail streamers. Tropicbirds fly with snappy, shallow wingbeats. **Similar species:** See Red-billed Tropicbird. **Voice:** Generally silent at sea. **Status:** Two records far offshore, both adults. One, May 20, 1970, five nautical miles off Westpunt, Curaçao. Second, June 23, 1970, 12 nautical miles northwest off Cape Malmok, Bonaire (Voous 1983). **Range:** Widespread in tropical and subtropical oceans worldwide (though not in eastern Pacific). In Caribbean, breeds in Greater Antilles, where Red-billed Tropicbird is absent, and northern Lesser Antilles, where it overlaps with Red-billed. Very pelagic, sometimes wandering great distances offshore, far from breeding areas.

Family Ciconiidae: Storks

Wood Stork (*Mycteria americana*)   **Plate 11**

Papiamento: *Ciguena americana, Garsa kabes chino*
Dutch: *Kaalkopooievaar*

Description: Distinctive. A tall, long-legged wader with dark, naked head in adult (feathered in first year) and large down-curved bill. White with black flight feathers and tail, though when not in flight appears all white, with only a black line visible on edge of wing. Black legs. **Similar species:** Unmistakable. **Status:** Only a single record for this species, which is known to wander widely and is an excellent long-distance flyer. The one record was at Bubali, Aruba; the individual arrived as an immature in Feb 1977 and apparently remained in residence there until at least Oct 1984. **Range:** Breeds in US, from coastal South Carolina to Florida, and from coastal Mexico south to Argentina. Rare resident on Cuba and Hispaniola.

Family Fregatidae: Frigatebirds

Magnificent Frigatebird (*Fregata magnificens*)   **Plate 5**

Papiamento: *Skerchi, macuacu, sherve, maniwá*
Dutch: *Amerikaanse Fregatvogel*

Description: An ummistakable bird of tropical seas. One of largest (certainly with the widest wingspread) and most distinctively shaped birds likely to be seen on the islands—and identifiable at a great distance. Males entirely black, with red throat pouch; females black except for white upper breast; immatures black, with white head and breast. Very long, thin, knife-like wings; slender body; long, forked tail; and small head with long, thin, hooked bill. **Similar species:** Only large, black seabird in the ABCs, unlikely to be confused with any other species. There is a report of Great Frigatebird (*Fregata minor*) from Sero Colorado, Aruba, on March 15, 2005 (S. Mlodinow in Prins et al. 2009), and another of two birds photographed from Bonaire on Jan 20, 2014 (C. Sheeter). If correct, this species is an exceptionally rare vagrant. There is much yet to be learned about changes in frigatebird plumage over the 8–10 years required for them to become breeders, and without such knowledge it is difficult to assess the accuracy of sightings of the Great Frigatebird. We urge interested observers to photographically document the range of plumage variation in frigatebirds on the islands. Howell et al. (2014) has an excellent summary and discussion of frigatebird identification. **Voice:** Generally silent; occasionally makes a guttural croak. **Status:** Commonly seen along coasts of all three islands, but nearest known breeding location is in coastal Venezuela. Often seen far out to sea and soaring very high, typically singly or in groups of 5–10. When fishermen discard fish guts on land or water, birds often swoop down very low and can be seen at close range; at the marina at Hadicurari, Aruba, 40 birds were counted in Apr 2000 (AJW). Roosts at night in mangroves, on reef islands off Oranjestad (Aruba), in Lac Bay (Bonaire), and on Isla Makuaku, in Sint Jorisbaai Bay, Curaçao. Largest numbers noted at Isla Makuaku, with counts of 120 in June 1990, 80 in July 1992, and 50 in Aug 1992 (Prins et al. 2009). Birds have been seen doing courtship displays, but no definite nesting documented on any of the three islands. Groups of 30–40 birds often seen soaring over Sero Colorado headland on Aruba. **Range:** Breeds in Caribbean and tropical Atlantic, from Dry Tortugas National Park, Florida, to Brazil, and on Cape Verde island off Africa; breeds in Pacific from Mexico to Peru.

Family Sulidae: Boobies

Masked Booby (*Sula dactylatra*)   Plate 6

Papiamento: *Bubi*
Dutch: *Maskerboebie*

Description: Relatively large seabird with long wings and big bill. In flight, appears pointed at both ends because of long, pointed bill and pointed, relatively long, tail. Adults strikingly white except for contrasting black tail and flight feathers, black face, and large yellow bill. Juveniles have brown back, upperwing, and tail, but with white rump. Brown head is separated from brown back by white collar; underside of juvenile white, including white underwing with contrasting narrow brown trailing edge. **Similar species:** Largest, bulkiest of the three booby species that occur in the ABCs; also has the shortest tail. Adults easily distinguished from all plumages of Brown Booby and from juvenile and brown-morph Red-footed Booby by white body and inner wings. Juveniles distinguished from juvenile Brown Booby and brown-morph Red-footed Booby by white collar, mostly white underwing, and white belly. Adult Masked Booby most similar in appearance to white-morph Red-footed Booby but that morph has longer, all-white tail (black in Masked); it also lacks black face and the black on rear of upperwing does not extend to body as in Masked. Red-footed much slimmer and more buoyant in flight. On Red-footed, portion of body behind wings longer than in front; on Masked, length of body behind wings approximately equal to length in front. **Voice:** Generally silent at sea. **Status:** Perhaps less unusual in waters of ABCs than previously thought and/or increasing in numbers. In 1983, Voous listed only five records for the ABCs; Prins et al. (2009) list seven additional records for the ABCs. More recently, Luksenburg and Sangster (2013) observed the species 44 times off Aruba in 2010 and 2011, 0.5–11.7 mi (0.8–18.8 km) from shore; about half of their sightings were in July and Sept, off the southwest and northwest coasts of Aruba. In recent years, reports of Northern Gannets (from visiting birders on arriving and departing cruise ships) are very likely misidentifications of this species (Ligon 2006). Few observers have attempted seabird watches with telescopes off northern and southern points of any of the islands. Following records, on or near shore, from Prins et al. (2009). <u>Aruba:</u> One adult observed from shore at California (northern tip of island) on Apr 27, 1980; one on (and near) San Nicolas Bay reef islands, May 21–22, 1987; one immature seen off south point of island in July, 1989; one adult seen near Malmok in May, 2003. <u>Bonaire:</u> One exhausted adult found on shoreline near Piedra Pretu, south of Lac, March 1, 1976; one immature among flock of Brown Boobies roosting on cliffs at Malmok, Oct 13, 1979; one immature seen off Punt Vierkant, Jan 24, 1988; one (reported by visiting ornithologist) from shore, near Boca Onima, Nov 8, 2002 (Ligon 2006); adult observed off "Invisibles" dive site, Oct 9, 2004. Curaçao: One bird seen from boat between Curaçao and Klein Curaçao, Nov 9, 2005 (Prins et al. 2009). **Range:** Worldwide in tropical oceans. Nests at 14 scattered locations in Caribbean, where total breeding population has been estimated at 550–650 breeding pairs (Schreiber and Lee 200). Has bred historically on Venezuela's Los Monjes, 56 mi (90 km) west of Aruba, and on Los Roques and Los Hermanos, to the east of Bonaire (Voous 1983).

Brown Booby (*Sula leucogaster*) Plate 6

Papiamento: *Bubi*
Dutch: *Bruine Boobie, witbuikboebie*

Description: Relatively large seabird, with long wings. In flight, appears pointed at both ends because of long, pointed bill and relatively long, pointed tail. In adults, chocolate-brown on upperside of body and wings and on upper breast contrasts sharply with white lower breast, belly, and undertail coverts. On underside of wing, white framed by brown border on leading and trailing edges; solid dark brown on outermost section. Bill and legs bright yellow (female) or dull yellowish-green (male). Juveniles chocolate brown overall, but note lighter brown basal two-thirds of underwing. Takes several years to reach adult plumage; immatures show similar pattern to adults but bright white on underside looks dirty-white. Under the windy conditions of the ABCs, flying birds have a characteristic arcing flight pattern, in which the bird hooks up from the ocean surface then turns and drops quickly back down toward surface, repeating pattern as it progresses, often with wings held stiffly or flapping near the top of the arc. Even at a distance, contrast between brown upperside on one side of the arc and white underside as the bird turns is readily apparent and an excellent key to identification. **Similar species:** The two other species of booby occur relatively regularly in the south Caribbean seas, but farther out to sea, at least 6 mi (10 km) from shore; but either could occur with Brown Boobies at night cliff roosts. No other booby has chocolate-brown upperside, throat, and upper breast constrasting with white belly and underwing (basal two-thirds). Brown-morph Red-footed Boobies always have pale brown head and underside with dark brown under-wing, upperwing, back, and upper tail. Juvenile Masked Booby superficially similar to adult Brown Booby but upperwing, body, and tail have white edges to brown feathers and white collar separating brown hood from brown back; white on under-side of wing extends farther to base of dark primaries; lacks brown leading edge of underwing. **Voice:** Not likely to be heard. **Status:** Stays close to shore, generally foraging within 6.2–12.4 mi (10–20 km) of land and roosting at night on cliffs. Not known to nest on any of the ABCs, though may have formerly nested on Klein Curaçao (Voous 1983) and does nest on a number of islands off Venezuela (Meyer de Schauensee and Phelps 1978). Largest numbers seen at traditional night roosting locations on northern windward cliffs on Bonaire (Malmok) and Curaçao (Watamula, Westpunt, and near Playa Grandi), where the birds were historically snared at night by local fishermen for food and use in black magic. Also roosts during the day on buoys, ship-rigging, and rusting shipwrecks, the most well-known site being near the so-called German shipwreck off Malmok, Aruba, where there were formerly sometimes 5–10 individuals. This location is also a popular snorkeling and diving location—birds can be approached quite closely while snorkeling, affording excellent close-up views. They can often be seen passing by the shore, especially in the morning and evening as they leave or return to roosts. Maximum numbers: 240, July 9, 2009, Malmok, Washington-Slagbaai NP, Bonaire (STINAPA fide eBird); 200, Malmok, Washington-Slagbaai NP, Bonaire (Voous 1983); 40, July 6, 2001, Malmok, Washington-Slagbaai NP, Bonaire (AJW); 35 on Jan 20 1971, Aruba (Voous 1983). On Bonaire, the largest numbers of birds roost at Malmok, within Washington-Slagbaai NP, but park closes before birds come to roost so special arrangements must be made with park officials. Historically, birds also roosted near Boca Bartol, also within the park, but that site hasn't been checked in recent years. Birds can be seen passing offshore from virtually anywhere on the island. Two adults, for example, were observed passing offshore Willemstoren Lighthouse in March, 2001 (AJW). On Curaçao, best location is the roost site near Westpunt and Watamula. Birds may have bred on Klein Curaçao based on extent of guano deposits (Voous 1983). **Range:** Occurs throughout the world's tropical seas, with breeding sites throughout Caribbean, including Venezuelan islands to the east of Bonaire (Las Aves, Los Orchila, Los Roques), but not currently breeding on the ABCs.

Red-footed Booby (*Sula sula*)   Plate 6

Papiamento: *Bubi bala, Bubi pia còrá, Bubi pia lòrá*
Dutch: *Roodpootboebie*

Description: Relatively large seabird, with long wings and big bill. In flight, looks pointed at both ends, having a pointed bill and pointed, relatively long, tail. Three color morphs in this species. Adults of white morph are all white except for black primaries, while adults of white-tailed brown morph are tannish-brown on body and wings but with white tail extending up onto lower back. Brown morph adults are all brown. Immatures (reach adult plumage in 2–3 years) are brownish overall, with dark underwing and lighter belly; tail with white tip (in some individuals, the white tip is worn away). **Similar species:** Slimmer and proportionately longer-tailed than either Brown Booby or Masked Booby. White-morph adult could be confused with adult Masked Booby but lacks black tail, and black on wings does not reach body. Young Red-footed Booby has dark underwing as compared to light underwing in Brown and Masked Booby. Immature Red-footed Booby always lacks the sharp contrast between darker head and lighter body as seen in Brown Booby and juvenile Masked Booby. **Voice:** Generally silent at sea. **Status:** Apparently sometimes relatively common offshore (10 mi [16 km] or more), with flocks containing as many as 50–80 birds; numbers increase west to east, so more common off Bonaire than Curaçao and least common off Aruba (Voous 1983). However, little recent information on offshore status and few observers have attempted seabird watches with telescopes off northern and southern points of any of the islands. Occasional onshore records (13 records by Voous), mostly of individual birds joining night roosts of Brown Boobies at ocean cliffs on northern shores of Bonaire and Curaçao. Onshore records essentially span all seasons. Aruba: One record of a bird captured on board a tanker 14 mi (22.5 km) west of Aruba, March 31, 1955 (Voous 1983); one adult at Sero Colorado, Sept 1978; one immature at Andicuri, Jan 1979; one immature at Malmok (at the German shipwreck), March 12, 2005 (all Prins et al. 2009). Bonaire: Ten records for the island (Prins et al. 2009) and has occurred uncommonly among roosts of Brown Boobies at Malmok (Voous 1983); an injured brown-morph adult was photographed on the beach at Sorobon, mid-July 2001 (Ligon 2006). Three reports of larger numbers mentioned without details in Prins et al. (2009) of 18–43 birds seen near Willemstoren, late June to early July. Said to be more common in offshore waters to the north and south of the island (Voous 1983). Curaçao: Has occurred uncommonly among roosts of Brown Boobies at Watamula, Westpunt, and one seen at Boka St. Marie on July 30, 1992 (Prins et al. 2009). Said to be quite common in offshore waters to the north and south of island (Voous 1983). **Range:** Worldwide in tropical oceans. In Caribbean, known to nest at 18 sites, with an estimated 8,200–10,000 breeding pairs (Schreiber and Lee 2000) as well as colonies on Venezuelan islands (at least historically) of Islas Las Aves, Los Roques, La Orchila, Los Hermanos, and Los Testigos (Hilty 2003).

Family Phalacrocoracidae: Cormorants

Neotropic Cormorant (*Phalacrocorax olivaceus*)   Plate 7

Papiamento: *Dekla*
Dutch: *Bigua Aalscholver*

Description: Black waterbird with long, yellow, hooked bill, relatively long tail (two-thirds length of body), serpentine neck, and black feet. Thin, white border to yellowish throat pouch, which is pointed at rear. Immature birds have drab, light brown head, neck, and underparts; has darker brown upper body and wings but usually has paler, thin border to throat pouch, as in adults. **Similar species:** Double-crested Cormorant is exceptionally rare (only a single record) but increasing in abundance in US so could occur again on the islands. Compared to Neotropic Cormorant, neck and body much larger and bulkier; tail shorter, perhaps one-half body length compared to two-thirds body length in Neotropic Cormorant. Double-crested Cormorant throat pouch is orange versus yellowish in Neotropic Cormorant, is rounded along rear edge instead of pointed, and lacks thin, white border. **Voice:** Generally silent; hoarse croaks occasionally heard at nesting colonies. **Status:** Originally, a regular nonbreeding visitor (probably from Venezuela) to all three islands, in varying numbers and typically March–Aug (Voous 1983), but birds began attempting to nest on Aruba, at Bubali, by 1978; established as breeder there by 1993. Numbers have remained high at Bubali, ranging from 60 to 230, Jan–Apr, from 1993 to 2000, with a small number of nests (10 or fewer) documented in Jan 1993, 1995, 1996, and 1999, and 22 active nests counted Apr 2000. Nesting documented in pond at Divi Links golf course near

Oranjestad in Nov 2011 (AJW). Birds are sometimes seen at other areas of the island; as many as 20 birds were seen flying by at Spanish Lagoon; 20 off Oranjestad, Jan 1999); and 2 or 3 at Savaneta, Jan 1999 (all AJW). On Bonaire small numbers may occasionally be found in bays in Washington-Slagbaai NP, Jan–Aug period, including at Boca Bartol, Slagbaai, Gotomeer, and Playa Frans. Single immature birds were seen at Boca Bartol and Slagbaai in Apr 2001 (AJW). Other recent records include a single adult seen at Goto, July 5, 2001 (AJW), and three at Boca Bartol, July 6, 2001 (AJW). Voous (1983) reports a maximum of 60 birds in Feb 1978 but does not note the specific location on Bonaire. On Curaçao occasional individuals seen drying wings and feeding in Schottegat, Jan–Aug, and small numbers (four and three) observed at sewage treatment plant ponds, at Klein Hofje, March 2000 (AJW). Apparently, in wet years larger numbers may occur at seasonal ponds, including approximately 25 seen at large pond at Muizenberg, March 2000 (AJW), during an exceptionally wet year. Maximum single location numbers: 300 in Apr and June, 1979, Bubali, Aruba (Voous 1983); 230 on Apr 14, 2000, Bubali, Aruba (AJW); 180 on Jan 12, 1993, Bubali, Aruba (AJW). **Range:** Range extends from coasts of Louisiana and Texas in southern US, south through Central America and throughout South America, to Tierra del Fuego. In West Indies, resident breeder on Cuba and vagrant elsewhere.

Double-crested Cormorant (*Phalacrocorax auritus*)   Plate 7

Papiamento: Deklá orea
Dutch: Geoorde Aalscholver

Description: Black waterbird with long, yellow, hooked bill, serpentine neck, black feet, and orangish throat patch (rounded at rear). Breeding adults have two small crests on the head but they are difficult to see. Immatures (more likely here than adults) have drab, light brown head, neck, and underparts; darker brown upper body and wings. **Similar species:** See Neotropical Cormorant. **Status:** One record, apparently with supporting photograph, of a single immature bird observed Sept 16–18, 1979, at Salina Martinus, Bonaire (Voous 1983). **Range:** Breeds across much of US and southern Canada and along Pacific and Atlantic coasts. Also winters on both coasts, south to Mexico. Populations have increased greatly in the US and Canada.

Family Pelecanidae: Pelicans

Brown Pelican (*Pelecanus occidentalis*)   Plate 7

Papiamento: *Pelicano, ganshi*
Dutch: *Bruine Pelikaan*

Description: Rather unmistakable large bird with characteristic long bill and large throat pouch, typically seen floating on—or plunge-diving into—the ocean, usually within 100 meters of shore, sometimes within a meter or so of people swimming. In flight, note very broad wings, short tail, and massive head and bill. Adults with dark brown underside, but pale on back, tail, and at base of upper surface of wings. Neck and head white during nonbreeding season, but note brown nape and golden crown in breeding season. Bill pouch dark brown; upper bill orange-yellow; feet dark. Young birds in first and second years have brownish drab upperparts, including head and neck; white belly. Underside of wings brown, with a white zig-zag mark running up center of each wing. **Similar species:** Unlikely to be confused with other species. **Voice:** Usually silent. **Status:** Commonly found throughout the year, on the leeward coasts of all three islands, but most numerous on Aruba. Slightly smaller Caribbean race augmented by larger northern migrants (indistinguishable in the field), Sept though Apr. Small numbers of Caribbean race breed on Aruba. Maximum numbers: 300 from 1977 to 1979, Bubali, Aruba (Voous 1983); 45 on Jan 11, 1994, evening count of birds flying toward roost near Oranjestad, Aruba (AJW); 50–100 said to have been counted along beaches on Aruba but no date given (Voous 1983). On Aruba, most common near beaches on southwest side of the island but occurs regularly along entire leeward coast. Typically encountered in groups of 5–10 but sometimes in groups of 20–30. Also occurs in small numbers at Savaneta salt pans and at Bubali, though numbers in late 1970s were vastly higher at Bubali than is the case today. Breeding first confirmed on Aruba in 1966 on the reef island just offshore from the airport, near Oranjestad, and birds apparently nested at that location annually, through the 1980s, with numbers varying from just a few to as many as about 20 pairs (Voous 1983), but no known confirmation of breeding in recent years. Breeding may also have occurred at Bubali, where three adults in breeding plumage observed in bushes with nesting cormorants and egrets, in Apr 2000 (AJW). On Bonaire small numbers occur year-round, usually 1–5 birds in and near the Pekelmeer and at Boca Slagbaai, Boca Bartol, and Gotomeer in Washington-Slagbaai NP (AJW), but individuals may be seen almost anywhere along the coast, especially on the leeward side of island. Although birds in breeding plumage have been observed on Bonaire in Apr and July, when migratory race is unlikely to be present, no evidence of breeding has been reported. On Curaçao, small numbers are seen scattered along leeward shore. **Range:** Breeds on Atlantic Coast of North America, from Delaware south to Florida, and across Gulf Coast to northeastern Mexico; on Pacific Coast, breeds from southern California south to Chile. In northern South America breeds along coasts of Colombia and Venezuela (estimated 17,500 birds in 1982), and apparently south to Brazil (Hilty 2003).

Family Ardeidae: Herons, Egrets, Bitterns

Pinnated Bittern (*Botaurus pinnatus*)   Plate 8

Papiamento: *Garabet di cana*, *Garabet di kana*
Dutch: *Zuid-Amerikaanse Roerdomp*

Description: Has characteristic bittern shape and habits. A medium-sized, thick-necked, brown-striped heron with rather short, yellowish legs. Whitish chin and underparts with brown streaking along sides of neck and upper breast. Slightly larger than night-herons, and quite similar in general coloration, but with longer bill and more upright posture, often with bill pointing upward. **Similar species:** Very similar in appearance to American Bittern, a species familiar to North American birders (there are no records of that species, however, from the ABCs), but lacking dark line down neck and with fine barring on cap, cheeks, and back of neck, as well as on edges of otherwise dark-centered feathers of back and wings. Most likely species to confuse with this are immature Yellow-crowned and Black-crowned night-herons, also with brown stripes. Both of these species have much shorter bills and squat-looking heads and necks, and lack any fine barring on body. Several species of tiger-herons occur in nearby Venezuela and Colombia and could be possible vagrants on the ABCs. The juveniles are very similar in appearance to Pinnated Bittern but show horizontal rather than vertical dark centers to wing and body feathers, which also lack fine horizontal barring. **Voice:** Call when startled is a low-pitched "gawk." In normal range, males have breeding song consisting of a repeated, low booming note. **Status:** Represented by only a single record, of a bird captured, photographed, and released in Oranjestad, Aruba, Jan 18, 1972 (Voous 1983). Note that this species is a secretive denizen of freshwater marshes, very local, and poorly known throughout its Central and South American breeding range. If it occurs in the future it would be most likely to occur in freshwater marsh habitats such as at Aruba's Bubali wetlands or on Curaçao, perhaps at Malpais or Muizenberg. **Range:** Patchily distributed from southern Mexico through Central and South America, to Argentina. Also breeds on Trinidad.

Least Bittern (*Ixobrychus exilis*) Plate 8

Papiamento: *Garabet enano*
Dutch: *Amerikaanse Woudaap*

Description: A tiny, often rather secretive, heron. Black cap and black (male) or brown (female) back contrast with rusty cheek and back of neck; buffy wing patches. Immature similar to female but streaked on back. **Similar species:** Green Heron and rare Striated Heron larger and lack dark patch on back and buffy wing patch. The related Stripe-backed Bittern (*I. involucris*) is similar in size and shape but dark patch on back replaced with dark streaking; lacks strongly contrasting buffy wing patch. **Voice:** Distinctive *cu-cu-cu* call, given on breeding grounds, has been heard at Bubali, Aruba, at dawn. When startled and in flight, can give a short *kuk*. **Status:** Two records up to Apr 2013; first sight record for Aruba was of two birds from Bubali, Apr 9, 2013 (S. Mlodinow fide eBird); first photographically documented record was at Bubali, Nov 18, 2013 (K. Hansen fide eBird). Now, apparently a resident at Bubali, where up to three adult birds have been seen at one time and the species has been sighted and photographed virtually continuously since Apr 2013 and continuing into 2016. Evidence of breeding should be watched for there. One record (sex and age not given) of this marsh-loving bird at Muizenberg, Curaçao, Oct 20, 2005 (Prins et al. 2009). **Range:** Extensive but patchy range extends from southern Canada south to Brazil and Paraguay.

Great Blue Heron (*Ardea herodias*)   Plate 8

Papiamento: *Garza blou grandi*
Dutch: *Amerikaanse Blauwe Reiger*

Description: As tall as an American Flamingo, this is the largest of the herons that regularly occur on the islands. Long, heavy, yellowish bill; long, thick, yellowish-greenish legs; long neck. Three color morphs; the dark morph, which is the most common, shows gray body. Adult has thin, black line down center of neck, with sides of neck brownish-gray; white crown; black patch behind eye extends to nape; rusty thighs and edge of wing and black sides. Immature dark morph with all-dark crown; compared to adult, neck is grayer, with more streaking. Rarer white morph is all-white, with yellow bill and legs. Intermediate "Wurdemann's" morph can show a mix of characters, but adults typically have white head and neck (retaining gray body). **Similar species:** Most similar to two species yet to be documented from the ABCs though likely to be found eventually. The Cocoi Heron, which is common in nearby Venezuela and throughout northern South America, is virtually identical in size and shape to Great Blue Heron but has entirely black crown that extends below the eye, white sides of neck, and white thighs, features that could make it appear similar to the "Wurdemann's" morph of Great Blue Heron except for white thighs and black cap. Immatures very similar but immature Cocoi Heron has lighter neck. Increasing numbers of another very similar species, the Gray Heron, continue to be documented in the nearby Lesser Antilles. Gray Heron is virtually identical in shape, size (perhaps slightly smaller), and coloration, but has white rather than rusty thighs. Rare white morph Great Blue Heron resembles Great Egret but is larger, with yellowish (not black) legs. **Voice:** Loud, low-pitched croaking *roark*. **Status:** Throughout the year, small numbers of nonbreeding birds visit the islands, frequenting edges of wetlands and mangroves. Typically 1–4 birds at any one location. Report of 40–50 birds at Bubali, Aruba, Feb 1977 (reported in Voous 1983), is dubious considering that nonbreeding birds are rarely found in such concentrations, even in areas where they are common. Most reports are of dark-morph birds, but there have been occasional records on all three islands of white-morph birds and "Wurdemann's" morph. **Range:** Breeding range extends from southern Canada to Greater Antilles, southern Mexico, and Belize, but there are isolated populations in Galapagos and islands off Venezuela. Wintering birds occur regularly as far south as Colombia and Venezuela.

Great Egret (*Ardea alba*)   Plate 10

Papiamento: *Garsa blanco grandi, Garsa blanku grandi*
Dutch: *Grote Zilverreiger*

Description: Tall, all-white heron with relatively long, yellow bill and black legs. **Similar species:** Only other bird with combination of black legs and yellow bill is the nonbreeding Cattle Egret, which is much smaller. **Voice:** Makes occasional low croaks. **Status:** Since the late 1980s, a regular breeder at Bubali wetlands, Aruba, where it can be seen year-round. Largest numbers documented on Aruba include:

50 at Bubali, Apr 2000; 30 at Spanish Lagoon, Jan 2003; and 25 at Bubali, Jan 1997 (AJW). Smaller numbers found at other Aruba locations, including Tierra del Sol wetlands and Savaneta wetlands. Decidedly uncommon on Bonaire and Curaçao, where one or two birds may be found at a few wetland locations; interestingly, there is some evidence that it may have nested at least once on Klein Bonaire. Great Egrets travel long distances beyond their breeding areas, and it is likely that many individuals seen on the islands have come from breeding areas in Venezuela. **Range:** From northern US to southern South America. Also occurs in southern Europe and Asia, Africa, and Australia.

Little Egret (*Egretta garzetta*)   Plate 10

Papiamento: *Garsa blanca oropeo, Garsa blanku oropeo*
Dutch: *Kleine Zilverreiger*

Description: Virtually identical to the Snowy Egret except for longer, slightly thicker bill, and flatter head; slightly larger, with thicker legs. Skin in front of eye usually gray or red (rarely yellow). In breeding plumage, has 2–3 long plumes extending from back of head, unlike scruffy head of Snowy Egret. **Similar species:** See Snowy Egret. **Voice:** Occasional harsh squawks. **Status:** Two records, both from Aruba: Tierra del Sol wetlands, March 25–20, 2003; Bubali, March 12, 2005 (Prins et al. 2009). Birders are urged to document any sightings with photos, as we are not aware of any published photographs. **Range:** Very widespread, from southern Europe through Africa, southeast Asia, and Australia. Breeds on Barbados and increasing reports in Caribbean and eastern US.

Snowy Egret (*Egretta thula*)   Plate 10

Papiamento: *Garsa blanco chikito, Garsa blanku chiki*
Dutch: *Amerikaanse Kleine Zilverreiger*

Description: Striking, slender all-white egret with black bill and yellow lores (skin in front of eyes). Adult has black legs with yellow feet. Immature birds have black legs (with yellow line running down back side of legs) and yellow feet. **Similar species:** Cattle Egret and Great Egret have yellow bills; adult white-morph Reddish Egret has bill with pink base. Immature white-morph Reddish Egret has dark bill but all-dark legs, and is much larger than Snowy Egret. Immature Little Blue Heron has pale bill, yellow legs, and some dark in plumage. Most similar is the very rare (but potentially increasing) Little Egret, which in breeding plumage has 2–3 long plumes extending from crown, unlike the many short plumes on Snowy Egret. Little Egret has a flatter head and longer, thicker bill, and its lores are often gray (sometimes yellow and other colors in breeding plumage). Although not recorded from the ABC islands, white-morph Western Reef-Heron has been recorded a number of times in the Caribbean; it is very similar but with a longer, thicker bill that has a more curved upper ridge, giving it a drooped look. See Howell et al. (2014) for more details about identification of white-morph Western Reef-Heron.

Voice: Occasional harsh squawks. **Status:** Common year-round resident and local breeder on all three islands. Largest numbers recorded at Bubali wetlands in 1970s and 1980s, with up to 200 birds at one time. In recent decades, counts at single wetland locations typically of 1–10 birds, but occasionally of 20–30. Have nested in past at a number of locations on each island (Voous 1983). **Range:** From US south through Central America and Caribbean, to southern South America.

Little Blue Heron (*Egretta caerulea*)   Plate 10

Papiamento: *Garsa blou chikito, Garsa blou chiki*
Dutch: *Kleine Blauwe Reiger*

Description: Adults have a slate-blue body; a dark reddish-purple head and neck in breeding pumage; dark legs and bill. Juvenile is all white, with yellow legs; bill with light base and dark tip; usually shows dusky-gray wingtips. As they age, immature birds, while still largely white, have increasing amounts of dark gray patches in body and wings, eventually molting into the completely dark adult plumage. **Similar species:** Adult is distinctive, perhaps only confused with larger and bulkier dark-morph Reddish Egret, which always shows striking pink bill with dark tip and has a much more reddish head and neck. Juvenile and immature Little Blue Heron most likely confused with Cattle Egret or Snowy Egret. Cattle Egret has rather stubby, bright yellow bill, while Snowy Egret has an all-dark bill and dark legs with yellow feet. **Voice:** Like other herons, makes low croaks. **Status:** Apparently quite uncommon, though numbers are said to have reached as many as 20 birds at Bubali, Aruba, prior to 1983 (Voous 1983). Breeding documented at Lac Bay, Bonaire, in 1983, and Spanish Lagoon, Aruba, in 1992 (Prins et al. 2009). In Jan surveys (1999–2003) on Aruba, we found the birds only at Savaneta, where they were regular, in groups of four or fewer. **Range:** From eastern US south to Argentina.

Tricolored Heron (*Egretta tricolor*)   Plate 9

Papiamento: *Garsa barica blanco, Garsa tres kolo, Gran gudjee, Garsa barika blanku*
Dutch: *Witbuikreiger*

Description: A medium-sized heron with rather slim neck and long, thin bill. Dark blue head, neck, back, and wings contrast with white belly and white line extending up front of neck to chin. Juvenile with rust-colored neck and wing coverts. **Similar species:** No other commonly occurring heron shows dark upperparts contrasting with white belly. **Voice:** Makes occasional low croaking sounds. **Status:** Relatively common year-round resident of the islands; documented historically as a breeder on all three ABC islands (Voous 1983). Occurs throughout the islands in lagoons, bays, salinas, and occasionally freshwater wetlands, usually in groups of 1–10. Largest numbers we have documented were up to 30 at Savaneta, Aruba, Jan 1996, but Voous (1982) reports that an observer in 1930 counted at least 50 nests on the Bucuti reef islands of Aruba. **Range:** Coastal regions, from southeastern US south through Mexico, Central America, and Caribbean, to Brazil and Peru.

Reddish Egret (*cambiar*)   **Plate 10**

Papiamento: *Garsa cora, Garsa kabes kane, Garsa kora*
Dutch: *Roodhalsreiger*

Description: A fairly large, stocky heron. Adults show distinctive black tip to pink bill, unlike any other common herons or egrets of the ABC islands. Two color morphs: one is entirely white; the other has a dark gray body and reddish head and neck. Both morphs have dark legs. Immature dark-morph birds have gray body; reddish-brown on head, neck, and part of wing; dark legs. Immature white-morph same as adult but with all-dark bill. **Similar species:** White-morph adult could be confused with other white egrets but they lack dark legs and bicolored bill. Immature white-morph Reddish Egret only white egret with dark bill and completely dark legs. Adult Little Blue Heron superficially similar to dark-morph birds but is much smaller; also, on its bill the contrast between the darkish tip and the rest of the bill is much less striking than on adult Reddish Egret. Immature dark-morph Reddish Egret is larger than adult Little Blue Heron, and has a heavier bill and thicker legs. **Voice:** Makes occasional low croaking sounds. **Status:** Breeding documented only on Bonaire (Pekelmeer, Lac Bay, and other sites), where regularly occurs in both color morphs. Often seen singly, in pairs, or groups of three, occasionally more. Largest numbers occur at Pekelmeer, but some also at salinas in Washington-Slagbaai NP and other locations. Small numbers occur on Curaçao and Aruba. On Aruba, our Jan surveys (1993–2003) turned up only one, a white-morph bird, at Savaneta; since 2011, groups of 1–3 mostly dark-morph birds have been seen regularly on Aruba, especially at salinas behind Palm Beach, near Rooi Santo (on the way to Malmok, close to the sky diving hut). **Range:** Coastal regions, from southern California south through Central America; also from the US Gulf Coast, through Caribbean, to coastal Venezuela and Colombia.

Cattle Egret (*Bubulcus ibis*)   **Plate 10**

Papiamento: *Garabet di baca, Garabet di baka*
Dutch: *Koereiger*

Description: Smallest of the all-white herons. In breeding season, shows orange-red to yellow legs and yellow or yellow-orange bill; in nonbreeding season legs are dark, bill yellow. In breeding season, adults have buff patches on breast, back, crown, and nape. Juveniles have black bill and legs. **Similar species:** Only the much larger Great Egret has a yellow bill. Juvenile and immature of other all-white herons often show a pale bill (black on juvenile Cattle Egret). No other white egrets or herons share combination of very short legs; small, stout size; and short bill. Reddish Egret and adult Snowy Egret have black or dark legs (like juvenile Cattle Egret) but both are larger and have longer bill, legs, and neck. Adult Snowy has yellow feet. **Voice:** Not often heard; gives occasional low, grunting calls around nesting sites. **Status:** Relatively common and widespread on

Curaçao and Aruba; infrequent and in small numbers on Bonaire. First found on Aruba in 1944. Has now become common year-round on Aruba and Curaçao; irregular on Bonaire, where there were only three records from 2001–2009 (Prins et al. 2009). First breeding record for Aruba from Bubali wetlands, in 1980; first breeding record for Curaçao from Isla Macuacu, Sint Joris Bay, 1967 (Voous 1983). Breeding continues on Aruba at Bubali, and breeding known from Schottegat on Curaçao, 2003. Maximum numbers recorded include 400 at Curaçao airport, Oct 1976 (Voous 1983); 175 on Bonaire at Salina Martinus, Jan 1975 (Voous 1983); 125–150 at Malpais, Curaçao, March–Apr 2000; 75 in Ruistenberg area of Curaçao, Nov 2003; 54 on Curaçao, at dump near Malpais, Nov 2003, and on same day, 51 at Klein Hoje sewage treatment plant; and 48 at Schottegat, Curaçao, Nov 2003 (AJW). **Range:** Cosmopolitan; found in Africa, Europe, Asia, Australia, and the Americas, where it occurs from southern Canada south to Argentina and Chile.

Green Heron (*Butorides virescens*)   Plate 8

Papiamento: *Galina di awa, Caw-caw*
Dutch: *Groene Reiger*

Description: Small, short-necked heron with long, thin bill. Adult with dark cap, chestnut neck, and dark, greenish back. Legs yellowish, bill dark. Immature with brownish, streaked neck and white edges to wing feathers. **Similar species:** Smaller than night-herons, with longer, thinner bill. Adult Green Heron has very different plumage color and pattern; immature very different, in size and shape, from night-herons; has dark cap (absent in immature night-herons). Although very rare, the most similar species is the closely related Striated Heron, a species widespread in South America. Striated Heron is virtually identical in size and shape, but adult has gray (not chestnut) neck. Immature Striated Heron almost indistinguishable from immature Green Heron but shows gray tones on sides and neck. **Voice:** Makes a sharp *skeeow*, often given by startled birds. **Status:** Widespread; seen in small numbers throughout the year, in freshwater and salt-water wetlands, where it often prefers to skulk in vegetation along edges. A resident subspecies occurs year-round; it has been recorded breeding in small numbers on all three ABC islands (Voous 1983). Birds are seen throughout the year and typically in groups of 1–4 (highest number: 16, Bubali, Feb 3, 2014 [S. Mlodinow fide eBird]), at Bubali wetlands, Spanish Lagoon, Tierra del Sol wetlands, and Savaneta on Aruba; Klein Hofje, Jan Thiel Bay, Malpais, Schottegat, and other bays and wetlands on Curaçao; and Goto, Lac Bay, Pekelmeer, Playa Tern, Put Bronswinkel, Salina Bartol, and Slagbaai on Bonaire. Note, however, that it could be encountered at any small freshwater wetland or coastal area. A subspecies that breeds in North America has been identified here from specimens and probably regularly augments the breeding population in winter but may be indistinguishable from resident birds in the field. **Range:** Breeding range extends from extreme southern Canada south through Caribbean and Central

America to Panama. Winter range extends from southern US to northern coast of South America, Colombia to Brazil.

Striated Heron (*Butorides striata*)   Plate 8

Papiamento: *Galina di awa strepia*
Dutch: *Mangrovereiger*

Description: Identical in shape and size to Green Heron but adults with gray instead of chestnut neck. **Similar species:** See Green Heron. **Voice:** See Green Heron. **Status:** A rare, but apparently fairly regular, visitor to the ABC islands from South America, where it is a common breeder (Green Heron only winters, in small numbers, in the northern part of the continent). Currently one record for Aruba; two for Curaçao (one a specimen), and five for Bonaire (one a specimen). This species and Green Heron have sometimes been considered forms of a single species; plumage variation within each species and hybridization leads to much confusion, especially near distributional boundaries of the two species. A photo taken along the Kaminda Sorobon road on Bonaire, Jan 21, 2014, was reported as a Striated Heron (C. Sheeter archived at eBird). The bird in the photo shows strong chestnut tones in the face, below the cap, and along the sides near the breast, suggesting that it is either a hybrid or an atypical Green Heron. Ross Wauben photographed a Striated Heron at Bubali, Aruba, on Feb 10, 2016 (archived at eBird), that does not appear to show any chestnut tones on the sides, neck, nape, or face. We are not aware of any other published photographs of possible or confirmed Striated Heron from the islands, although we and others have photographed a great many Green Herons in the islands over the years. We urge birders to photograph any Striated Heron or Green Heron to assist in learning more about variation in these species. **Range:** Southern Panama through South America, to Argentina and Chile.

Whistling Heron (*Syrigma sibilatrix*)   Plate 9

Papiamento: *Garsa fluitdo*
Dutch: *Fluitreiger*

Description: This striking, medium-sized heron is unlikely to be confused with any regularly occurring species. Has black cap and sky-blue skin around yellow eye; pink bill has a black tip; tan on neck and upper breast (also note tan wing patch); back and outer wings are pale gray. White rump and tail obvious in flight. **Similar species:** No regularly occurring species has similar characteristics. Not yet recorded from the ABC islands, the Capped Heron (*Pilherodius pileatus*) has black cap and blue around eye but is smaller, with white forehead and white body. **Voice:** "High reedy complaining whistle," (Hilty 2003). **Status:** Five or six records of this rare vistor. One bird was photographed at Salina Boca Slagbaai, Bonaire, Jan 8, 2003 (Ligon 2006). Two birds were photographed together at Palm Beach, Aruba, June 2010 (B. Daly, photos available at www.arubabirds. com). One was photographed by Ross Wauben at Divi Links, Aruba, on Nov 13,

2013 (fide eBird). One (perhaps the same bird) was seen at Divi Links, Aruba, on Feb 13, 2014 (R. Wauben fide eBird). One bird was seen at Bubali, Aruba, from Feb–Apr, 2016 (photograph by P. Sprockel, archived in eBird). One was photographed by Sipke Stapert near the airport on Bonaire, Feb 15, 2016 (personal comm.). **Range:** Occurs in two distinct populations, one extending from Colombia to Venezuela, another from Brazil to Argentina.

Black-crowned Night-Heron (*Nycticorax nycticorax*) Plate 9

Papiamento: *Garza di anochi, Krabechi bachi preto, Krabechi bachi pretu*
Dutch: *Kwak*

Description: A medium-sized, stocky heron with fairly short legs. Adult has black cap, large, black patch on back, and black bill; gray wings; white below and on sides of neck; yellow legs. Immature has brown back; wings with bold, white streaks and spots; streaked underparts; bill with yellow base. **Similar species:** Closely related Yellow-crowned Night-Heron, which is common, is the species most likely to cause confusion. Adult Yellow-crowned Night-Heron has all-gray body and wings and lacks black patch on back; black stripe across eyebrow and cheek join together on nape to frame white cheek and create masked appearance. Immatures of two species are very similar. On immature Yellow-crowned, bill lacks yellow base; also with less prominent spotting on wings and back; longer legs and neck. **Voice:** Call is a loud *phawk*, sometimes heard at night when birds are foraging. **Status:** Thought to breed in small numbers but apparently never confirmed on any of the islands. Occurs throughout the year, generally in small numbers (except at Bubali wetlands), likely augmented by post-breeding dispersers from Venezuelan mainland. In 1970s, up to 200 reported from Bubali wetlands, Aruba. Largest number we recorded over 10 years at Bubali was 22 on Apr 14, 2000. Occasional birds have been recorded from Tierra del Sol wetlands and from Spanish Lagoon, Aruba. Also recorded on Bonaire, from entrance to Gotomeer and Salina Tam (pers. obs. and Prins et al. 2009). Formerly seen at Sint Jan and Santa Cruz, Curaçao (Voous 1983), and apparently other locations as well (Prins et al. 2009). **Range:** One of the world's most cosmopolitan bird species, occurring on all continents except Australia and Antarctica.

Yellow-crowned Night-Heron (*Nycticorax violoacea*) Plate 9

Papiamento: *Krebechi Cu, Krabechi korona hel*
Dutch: *Geelkruinkwak*

Description: A medium-sized, stocky heron with fairly short legs. Adult has white crown and dramatic black mask. Body and wings blue-gray, legs yellow, bill black. Immature has brown back; wings with fine, white streaks and spots; streaked underparts; all-black bill. **Similar species:** See Black-crowned Night-Heron. **Voice:** Call similar to that of Black-crowned Night-Heron, but *phawk* made at slightly higher pitch. **Status:** Typically seen in small numbers (1–4) at coastal locations, most regularly on Curaçao and Bonaire; apparently rare on Aruba, where not observed in 150+ field observation days, across 10 years (AJW). Breeding documented on Bonaire in 1970s, near Slagbaai, where 10–15 nests were found. Historically, other breeding locations documented on Bonaire include Goto and Klein Bonaire, and, on Curaçao, at Sint Joris Bay and Isla Macuacu (Voous 1983), but no recent confirmed breeding records found from any of the islands. **Range:** Eastern US south through Caribbean and Central America to coastal South America, as far south as Peru on Pacific side and Brazil on Atlantic side.

Boat-billed Heron (*Cochlearius cochlearius*)   Plate 9

Papiamento: *Garabet boca di lancha, Garabet bok'i lancha*
Dutch: *Schuitbekreiger*

Description: Short, stocky heron with large eye, black cap and crest, and gray back with black crescent on upper back; also note white throat, breast, and hind neck, chestnut belly, yellow legs, and huge, black bill. Immature has brown back and lacks black back crescent and chestnut on belly. Apparently almost entirely nocturnal. **Similar species:** Virtually unmistakable but superficially similar to Black-crowned Night-Heron; immature Boat-billed has much larger, thicker bill, and lacks streaking and spotting. Adult Black-crowned Night-Heron lacks chestnut belly and has much more extensive black patch on back. **Voice:** When startled, makes a low pitched *quawk* similar to call of Black-crowned Night-Heron. **Status:** One record of immature bird seen in Pekelmeer, Bonaire, Oct 1972 (Voous 1983). Given widespread range in South America and the heron family's well known propensity for long, post-breeding movements, this is a species to be expected again. **Range:** Mexico south to Argentina. Also Trinidad.

Family Threskiornithidae: Ibises, Spoonbills

White Ibis (*Eudocimus albus*)   Plate 12

Papiamento: *Ibis blanco, Ibis blanku*
Dutch: *Witte Ibis*

Description: Adult pure white except for black wingtips, bright red legs, and red, down-curved bill. Juvenile has brown back, wings, and neck but with white rump, belly, and underwing (visible in flight), and with more orangish bill. As juvenile matures, brown feathers began to be replaced by white, starting with back feathers, and for a period looks patchy and mottled. **Similar species:** Adult is unmistakable. Immature could be confused with White-faced and Glossy ibises but neither has white belly, rump, and underwing. Juvenile Scarlet Ibis virtually identical to juvenile White Ibis and probably not distinguishable in field; immature Scarlet Ibis eventually takes on pink-tinged plumage (white on White Ibis). **Status:** From 1977 to 1984, this species occurred regularly in small numbers at Bubali, Aruba, with at least 30 records of 1–3 individuals (Voous 1983), but it has not been recorded there since 1984. A bird reported as an immature White Ibis (Jan–May, 2015, at various locations on Aruba), of which several photos have been submitted to eBird, appears to show the typical but often overlooked pinkish sheen of an immature Scarlet Ibis. There is one record for Curaçao of an immature bird captured and photographed in 1925 and a single adult photographed on May 9 and May 28, 2016, at the LVV site in Willemstad, near Seru Loraweg (Carel de Haseth, Rob Wellens). No records for Bonaire (Prins et al. 2009). **Range:** From southeastern US to northern South America.

Scarlet Ibis (*Eudocimus ruber*)   Plate 12

Papiamento: *Ibis cora, Ibis kora*
Dutch: *Rode Ibis*

Description: Adult reddish-pink with black wingtips. Juvenile has brown back, wings, and neck, but with white rump, belly, and underwing (visible in flight), like White Ibis. By second year, immature begins to acquire pink tones in rump, underside, and underwing. **Similar species:** Adult is unmistakable. Only other pink-red birds are American Flamingo and Roseate Spoonbill, both of which are larger and lack long, down-curved bill. Juvenile is probably indistinguishable from juvenile White Ibis. **Status:** Thirteen records at Bubali, Aruba, between 1977 and 1980, though sightings of only 1–2 birds at a time (Voous 1983; Prins et al. 2009). An adult was photographed at Bubali, Aruba, on Apr 7, 2013 (S. Mlodinow); likely the same bird was seen at Tierra del Sol, Aruba, through Aug 2013 (fide eBird). We and others, documented (still photos and video) an immature bird near and at Bubali, Aruba, from at least Nov 2013 through at least March 2014 (AJW, J. Wierda, S. Mlodinow). A bird that was seen at Spanish Lagoon, Divi Links, and Bubali, from May 2015 through at least May 2016, started off in a subadult plumage and eventually molted into full adult plumage (fide eBird). A bird reported as an immature White Ibis, Jan–May, 2015, at various locations on Aruba and of which several photos have been submitted to eBird, appears to show the typical but often overlooked pinkish sheen of an immature Scarlet Ibis. On Curaçao an immature bird was seen at Klein Hofje in Apr 2007 (Prins et al. 2009) and an adult at Schottegat in May 2010 (Eric Newton, pers. comm). Two records from Bonaire, perhaps pertaining to the same individual: one photographed at Lac Bay, Apr 26–27, 2011; one at Lac Bay on July 12, 2011 (fide eBird). **Range:** Northern South America.

Glossy Ibis (*Plegadis falcinellus*)   Plate 12

Papiamento: *Ibis preto, Ibis pretu*
Dutch: *Zwarte Ibis*

Description: In all plumages, often appears very dark in field. Has reddish legs and long, brownish, down-curved bill; a thin, white line extends from both the top and bottom of the eye to the bill. Dark eye. Breeding plumaged adult shows chestnut colors on body. **Similar species:** Immature White and Scarlet Ibises show white on belly, rump, and underwing. Glossy Ibis almost identical to very rare White-faced Ibis, which is distinguished in adult plumage by red eye and in breeding plumage (usually) by brighter and wider white line that extends from top and bottom of eye to bill. Nonbreeding White-faced Ibis lacks pale lines between eye and bill. Facial skin typically pink compared to dark or gray facial skin in Glossy Ibis. Legs of White-faced Ibis usually brighter red. Immature birds probably indistinguishable from White-faced Ibis. There are several species of all-dark ibises that occur in Venezuela; all are theoretically possible but unlikely, as very rare vagrants. Two of these show bright red facial skin that would make them unmistakably different. Most similar would be Green Ibis but note lack of white lines on face, shorter and thicker build, and hackles on back of neck. **Status:** Before 1977 there were only two records, one from Aruba in 1965 and a second from Malpais, Curaçao, in 1971 (Voous 1983). From 1977–1983, there were at least 20 records from Bubali, Aruba, with most sightings of 1-3 birds and an exceptional record of 8 individuals (Voous 1983). Since that time, there have been at least 7 records, 5 of them seemingly referring to single individuals on Aruba: one at Bubali, July–Aug 1989 and Aug 2003; one at Tierra del Sol wetlands, March 2004 (Prins et al. 2009); one in wetlands across from Bubali, near Windmill Resort, Feb 26, 2012 (J. Wierda, as reported on www.arubabirds.com); three from the same wetlands on Aug 10, 2014 (fide eBird); one regularly seen and photographed at Bubali, behind Windmill Resort, at Divi Links, and other places, Jan 2015 through Aug 2016 (all fide eBird); two seen from wetlands across from Bubali, Aug 26, 2016 (fide eBird). On Curaçao, single individuals were seen at Klein Hofje, in Nov 1992, Jan and Apr 1993, Nov 1997, and March 1998; and at Malpais in Nov 1992. Single adults (perhaps the same bird?) were seen and photographed at the LLV site in Willemstad, near Seru Loraweg, Curaçao, multiple times from 2013 to 2016 (Carel de Haseth, Rob Wellens). Six records from Bonaire of single birds at Lac, Sept–Oct 1980; at a wetland near the airport in Jan 2005 (both Prins et al. 2009); one at the Echo Dos Pos Conservation Center on May 18, 2008; one near the Bonaire Sewage Treatment Plant on June 20, 2014; one at Lac Bay on March 22, 2015; and one at the Bonaire Sewage Treatment Plant on June 2, 2016 (all fide eBird). **Range:** Occurs in tropical and subtropical regions around the world. In Western Hemisphere, ranges from Atlantic Coast of US, southern Maine to Florida, and along Gulf Coast to Mexico, Caribbean, Central America, and northern South America.

White-faced Ibis (*Plegadis chihi*)   Plate 12

Papiamento: *Ibis cara blanco, Ibis kara blanku*
Dutch: *Witmaskeribis*

Description: Virtually identical to Glossy Ibis except for red eye. Has pink lores; bold, thick, white line encircling face and eye in breeding plumage, and pinker legs. **Similar species:** See Glossy Ibis. **Status:** Voous (1983) reported studying a photo of one nearly adult bird in nonbreeding plumage with distinct white lines above and below eye. This description sounds more similar to nonbreeding Glossy Ibis, making the record uncertain. The species is known to wander widely, so although it does not occur in northern South America, it is not inconceiveable that the species will eventually be documented on the islands. **Range:** Split distribution, with northern population from western US to northern Central America and southern population in southern half of South America.

Roseate Spoonbill (*Ajaia ajaja*)   Plate 12

Papiamento: *Cucharon cora, Kucharon kora*
Dutch: *Rode Lepelaar*

Description: Medium-sized wading bird with large, flattened, spoon-shaped bill. Adult with pink body and wings, white neck, unfeathered gray head, and reddish legs. First-year birds are white with pinkish tinge, darker wingtips, dark legs, and feathered head. **Similar species:** Bill shape is characteristic. Scarlet Ibis is smaller, with down-curved bill and black wingtips. American Flamingo much larger, with distinctive bill. **Status:** At least 11 records for ABC islands up to Feb 2013 (mostly from Aruba). After that date, numbers showed a marked increase on Aruba, where it is now resident, and in small numbers; single birds and groups of 2–9 regularly seen at several locations, including Spanish Lagoon, Divi Links, Bubali, and the salinas behind Eagle Beach. Two records for Bonaire: one immature at Slagbaai, Apr 1960; one at Salina Boca Slagbaai, May 1998 (Prins et al. 2009). One record for Curaçao, of single bird at Santa Krus, May 1998 (Prins et al. 2009). **Range:** Southern US to southern South America, and including Greater Antilles.

Family Cathartidae: Vultures

Black Vulture (*Coragyps atratus*)   Plate 15

Papiamento: *Zamuro preto, Zamuro pretu*
Dutch: *Zwarte Gier*

Description: When perched, all black with unfeathered grayish head; short legs. In flight, underwing shows silvery-white wingtips; also note short tail. **Similar species:** Should be easy to distinguish from all species regularly occurring on the islands. **Status:** In the 1970s, two captive birds were reportedly brought from Colombia and released on Aruba. The three sight records known from Aruba are likely of these released birds: one bird at Savaneta, Jan 1973; two at Savaneta, July 1977; one at Savaneta, 1982 (Prins et al. 2009). At least two were photographed on Curaçao in Nov 2016 (Carel de Haseth, pers. comm.). No records from Bonaire. This species has occurred a number of times as a vagrant in the Caribbean, so it plausibly could occur as a rare vagrant on the islands. **Range:** Eastern US south to southern South America.

Turkey Vulture (*Cathartes aura*)   Plate 15

Papiamento: *Zamuro cabez cora*
Dutch: *Kalkoengier*

Description: Very large, similar in size to Crested Caracara, but note very distinctive head. When perched, all black, with unfeathered head either red (adult or first year) or dark (juvenile); short legs, long wings and tail. In flight, this very large bird is not likely mistaken; flies with its wings held in a shallow V, showing silvery flight feathers that contrast with black wing linings (underwing coverts). **Similar species:** Easily distinguished from all regularly occurring species on the islands. Vultures that commonly occur on mainland Venezuela have yellow heads and different pattern in flight feathers. **Status:** Two records of birds photographed on Aruba. One was photographed at 6:50 am, gliding 200 meters above the area between Pos Chiquito and Savaneta, June 17, 2010 (G. Peterson). Another photographed in flight at Sero Colorado, Feb 8, 2011 (R. van der Zwan). **Range:** Southern Canada south throughout South America.

Family Pandionidae: Osprey

Osprey (*Pandion haliaetus*)   Plate 13

Papiamento: *Gabilan piscador, Gabilan piskadó, Gabilan di laman, Patalewa*
Dutch: *Visarend*

Description: Large, long-winged, fish-eating hawk with brown upperparts, variable amount of white on crown, brown eye stripe. White underside, with some streaking on upper breast in female. In flight, note long wings held in bent M-shape, white wing linings contrasting with dark wrist patch, and mottled flight feathers; narrow dark bands on tail. Immature similar but with light edging to feathers of back and upperwing. Blue-gray legs. **Similar species:** On the islands, only birds of similar size are the Crested Caracara and White-tailed Hawk. Crested Caracara shows dark belly, dark wings with white patches in primaries, red facial skin, and long yellow legs. White-tailed Hawk much smaller, with dark head, chestnut shoulders in adult, much shorter wings and tail. **Voice:** Generally silent. Makes a high-pitched shrieking *heep-heep-heep-heep*, sometimes heard during aggressive interactions between wintering individuals. **Status:** Although it doesn't breed in the islands, individuals occur throughout the year, though more numerous from Sept through Apr. Can be found near reef islands, bays, lagoons, ponds, and deeper salinas, where they hover before plunging into the water feetfirst to capture fish; also seen perched on exposed poles, dead trees, and other perches. **Range:** Occurs throughout much of the world. Winterers on the islands are from North American breeding population, which winters from southern US to South America. Banded birds from Chesapeake Bay, US, have been recovered in Aruba and Bonaire.

Family Accipitridae: Hawks, Kites

White-tailed Kite (*Elanus leucurus*)   Plate 13

Papiamento: *Milano rabo blanco, Milano rabu blanku*
Dutch: *Amerikaanse Grijze Wouw*

Description: A small hawk. Adult with white head and underparts; light gray back; and wings with black bar on shoulder. Underwing white, with black primaries and black spot at bend of wing. White tail is square-tipped. Black patch surrounds red eye in adult (eye yellowish in immature). When hunting, often hovers; in soaring flight, holds wings in shallow V-shape (dihedral). At rest, wings do not reach end of tail. **Similar species:** A distant falcon seen in silhouette or a distant hovering American Kestrel might look similar, but with decent views the very different plumage pattern is generally unmistakable. **Voice:** Weak whistles and harsh notes, but unlikely to be heard. **Status:** One record of immature bird on Aruba, from early June to at least June 13, 1980, observed near Bubali and various interior locations (Voous 1983). It is a common resident of dry grasslands in Venezuela, so future records on the islands not unlikely. **Range:** Breeds from southern US through Central America and South America, to northern Argentina and Chile.

Swallow-tailed Kite (*Elaniodes forficatus*)   **Plate 13**

Papiamento: *Milano rab'i swalchi, Milani rab'I souchi*
Dutch: *Zwaluwstaartwouw*

Description: A slim, medium-sized hawk. Virtually unmistakable in flight, with black, deeply forked tail; white head, underside, and wing linings; and black flight feathers. **Similar species:** Only other large bird with deeply forked tail and white head that occurs on the islands is immature Magnificent Frigatebird, which is larger and longer-winged, with all-dark underwing, dark undertail coverts, and a massive, long bill. **Voice:** High-whistled notes, unlikely to be heard on the islands. **Status:** Two records. Two independent sightings of what was likely the same bird at Bubali, Aruba, March 25–30 and Apr 13, 2003. Another sighting of a single bird at Noord Salina, Kralendijk, Bonaire, Apr 26, 2002 (Prins et al. 2009). Although Prins et al. (2009) speculate that the birds that have appeared in the islands are likely of South American origin, satellite tracking of Swallow-tailed Kites from US breeding grounds has registered birds in movement from the Dominican Republic south, on a flight path that could pass over, or near, Aruba, Bonaire, and Curaçao. Spring sightings in the islands roughly coincide with spring migratory movements of Swallow-tailed Kites moving north to breeding grounds. **Range:** Widespread across South and Central America, with a disjunct breeding population in the southeastern US. Northernmost birds known to migrate to South America, as far south as Brazil. Southern South American populations thought to migrate north to unknown location, and other South American populations may have as yet undocumented migratory movements as well.

Long-winged Harrier (*Circus buffoni*)

Plate 13

Papiamento: *None yet proposed*
Dutch: *None yet proposed*

Description: Long-winged, long-tailed hawk with a bold, white rump in all plumages and flattened, owl-like facial disk. Adult male of pale morph is charcoal black above with contrasting lighter gray on inner primaries and secondaries of upperwing. One of the most striking features is the very dark head, nape, and upper chest band that contrast with the light eyebrow line and thin light "necklace" line that reaches up to posterior end of eyebrown line. Underwing heavily barred. Underparts light, with some limited teardrop streaking on upper breast and sides. Tail with alternating black and gray (appears white from below) bands. Female similar to male but more brownish above. Immature browner and more heavily streaked below and on underwing coverts. Dark morph of Long-winged Harrier has not been observed on the islands; entirely dark with contrasting lighter gray flight feathers. Flies with wings held in shallow V-shape, and, when hunting, glides low to the ground. **Similar species:** No other commonly occurring species looks similar. Adult male Northerrn Harrier lacks dark head and upperparts and has unbarred underwing with extensive black wingtips. Immature and female Northern Harrier can appear very similar but note the more contrasting head pattern in Long-winged Harrier and darker upperwing coverts in Long-wing Harrier that show more contrast with flight feathers on upperwing. Immature Northern Harrier usually also shows strong rusty-orange underside (typically lacking in Long-winged Harrier). Any harrier observed on the islands should be carefully studied and documented with photos and/or video. **Voice:** Whistles and barking notes, but unlikely to be heard. **Status:** One photographed July 6, 2015, by Elly Albers at Lac Bay, Bonaire, and seen at least until July 20, 2015 (fide eBird). One individual extensively photographed by Carel de Haseth at the LVV site in Willemstad, near Seru Loraweg, and present from Sept 20, 2015, through Dec 26, 2015 (personal communication). An individual that was likely the same bird was spotted at Malpais in Jan 2016. **Range:** Northern Colombia and Venezuela southeast through Guianas to Brazil and south to northern Argentina and Chile.

Northern Harrier (*Circus cyaneus*)

Plate 13

Papiamento: *Gabilan americano, Gabilan merikano*
Dutch: *Blauwe Kiekendief*

Description: Long-winged, long-tailed hawk with a bold, white rump in all plumages and flattened, owl-like facial disk. Adult male is blue-gray on upperparts, throat, and upper chest; white underparts with variable amount of reddish barring. Tail barred. Underside of wings white with black wingtips and thin black trailing edge, thickest near body. Adult female brownish with barred underwing and streaked chest and belly, but white rump obvious. Immature like female except chest, belly, and inner wing linings rusty-orange. Flies with wings held in shallow

V-shape, and, when hunting, glides low to the ground. **Similar species:** No other commonly occurring species looks similar. Immature Long-winged Harrier and Cinerous Harrier (not described) generally lack strong rusty-orange underside. Cinerous Harrier probably least likely to occur but is very similar to Northern Harrier in all plumages. Any harrier observed on the islands should be carefully studied and documented with photos and/or video. **Voice:** Whistles and barking notes, but unlikely to be heard. **Status:** One sight record from Klein Hofje, Curaçao, Oct 31 through Nov 30, 1997 (Prins et al. 2009). **Range:** Sometimes considered conspecific with Hen Harrier of Old World but range here given only for New World form. Breeding range extends from Alaska east to Quebec and south to Mid-Atlantic, Midwest, and California in United States. Winter range extends south regularly to Panama and Greater Antilles but with a handful of records in northern South America and Lesser Antilles. One record for Venezuela.

White-tailed Hawk (*Geranoaetus albicaudatus*)　　　Plate 13

Papiamento: *Falki, Partawela, Gabilan di seru*
Dutch: *Witstaartbuizerd*

Description: Large, broad-winged hawk with a short tail. Adult white underneath, with pale gray back and head. Upperwing dark gray, with prominent rusty shoulders. Upper tail coverts and tail white with single black subterminal band. Underwing linings white, flight feathers gray with darker wingtips and with black trailing edge of wing. When perched, wingtips extend well beyond tail. Legs bright yellow. In first year, young birds are very dark chocolate-brown on uppersides and tail, with some white patches on sides of face; white or yellowish-white patch on chest with dark band across belly; shows white upper tail coverts, which can look like a white U on the rump when seen in flight from above. By second year, more closely resembles adult but may continue to show dark breast band and dark underwing. (A dark morph of the species occurs in many populations but we find no documentation that it has occurred in the islands.) **Similar species:** Only regularly occurring species on the islands with which White-tailed Hawk likely to be confused are Crested Caracara and Osprey. Both much larger and with several plumage differences. Crested Caracara has distinctive, colorful, large, hooked bill and capped appearance. Osprey has largely white head, not dark as in White-tailed Hawk; has much longer wings; and tail lacks black subterminal band. **Voice:** Call described by Voous (1982) as a "penetrating 'kee-weet, kee-weet.'" **Status:** Regular in small numbers on Curaçao, where it apparently still breeds, with a population estimated at 15–20 pairs by Nijman et al. (2009), based on fieldwork in 2005 and 2006. In more than 23 years of visiting Aruba, we have never encountered the species, though there continue to be occasional unverified reports. Two reported sightings in 1988 and one sighting in 1997 on Aruba may have been the last known sightings there (Prins et al. 2009). Ligon, a resident birder on Bonaire from 2001–2015, noted that he never observed the species on Bonaire and suggested that the last observation of the species there may have been in 1998 (Ligon 2006), but Nijman et al. (2009) reported seeing three individuals in the northwestern part of the island in 2003. **Range:** Patchy distribution. The northernmost population is in coastal Texas and northern Mexico, with a series of apparently disjunct populations extending south to Argentina.

Family Rallidae: Rails, Gallinules

Sora (*Porzana carolina*)   **Plate 16**

Papiamento: *Gaito sora, Gaitu sora*
Dutch: *Soraral*

Description: Small marsh bird, with big feet and small bill; it almost looks like a miniature chicken (though unrelated to to that bird). Typically seen walking along edge of marshy wetland. Note stubby yellow bill, yellow-green legs, short tail. Grayish throat with varying amounts of black. Belly shows white barring against darker brown background. Upperparts dark brownish with light feather edgings on back. Immature birds show a light brown throat (gray and black on adult). **Similar species:** Common Gallinule, Caribbean Coot, and Purple Gallinule have similar shape and display similar behavior, as they are all members of the same family, but each is twice as big as Sora and either uniformly dark (coot and Common Gallinule), uniformly purple (adult Purple Gallinule), or tannish-green (immature Purple Gallinule). None have yellow bill with white barring on belly, as shown in Sora. A number of rails are theoretically possible as vagrants from both North America (perhaps Virginia Rail) and South America. **Voice:** Birds wintering on the islands often make a high, piercing *keek* but also give well-known descending whistled winny, commonly heard on breeding grounds, as well as loud, high whistled *sur whee*. **Status:** Uncommon but regular nonbreeding winter resident in the limited freshwater wetland habitats on the islands, occasionally also in mangroves. Seen very regularly from Oct to Apr in the Bubali wetlands on Aruba; in wet years in Malpais and Muizenberg wetlands on Curaçao, and scattered seasonal wetlands on Bonaire, as well as at the newly constructed sewage treatment plant grounds. Although 4–5 individuals are occasionally seen at one time, it is more common to see 1–2 birds at a given location. Birds have been observed as early as Aug, and there are apparently two July sight records from Bonaire (Prins et al. 2009), though the original source appears to only list a single July record (Ligon 2006). **Range:** Breeding range extends across southern half of Canada and most of US (not in the Southeast and southern part of Midwest). Winter range extends from southern US to northern South America.

Purple Gallinule (*Porphyrio martinicus*)   **Plate 16**

Papiamento: *Gaito biña, Gaitu biña*
Dutch: *Amerikaans Purperhoen*

Description: Adult unmistakable. A chickenlike (unrelated to chickens) marshbird with purple head and underparts, greenish back, bright yellow legs with long toes, bright reddish-orange stubby bill, and very short tail. On immatures, head and underparts are tan, back a dull greenish-blue. **Similar species:** Common Gallinule and Caribbean Coot are similar in shape and behavior but are uniformly dark. **Voice:** Makes harsh, often loud, clucks, chortles, and trumpeting sounds. **Status:** Rare to uncommon. Occurs irregularly in freshwater wetlands. Has bred at least once at Bubali, Aruba, in 2016 (two young fledged). Most records have been of immature birds, though adults have occasionally been seen. Typically a single bird is discovered every 2–5 years; single birds can sometimes be seen in the same wetland for months before disappearing. Perhaps most regular sightings have been at Bubali wetlands on Aruba, but birds have also been seen at Klein Hofje on Curaçao and, on Bonaire, at a pond near the entrance to Washington-Slagbaai NP, near Dos Pos, and at the sewage treatment plant. **Range:** Breeds from southeastern US south through Caribbean, Central America, and South America, to Chile and Argentina. Birds in most of US migrate south in winter; migratory movements of other populations poorly understood.

Common Gallinule (*Gallinula galeata*)   Plate 16

Papiamento: *Gaito pico còrá, Gaitu pik kòrá*
Dutch: *Waterhoen*

Description: A chickenlike marsh bird (unrelated to chickens). Black head and underparts; dark brown back with narrow white line along flank. Very short tail. White outer undertail coverts. Bright yellow legs and long toes. On breeding birds, forehead shield and bill red, with yellow tip to bill. On nonbreeding birds, bill and forehead shield brownish. Immature dull brown overall but still shows white flank line. **Similar species:** Most similar to Caribbean Coot but latter species has white, not red, forehead shield and bill, and lacks white line along flank. **Voice:** Makes harsh, often loud, clucks, chortles, and trumpeting sounds. **Status:** Common year-round breeding resident at Bubali wetlands and Tierra del Sol ponds on Aruba (and small numbers seen regularly at Savaneta and pond at Divi Links); smaller numbers at Klein Hofje sewage ponds on Curaçao. In rainy years, common breeder in wetlands on Curaçao (Malpais, Muizenberg, for example) and Bonaire. Maximum counts we have recorded include 87 adults on Jan 8, 1994, and 78 adults on Jan 12, 1993, both at Bubali, Aruba. **Range:** Very widespread, occurring across North and South America, Europe, Asia, Africa, and the Pacific.

Caribbean Coot (*Fulica caribaeae*)   Plate 16

Papiamento: *Gaito frente blancu*, Kùt Cu, *Gaitu frente blanku*, Kùt
Dutch: *Caribische Koet*

Description: Chickenlike marsh bird (unrelated to chickens). Entirely black, with white outer undertail coverts. White forehead shield; white bill with black ring. Yellow legs and toes. Immature similar but grayer overall, with dusky bill and legs. **Similar species:** Most similar to Common Gallinule but latter species has red (not white) forehead shield and bill, and has white line along flank. Still some question as to whether or not Caribbean Coot and American Coot should be considered separate species, as small numbers of American Coots in the US and Canada show all-white or largely white forehead shields rather than the extensive dark red forehead calluses normally seen on American Coots. Apparently the higher the testosterone level in a male American Coot, the more likely it is that the red callus will be reduced or absent. Individuals showing varying amounts of red callus have been documented on all three islands, even breeding on at least one occasion. This would seem to indicate that birds with red calluses are unlikely to have originated as migrants or vagrants from North American populations of American Coot, but rather are variants of the form (Caribbean Coot) found locally. **Voice:** Makes harsh, often loud, clucks, chortles, and trumpeting sounds. **Status:** Common year-round breeding bird in permanent wetlands at Bubali and Tierra del Sol on Aruba; Klein Hofje sewage ponds on Curaçao; and likely at the sewage treatment ponds on Bonaire. With sufficient rain, large numbers can build up at wetlands like those at Muizenberg and Malpais on Curaçao and at various temporary ponds, including those on Bonaire, near Onima, and at the pond near the entrance to Washington-Slagbaai NP, where birds often breed until the wetlands dry up again. Exceptional counts of 800 (1997) and 600 (2005) were documented at Muizenberg, but counts of 10–100 individuals are more typical. **Range:** Caribbean and northwest Venezuela.

American Coot (*Fulica americana*)   not illustrated

Papiamento: *Gaito pico blancu, Gaitu pik blanku,* Kùt
Dutch: *Amerikanse Meerkoet*

Description: Chickenlike marsh bird (unrelated to chickens). All black, with white outer undertail coverts. White bill with red forehead and black ring on bill. Yellow legs and toes. Immature similar but grayer overall, with dusky bill and legs. **Similar species:** See Caribbean Coot. The one characteristic used to differentiate American Coot from Caribbean Coot is the color of the forehead: red in American and white in Caribbean; however, there is evidence that either color may appear on both species. Nevertheless, birds with white foreheads are rare in American Coot populations and those with red foreheads are rare in Caribbean Coot populations. **Voice:** Makes harsh, often loud, clucks, chortles, and trumpeting sounds. **Status:** At least 8 birds with red foreheads have been found on Aruba; 4 on Curaçao (including one sharing nest incubation duties with a typical white-shielded Caribbean Coot) and 10 on Bonaire (Voous 1983; Prins et al. 2009). This and other evidence has led some avian classifications to lump Caribbean Coot and American Coot into a single species that they designate as American Coot. American Coots breeding in the US and Canada are known, from band returns, to winter in the Greater Antilles and south to Costa Rica. It is conceivable that a migrant North American bird could occur on Aruba, Bonaire, or Curaçao, although such a bird would presumably be impossible to differentiate from a rare resident Caribbean Coots with a red forehead. We have provisionally included both forms here as separate species to encourage observers to continue to document any red-shielded individuals in order to learn more about their status. **Range:** Breeding range extends from Canada south to Central America and Greater Antilles. Wintering birds are known as far south as Panama. A disjunct resident breeding population occurs in Colombia; formerly a breeding population also occurred in Ecuador (Hilty and Brown 1986; Ridgely and Greenfield 2001).

Family Heliornithidae: Sungrebe

Sungrebe (*Heliornis fulica*) **Plate 15**

Description: A small waterbird, somewhat ducklike. Note brown back, bold black stripes on head and neck, white chin and throat, and slim, pointed, often reddish, bill. **Similar species:** Nothing similar in the islands, although young Pied-billed and Least Grebes show striped heads; they are smaller, darker overall, with shorter, smaller bills. Any sighting of this species should be carefully documented with photos and/or video. **Voice:** Unlikely to be heard in the islands. **Status:** One photo taken on Nov 15, 2010, of a male or nonbreeding female, in a pond near the entrance to Washington-Slagbaai NP (Rozemeijer 2011). **Range:** Year-round resident from Mexico south through Central America and South America, to Brazil.

Family Aramidae: Limpkin

Limpkin (*Aramus guarauna*) **Plate 11**

Papiamento: *Garao*
Dutch: *Koerlan*

Description: Heron-sized brown wading bird with dark legs and rather thick, mostly yellowish, slightly down-curved bill. South American subspecies shows thin, white streaking only on the head and neck (in the northern subspecies, the white streaking and spotting also occurs on the back and wings). **Similar species:** Brown plumage, with white streaks and spots, could suggest immature Black-crowned and Yellow-crowned night-herons but note larger size, dark (not yellow) legs, and longer, down-curved bill. Immature night-herons also show extensive white spotting and streaking on back and wings but have white undertail coverts (dark in Limpkin). Immature and nonbreeding White-faced and Glossy ibises also have dark plumage with white streaking on head and neck, dark legs, and a bill that (sometimes) appears yellowish. However, Limpkin is taller, with thicker, less down-curved bill, and with more extensive and obvious white streaking on head. Ibises also usually show bare facial skin between base of bill and eye—this area is feathered in Limpkin. **Voice:** Within its normal range, the screaming cry of the Limpkin often emanates from wetlands, usually at night. Unlikely to be heard on ABC islands, but its call has been likened to *keEEEuur* (Crossley 2011) or "carrao" (Voous 1983). **Status:** One record confirmed with photos from Sero Colorado, Aruba, for 10 days in Feb, 1975 (Voous 1983). Recent sight record of a bird reported from Tierra del Sol ponds, Aruba, on March 12, 2005 (Mlodinow 2006). **Range:** There is a disjunct population in Florida and neighboring southern Georgia. The bird also occurs from southern Mexico through Central America, to northern Argentina, and in Greater Antilles and Bahama.

Family Burhinidae: Thick-knees

Double-striped Thick-knee (*Burhinus bistriatus*)   Plate 17

Papiamento: *Snepi di mondi*
Dutch: *Caribische Griel*

Description: Medium-sized shorebird that prefers dry, open habitat; active at night, typically sits motionless during the day. Has long, yellow legs; nondescript brownish upperparts; short tail. Also note stout bill, bold black "eyebrow" line above white eye line, and yellow eye. **Similar species:** Larger than plovers or sandpipers; perhaps most similar in size and general appearance to a nonbreeding Black-bellied Plover, American Golden-Plover, or Upland Sandpiper. Of these, only the Upland Sandpiper has yellow legs, but it is much smaller and lacks black "eyebrow" and white eye line. **Voice:** Not often heard during the day; at night gives a variety of loud calls, including one described as *pip-prri-pippridipip-pip-pip-pridipip* (Garrigues and Dean 2007). **Status:** One confirmed record of a bird captured and photographed at Piscadera Bay, Curaçao, in July 1934 (Voous 1983). Some historical references seem to suggest that the species once occurred regularly on the islands. Apparently relatively widespread and common on Venezuelan mainland, including on Isla Margarita (Hilty 2003), and it shows some propensity for movement, as there are five records of vagrants on Trinidad and Tobago (Kenefick 2007). So, it should be looked for on the ABC islands. **Range:** Southern Mexico south through Central America and northern Colombia, to northern edge of Brazil. Also on Hispaniola.

Family Recurvirostridae: Stilts, Avocets

Black-necked Stilt (*Himantopus mexicanus*)   Plate 17

Papiamento: *Snepi hudiu, Macamba, Kaweta di patu, Redadó*
Dutch: *Steltkluut*

Description: Striking bird with long, pink legs; black upperparts that contrast with white underparts; and long, thin bill. The overall effect is that of a bird walking on stilts. **Similar species:** Unmistakable. **Voice:** Makes a loud, strident *weep-weep-weep* (Voous 1983); also produces a *kek-kek-kek-kek* (Paulson 2005). Flocks of birds call incessantly when disturbed, and adults with young give alarm calls with great insistence. **Status:** Nesting has been documented on all three islands, in seasonal wetlands as well as permanent wetlands such as coastal salinas and mangroves. Nonbreeding birds are commonly found in flocks ranging from dozens to occasionally hundreds of individuals. Maximum number reported: 500 near Sorobon, Bonaire, on Nov 1, 2001 (Ligon 2006). **Range:** Breeding range extends from northwestern US and Mid-Atlantic region in eastern US south to northern South America and the Caribbean.

American Avocet (*Recurvirostra americana*)   Plate 17

Papiamento: *Kaweta di pato boka lantá, Kaweta di patu boka lantá*
Dutch: *Amerikaanse Kluut*

Description: A large, long-legged shorebird with distinctive long, thin, upturned bill. White underparts and upper back. Black lower back and wings show a broad, white scapular bar. Head is tannish in breeding plumage, grayish in nonbreeding plumage. Dark legs. **Similar species:** Unmistakable. **Voice:** Makes single note calls similar to that of Black-necked Stilt but slightly higher pitched. **Status:** One record of two birds together with Black-necked Stilts at Lac, Bonaire, on March 7, 1979 (Voous 1983). **Range:** Breeds in southwestern Canada and western US south to interior Mexico. Winters from southern US to Central America.

Family Haematopodidae: Oystercatchers

American Oystercatcher (*Haematopus palliatus*)   Plate 17

Papiamento: *Kibra kokolishi, Shon Piet*
Dutch: *Amerikaanse Bonte Scholekster*

Description: A large shorebird boldly patterned with black head, dark brown upperparts, and white underparts. Massive orange bill and pink legs. In flight shows bold, white wingstripe. **Similar species:** Virtually unmistakable. Willet, a relatively large shorebird, also shows a bold, white wingstripe in flight, but it lacks bright orange bill and sharply contrasting plumage. **Voice:** Loud, insistent calls described by Voous (1982) as *peet-a-peet-a-peet*, and said to be the origin of the Papiamento name *Shon Piet*, which means "Mr. Peter" (Voous 1983). **Status:** Throughout the year, small numbers occur on rocky sections of the coast of all three islands. Though birds logically must breed on all three islands, they have only been confirmed on Aruba (San Nicolas reef islands in 1987 and 1989) and Bonaire (near Pekelmeer, 1968, Boca Onima, 1971, and Playa Chiquitu, 1975); breeding has apparently not been confirmed on Curaçao (Prins et al. 2009). Breeding activity seems to be concentrated in the spring and early summer. In five days of birding on Bonaire, in March 2001, we found 6 birds; in four days of birding on Bonaire, in July 2001, we counted 7–8 individuals. On Aruba, we have found that the species occurs more regularly on the island itself than on the islets off Oranjestad, at Malmok, and Boca Prins, and invariably in pairs. We have received reports from other observers of sightings on Aruba, from near Savaneta, Arashi, the coast near

California Lighthouse, and Dos Playas. **Range:** Occurs exclusively on marine coasts. Found on both Pacific and Atlantic coasts of North America, Central America, and South America. Throughout Carribean, generally less common, and with spotty distribution.

Family Charadriidae: Plovers, Lapwings

Black-bellied Plover, Gray Plover (*Pluvialis squatarola*)   **Plate 18**

Papiamento: *Lopi gris, Lopi shinishi*
Dutch: *Zilverplevier*

Description: As on the American Golden-Plover, the large, dark eye is often a particularly striking feature. In nonbreeding or juvenile plumage (the plumage state most likely to be seen on the islands), rather nondescript grayish-brown on upperparts. Light brown mottling underneath does not extend beyond legs. Dark bill and thick, stubby, dark legs. In flight, note black "armpits," white rump and upper tail, and white wingstripe. In breeding plumage, black underneath does not extend to undertail coverts, sometimes not even beyond legs. In all plumages, wingtips are even with or extend only slightly beyond tail tip. **Similar species:** See American Golden-Plover. **Voice:** Long, plaintive whistled *tur-eeee* or *tee-urr-eee*, upslurred at the end. **Status:** Has occurred throughout the year on all three islands, typically at coastal areas or sometimes around edges of ponds and salinas, numbering from single individuals to flocks of 5–15. During spring and fall migration periods, numbers can increase as suggested by our observation of a group of 120 on March 24, 2001, at the Pekelmeer on Bonaire. **Range:** Breeds in high Arctic of Canada and Alaska. Winters in coastal areas from US south to southern South America.

American Golden-Plover (*Pluvialis dominica*)   **Plate 18**

Papiamento: *Lopi dorado, Lopi dorá*
Dutch: *Amerikaanse Goudplevier*

Description: Although all plovers have large, dark eyes, this feature is particularly striking in the American Golden-Plover. In nonbreeding or juvenile plumage (the plumage most likely to be seen on the islands), note nondescript golden-brown on upperparts. Light brown mottling underneath extends along flanks to well beyond legs. Dark, rather thin, bill and dark legs. In flight, note lack of black "armpits" and all-dark uppersides. In breeding plumage, black underneath extends from chin to undertail coverts. In all plumages, wingtips extend well beyond tail. **Similar species:** Most likely to be confused with nonbreeding or juvenile Black-bellied Plover. Nonbreeding and juvenile American Golden-Plover lack black "armpits," white rump, and bold, white stripe on upperwing seen on Black-bellied Plover; on American Golden-Plover, white stripe over eye is more prominent and barring on chest extends along the flanks (on Black-bellied Plover, barring stops before the legs). American Golden-Plover has longer wings that extend well beyond tail tip. In Black-bellied, wings are even with tail or barely extend beyond it. In breeding plumage, American Golden-Plover shows more extensive black on underside extending all the way to undertail coverts. **Voice:** High, nasal "weedle" or "queedle," most likely made by birds in flight or disturbed birds in a feeding or roosting area. **Status:** Occurs with most frequency and in highest numbers Sept–Nov, but generally rather rare and unpredictable. Numbers typically range from 1–3 birds, but larger flocks occasionally occur, probably when migrating birds are blown westward or forced down by storms, as they make their way south to southern South America, where they normally winter. For example, about 100 were seen at Malpais, Curaçao, on Nov 4, 1962 (Voous 1983); flocks of 25 and 15 were again seen on Curaçao, in Oct 2001 (Prins et al. 2009). We have encountered the species once (single bird) on Curaçao, at Blue Bay Golf Course, in Nov 2003, and once (two birds) on Aruba, at Tierra del Sol ponds, in Nov 2011, but in recent years resident birders have been finding the species with increasing regularity during fall migration. Only a single encounter with the species reported on Bonaire—a single bird at Slagbaai, in Sept 1997 (Ligon 2006). There are only a few spring records; during spring migration, flocks pass quickly through Central America to reach northern breeding grounds. **Range:** Breeding range extends across northern North American tundra zone, from Alaska to Nunavut. Winters on grasslands of southern South America, largely in Argentina, Uruguay, Bolivia, and southern Brazil.

Southern Lapwing (*Vanellus chilensis*) Plate 18

Papiamento: *Kivit sur-amerikano*
Dutch: *Chileense Kievit*

Description: A medium-sized plover with pink legs and distinctive black, wispy crest; bill pink with dark tip; also note red eye-ring and eye. Gray head separated from black face and chin by narrow white border. Gray back, black breast, white belly. Bronzy feathers above dark green at bend of wing. In flight, black flight feathers on wing contrast with white bar on inner wing; black tail with white rump and base of tail. **Similar species:** With a good view, unmistakable. **Voice:** Loud, persistent *kep-kep-kep* or *keeah-keeah-keeah*. **Status:** First recorded from Bubali, Aruba, June 6, 1979, but second record not until May 30, 2001 (also at Bubali). After 2001, sightings became more frequent, with regular sightings from Tierra del Sol, where breeding was suspected in 2004 and 2005 and unequivocally confirmed in 2011. Birds are now regularly sighted at 4–5 wetland locations on Aruba and breed at Tierra del Sol, Divi Links, and, possibly, at Bubali and Savaneta. We observed 10 at Tierra del Sol golf course, Aruba, on Oct 9, 2016, and 5 together at Divi Links near Oranjestad, Aruba, in Nov 2011. On Curaçao, first record of a single bird at Klein Hofje on May 17, 2004 (Prins et al. 2009); pair of birds observed and photographed at Blue Bay Golf Course, Dec 2010 (Leo Spoormakers, Observado) and a single bird photographed at Muizenberg, Apr 2012 (John van der Woude, Observado); now confirmed nesting at Blue Bay Golf Course, Santa Barbara Golf Course, Malpais, and Curaçao Golf and Squash Club, and seen at multiple locations on the island (fide eBird and Observado). On Bonaire, 2–3 birds were seen multiple times between May and Oct 2006 at the ponds near the entrance to Washington-Slagbaai NP (Ligon 2006), and up to 5 have been present at the Bonaire Sewage Treatment Plant since at least 2014 and are now likely breeding there (fide eBird and Observado). **Range:** Occurs across South America, at low elevations, and its range has been expanding northward over the last 15–20 years. Now occurs as far north as Costa Rica and, in Caribbean, in Barbados.

Collared Plover (*Charadrius collaris*)   Plate 19

Papiamento: *Lopi coyar, Lopi koyar*
Dutch: *Kraagplevier*

Description: A small, elegant plover, with mud-brown on upperparts, white forehead bordered by black, orange on crown and at base of complete black breast band. Immatures lack orange tones and have incomplete breast band. Although often said to lack the white collar seen in other similar plovers, we have observed birds with a distinctive white collar. Legs pink. Bill dark. **Similar species:** Semipalmated Plover is slighty darker on back, lacks cinnamon tones on head, and has shorter, stubbier bill. Snowy Plover is sandy-colored on upperparts, has black legs, and lacks full breast band. Wilson's Plover also usually shows colorful orangish tones in its head, but is larger and has a larger, longer bill. **Voice:** Call is a sharp *peep* or *whit*. **Status:** Apparently an occasional nonbreeding visitor, mostly Apr to Aug, but has occurred in all seasons. The species may be more regular than historically documented. We counted six birds on Bonaire during surveys from July 4–7, 2001, and photographed several individuals. Since about 2010, birders on Aruba have documented (often photographed) many occurrences of the species at many locations, including some remarkable concentrations. Maximums reported include: 40 near California Lighthouse, Aruba, Oct 16, 2016; 30 near California Lighthouse, Aruba, Oct 13, 2014; and 18 near California Lighthouse, Aruba, Oct 4, 2015 (all M. Oversteegen fide eBird). **Range:** Occurs along coasts of Mexico, south through Central America and most of South America, where it also occurs in interior regions.

Snowy Plover (*Charadrius nivosus*)   Plate 19

Papiamento: *Lopi blancu, Lopi blanku*
Dutch: *Strandplevier*

Description: A small plover, with pale sandy upperparts with white forehead and face; sometimes has a small, dark area in front of eye. In nonbreeding plumage, shows a single, partial gray breast band; in breeding plumage, shows a single, partial black breast band, black line behind eye, and black border to white forehead. In all plumages, shows white collar, all-black bill, and dark legs. **Similar species:** Only small plover with dark legs, although legs can appear lighter and reddish when strongly backlit. Smaller and lighter colored than Wilson's Plover, with dark (not pink) legs and much smaller, slimmer bill. Semipalmated Plover is darker-backed, with full breast band and orangish legs. Very rare Piping Plover has orangish legs and thicker, shorter bill, and always lacks black behind eye. **Voice:** A soft whistled *puwee*. **Status:** Occurs along beaches and salt pans, often in hot, dry areas. Most regular on Bonaire, where during surveys we counted 40 individuals, from July 4–7, 2001. At least seven pairs showed evidence of nesting. Smaller numbers occur through winter period. Said to be a scarce breeding resident on Curaçao (Prins et al. 2009) and rare on Aruba, where we have not encountered the species ourselves. **Range:** A complex of subspecies, some considered distinct species, occur globally in tropical regions. In the Americas, breeds in western US and Gulf Coast south to Mexico. Also breeds in Greater Antilles, a section of the Pacific Coast of South America, on ABC islands and adjacent coast of Venezuela, and possibly Colombia.

Wilson's Plover (*Charadrius wilsonia*)   Plate 19

Papiamento: *Lopi pico diki, Lopi pik diki*
Dutch: *Dikbekplevier*

Description: A medium-sized plover, with mud-brown on upperparts and white on forehead that extends to behind eye. White collar and single black (breeding plumage) or brown (nonbreeding) breast band against white underparts. Legs pink. Stout, black bill. The resident subspecies shows cinnamon tones around head and where the breast band joins brown upperparts. **Similar species:** Killdeer is much larger, with two breast bands and orangish rump and upper tail in flight. Semipalmated Plover is much smaller, with shorter, thinner, all-black bill and pink legs; it lacks cinnamon tones on head. Snowy Plover is sandy-colored on upperparts, has black legs, and lacks full breast band. In nonbreeding plumage, very rare Piping Plover lacks full breast band. Collared Plover has pink legs, longer bill (finer than in Wilson's), but also usually shows some colorful orangish tones in head, as in resident subspecies of Wilson's. **Voice:** Normal call is a sharp *whit* or *pip*; alarm call a more strident *weet*. **Status:** Year-round breeding resident on Bonaire; probably used to breed on Aruba and Curaçao. Occasionally observed on Curaçao, rarely on Aruba. (In surveys in July 2001, we counted 16–18 pairs on Bonaire, including 8–10 birds at Lac and a single chick.) Three subspecies of Wilson's Plover have been described. The resident subspecies (*cinnamominus*) on Aruba, Bonaire, and Curaçao, which occurs along the northern South American coast from Colombia to Brazil, is the subspecies normally seen. But Voous (1983) mentions one specimen record and one possible sight record of a wintering bird from a subspecies (*wilsonia*) that breeds from the Atlantic and Gulf Coasts of the US south into Greater Antilles and the Atlantic coast of Central America. Prins et al. (2009) mentions two possible sight records of subspecies that breeds along the Pacific Coast of Mexico south to Peru (*beldingi*). Northern populations of *wilsonia* migrate south, though the exact boundaries of their "normal" wintering range are poorly understood; it is conceivable that they could occur in the ABCs during the winter months. Northern birds lack the rusty, cinnamon tones in the head and breast band and apparently have a larger bill than the resident subspecies, but caution should be exercised in subspecific identification as immatures of the resident subspecies may show less of the rusty coloration. We are skeptical of claims of the *beldingi* subspecies, especially as we suspect that there has been insufficient documentation of plumage variation across the range of Central and South American populations of Wilson's Plover. The *beldingi* subspecies is said to be darker than *wilsonia*, with broader dark lores and reduced white eye line (O'Brien et al. 2006), and sometimes with rusty tones at back of sides of head (Howell and Webb 1995), but this may vary across the range. **Range:** On Pacific Coast from Mexico south to Peru. On Atlantic Coast from mid-Atlantic of US south to northern South America, and through Caribbean.

Semipalmated Plover (*Charadrius semipalmatus*) Plate 19

Papiamento: *Lopi semipalmado*
Dutch: *Amerikaanse Bontbekplevier*

Description: A small plover, with mud-brown upperparts, white forehead bordered by black, and black mask through eye (mask brown in nonbreeding plumage). Also note white collar and single black (breeding plumage) or brown (nonbreeding) breast band against white underparts. Legs orange. In breeding plumage, shows orange bill with black tip and narrow yellow ring around eye. Bill dark in nonbreeding plumage. **Similar species:** Killdeer is much larger, with two breast bands and orangish rump and upper tail in flight. Wilson's Plover is larger, with larger, thicker bill and pink legs; subspecies seen on islands shows rusty cinnamon tones around head. Snowy Plover is sandy-colored on upperparts, has black legs, and lacks full breast band. Like Semipalmated, very rare Piping Plover has orangish legs, but in nonbreeding plumage it lacks the full breast band. Collared Plover has pink legs, longer bill (finer than in Wilson's), and usually shows some colorful orangish tones in head. **Voice:** Makes a soft, whistled *chu-wee*. **Status:** Regular throughout the year, in small numbers at coastal locations and around edges of freshwater ponds and salinas. Probably more numerous during migration periods, when flocks of 50–100 have been counted. The Pekelmeer, Bonaire, is a particularly good area to see larger numbers of this species. An individual banded and dyed on the coast of James Bay in Ontario, Canada, in mid-Aug, 1977, was photographed later that month on Bonaire, Aug 29–31, 1977 (Voous 1983). **Range:** Breeds in northern Canada and Alaska. Winters from southern US south to southern South America.

Piping Plover (*Charadrius melodus*) Plate 19

Papiamento: *Lopi melódic, Lopi melódiko*
Dutch: *Dwergplevier*

Description: A small plover, with pale sandy upperparts and white forehead and face. Also note white collar and, in nonbreeding plumage, a single, partial gray collar and white underparts. In breeding plumage (less likely here on ABCs), shows a single black (complete or partial) breast band and black band above white forehead. In nonbreeding plumage, shows all-black bill, which has orange base in breeding plumage. Legs orange. **Similar species:** Snowy Plover is most similar, but it always shows dark (not orangish or pinkish) legs and has thinner, longer bill. Use caution, as a Snowy Plover observed with strong backlighting can appear to show reddish legs. Semipalmated Plover has orangish legs and, in bright light, its back could look as pale as that of Piping Plover, but note dark area between bill and eye (white or gray in Piping Plover); also note that one subspecies of Piping Plover has darker lores than the other. **Voice:** Makes a gentle *peep-lo*. **Status:** The only well documented record that we know of is a bird photographed by Sipke Stapert at Pekelmeer, Bonaire, on March 17 and 20, 2016 (fide eBird). Three undocumented reports, all sight records, two from near Pekelmeer, Bonaire, respectively on Jan 1977 and June 1982 (Voous 1983; Ligon 2006; Prins et al. 2009,). Third sight record from Salina Slagbaai, Washington NP, Bonaire, on Jan 15, 2010 (fide eBird). Piping Plover would be exceedingly rare so identification should be made carefully and documented with photos. **Range:** Breeds on sandy beaches, from Canadian Maritimes south to mid-Atlantic states of US, with disjunct inland breeding populations extending from the Canadian prairies south to Nebraska and eastern Colorado. Birds winter along Atlantic and Gulf coasts of southern US and Mexico, in northern Bahamas, and locally in Cuba. Birds also winter irregularly on Pacific Coast of Mexico. Single specimen taken in 1955 in Ecuador (Ridgely and Greenfield 2001) but apparently no documented records in Central America and very few records (even fewer well-documented) in Caribbean outside Bahamas and Cuba.

Killdeer (*Charadrius vociferus*)   **Plate 19**

Papiamento: *Lopi killdeer*
Dutch: *Killdeerplevier*

Description: A medium-sized plover, with mud-brown on upperparts. White forehead bordered by black line; white eye line and collar; and note two black breast bands against white underparts. Legs pink. Stout, black bill. Rump and much of tail bright orange. **Similar species:** Double black breast bands are distinctive, as all other similar plovers have only a single breast band. **Voice:** Says its own name with a loud strident *kill-deer, kill-deer, kill-deer*. **Status:** Formerly only a wintering bird, in small numbers along coasts and wetland areas generally, from Aug through Apr; wintering birds presumably migrate from northern part of breeding range. Since 1979, a documented breeder and year-round resident on Aruba, at Bubali and Tierra del Sol, and on Curaçao, at Klein Hofe (Prins et al. 2009). Seemingly least common on Bonaire, where breeding has not been documented and where a resident birder encountered the species only four times (Ligon 2006). It is unclear if birds breeding on Aruba and Curaçao originated from Central or South American breeding populations or from North American populations. Similarly, the origin of wintering birds is unclear. It is interesting that nearby Venezuela had fewer than 10 records of Killdeer as of 2003 (Hilty 2003; Prins et al. 2009). On Aruba, we have found the species regularly in winter at numbers normally ranging from 1–5 birds at Bubali, Savaneta, Spanish Lagoon, and Tierra del

Sol and occasionally at other locations. Our largest count was of 13 on Jan 12, 1998, at Bubali, Aruba, where 14 were recorded May–June 1984 (Prins et al. 2009). **Range:** The species has an interesting distribution; the North American breeding range extends from Canada south to Mexico, with birds migrating south to vacate the northern half of the breeding range in winter. There is a resident breeding population in the Bahamas and Greater Antilles, Caribbean, and on the Pacific Coast of South America, from Ecuador to Peru. Along with those found on Aruba and Curaçao, there are small populations breeding in Costa Rica, occasionally in Panama, and perhaps other parts of Central America.

Family Jacanidae: Jacanas

Wattled Jacana (*Jacana jacana*)   **Plate 16**

Papiamento: *Jacana*
Dutch: *Leljacana*

Description: Combination of long, grayish-green legs, very long toes, and bright red shields and wattles at base of yellow bill makes this species unmistakable. Head and underparts black; in subspecies expected here, normally shows rusty inner wings and "saddle" on back. Flight feathers bright yellow; when wing is folded, yellow shows only as a narrow yellow line or not at all. Immature bird is quite different, with white underparts, brownish upperparts, and white line over eye; still shows yellow in wing, some rusty color on inner wing, and pinkish or light purple rudimentary facial shield at base of bill. **Similar species:** Common Gallinule and Caribbean Coot are both black-bodied, marsh-loving birds with relatively long legs (not as long-legged and certainly not as long-toed as Wattled Jacana) but lack yellow in wings and rusty colors on lower body and inner wings. Common Gallinule does have a red facial shield that could suggest a Wattled Jacana, but note the lack of yellow in wings in Common Gallinule and more yellow legs. As Sora and immature Purple Gallinule both have yellowish bills, they could be mistaken for immature jacana, but both lack bold eye stripe and yellow in wings. Northern Jacana occurs south to northern Panama and is a plausible rare vagrant to the islands; adult lacks red wattles and bill shield is normally yellow. Immature is very similar but has yellow (not pink or purplish) rudimentary bill shield. Any jacana species observed here should be carefully documented with photos. **Voice:** Harsh, raspy cackling calls. Described by Hilty (2003) as "a loud complaining *kee-kick, kee-kick*." **Status:** Four records. One dead adult bird found on Bonaire in 1971 preserved as a specimen (Voous 1983). Two sight records from Curaçao: one at Piscadera Baai, Aug 21, 1980, and one at Klein Hofje sewage ponds on March 15, 1992 (Prins et al. 2009). One adult photographed at Bubali, Aruba, present from June 20 to Aug 11, 2012 (A. and J. King, photo on www.arubabirds.com and from unnamed source on eBird.org). **Range:** Occurs from southern Panama south through most of South America to Argentina. Several subspecies have been described.

Family Scolopacidae: Sandpipers and Allies

Upland Sandpiper (*Bartramia longicauda*)   Plate 20

Papiamento: *Snepi Bartram*
Dutch: *Bartrams Ruiter*

Description: Medium-sized, brown-backed shorebird with yellow legs; small head, long, thin neck; large eyes; and rather short, delicate bill that is yellowish on inner two-thirds of lower mandible. Streaked on neck, breast, and flanks. At rest, wingtips reach only part way down tail. In flight, lacks any obvious wing-stripes or other markings above. In contrast to other shorebirds, this species is unlikely to be found wading in water, preferring instead grasslands, both on breeding and wintering grounds. **Similar species:** Combination of small, dove-like head and neck, large eyes, and short, fine bill is distinctive. Greater and Lesser yellowlegs are larger, with longer bills, and show white rump and lower back. Solitary Sandpiper is grayer and more spotted on back, with an obvious white eye-ring, white lores, and generally dark bill. In flight, Solitary Sandpiper has white and black barring on sides of tail. Stilt Sandpiper has longer down-curved bill and shows white-rump in flight. At rest, all four of these species show wingtips that extend to end of tail or beyond. Buff-breasted Sandpiper, which, like Upland Sandpiper, prefers drier, grassy habitats, is distinctly buffy underneath, with an all-black bill, and it lacks streaking on neck and breast. **Voice:** Call is a rapid, liquid chatter, but unlikely to be heard on the islands. **Status:** Rare migrant visitor. Eight records, three of them in Aug, when birds migrate south to wintering grounds in southern South America. In chronological order, records are as follows: One collected at Rif, Curaçao, Aug 30, 1956; one at Pekelmeer, Bonaire, Aug 25, 1977; one at Bubali, Aruba, Nov 1978; one at Zuidweb, Bonaire, Oct 22, 1984; one at Klein Hofje, on Feb 12, 1993, a very odd date, since they presumably would normally be on wintering grounds thousands of miles away at that time; one (of two seen) photographed by Greg Peterson at San Nicolas, Aruba, on Oct 2, 2011 (Voous 1983; Prins et al. 2009; Greg Peterson 2011); one photographed by Ross Wauben at Divi Links, Aruba, on Oct 18, 2015 (fide eBird). **Range:** Breeds from interior Alaska south—through the prairies of western Canada and the midwestern US—and east, spottily to southern Quebec and New Brunswick. Winters in grasslands, from southern Brazil to northern Argentina.

Whimbrel (*Numenius phaeopus*)   **Plate 20**

Papiamento: *Lopi piku doblá, Lopi pik doblá*
Dutch: *Regunwulp*

Description: Large, brown shorebird with a long, down-curved bill and dark brown head stripes. **Similar species:** Long, down-curved bill should make it virtually unmistakable. Although never recorded from these islands, the

Long-billed Curlew, which also has a down-curved (but longer) bill, lacks head stripes, has cinnamon colored underwing, and is larger. **Voice:** Makes a rapid series of high notes, all on the same pitch: *pip-pip-pip-pip-pip-pip*. **Status:** On all three islands, a regular migrant and winterer in small numbers, at coastal wetlands, occasionally in summer (we recorded two on Bonaire in July 2001). Voous (1983) wrote that the species "Occurs singly or in flocks of up to 10, rarely as many as 15–20." Most of our counts have been of 1–3 birds, with highest counts of 9 at Malmok, Aruba, in Jan 1998, and 8 at Savaneta, Aruba, in Jan 1996. Although usually seen along the shore or in wetlands, we noted two individuals foraging on the grass of a section of the Divi Links golf course near Oranjestad, Aruba, in Nov 2011; in North America, migrating birds often stop at golf courses. **Range:** Breeds from Alaska to northern coast of Northwest Territories; also breeds in a disjunct breeding territory, along western coast of Hudson Bay. Winters in coastal regions, from southern US to southern South America.

Hudsonian Godwit (*Limosa haemastica*)   Plate 20

Papiamento: *Lopi còrá, Lopi kòrá*
Dutch: *Rode Grutto*

Description: A large, long-legged shorebird with long, slightly upturned bill. Juveniles and nonbreeding adults are grayish on head and upperparts (juvenile buffier on breast, with checkered appearance on back). Dark legs. Bill pink at base and dark on outer two-thirds. Shows fairly distinct, light eye line. In flight, black underwing, narrow white wingstripe, black terminal tail band, and white rump are distinctive. Adult in breeding plumage has rufous underparts and darker markings on back. **Similar species:** Willet is smaller and has a straight, short, thick bill that lacks pink base. In flight, Willet shows gray (not black) tail band and has broad white wing strip. **Voice:** A variety of high, squeaky calls. **Status:** Uncommon fall migrant (Sept–Dec), with 11 records of single birds across all three islands as of 2009 (Prins et al. 2009). Three records each for Aruba and Curaçao, five for Bonaire. **Range:** Breeding populations in coastal Alaska, Mackenzie Delta, and Hudson Bay. Winters in southern South America. Famous for making exceptionally long non-stop flights, with this bird often bypassing regions between breeding and wintering grounds, except when forced to stop by bad weather or hunger.

Ruddy Turnstone (*Arenaria interpres*)   Plate 21

Papiamento: *Totolica di awa, Totolika di awa, Giripitu, Verfdo di boto*
Dutch: *Steenloper*

Description: One of the most familiar shorebirds on the islands, at least during winter. Small, stocky shorebird with a fairly stout, short, black bill, slightly upturned at the tip; orangish legs. Very distinctive in any plumage, with an elegant black pattern of lines running from face to upper breast. In breeding plumage, shows a bright rust color on back and wings; duller in nonbreeding and juvenile plumages. In flight (not shown), shows a striking white lower back and rump, with a black V-mark; white tail with a black terminal band; and white shoulder and wingstripes. **Similar species:** No other shorebird has the distinctive black pattern on head and breast. **Voice:** Calls most often heard made by birds fighting for food: a high, full-bodied *kip-kip-kip* and various rattle sounds. **Status:** Common nonbreeding migrant and winterer, with some birds staying throughout the summer. Certainly the most commonly seen shorebird on the ABCs, due in part to its habit of foraging for food scraps around open-air restaurants and bars. Some flocks also feed in natural settings along marine shores, often frequenting rocky areas or the sharp, limestone ancient reefs, but also beaches with rocky substrate (or restaurants and bars) nearby. Often also seen roosting on docks and on small boats in harbors or at anchor. Flocks of 10–40 are frequently seen, and flocks of up to 80 birds have been reported (Voous 1983). **Range:** North American subspecies breeds in Arctic of Canada and Alaska. Winters along marine coasts, from US to southern South America. Birds wintering on ABCs are thought to be from eastern Alaskan and low-Arctic Canadian breeding populations.

Red Knot (*Calidris canutus*) Plate 21

Papiamento: *Snepi còrá, Snepi kòrá*
Dutch: *Kanoet*

Description: A small shorebird, slightly larger and stockier than Ruddy Turnstone. In nonbreeding plumage, rather dull gray on upperparts and upper breast, with some barring on flanks. Bill black, fairly short and stout. Short, yellowish-green legs in nonbreeding season; these become darker, even black, in breeding season. In breeding plumage, brick-red face and underparts contrast with dark brown to black upperparts. White lower belly and undertail. In flight (not shown) generally rather nondescript and lacking any major white patches; tail, rump, and underwing are pale gray. **Similar species:** Can look remarkably similar to a dowitcher, until you see the much shorter bill and legs. In flight, dowitcher shows white wedge on lower back. Sanderling in nonbreeding plumage smaller and paler, with black legs; also note bold, white wingstripe, white underwing, and absence of barring on flanks. Breeding Sanderling has less extensive reddish color, restricted to upper breast and face. Dunlin in nonbreeding plumage is smaller; has a long, black bill that droops at the tip; and, in flight, shows prominent wingstripe and white sides (white extends to upperpart of tail). Concerns about recent major declines in Red Knots and the poverty of data about where some populations winter make it important to photographically document any sightings of Red Knots on the islands. **Voice:** Generally silent, but flight call is a "soft, melodius *cur-ret*" (Paulson 2005). **Status:** Rare migrant on Aruba and Curaçao. Apparently regular winterer and/or migrant on Bonaire. Three documented occurrences on Aruba: one in Nov 1971, at Coral Strand; two in Aug 1984, at Savaneta; one in Sept 1988, at Bubali. One record for Curaçao of a single individual in Nov 1962, at Malpais (Voous 1983; Prins et al. 2009). In contrast, there are numerous records of the species from Bonaire (Voous 1983), including of relatively large numbers: over 40 on Jan 8, 1973, at Lagoon (exact location unclear to us); over 80 on March 5, 1977, at the Pekelmeer; 160 on Feb 25, 1978, at the Pekelmeer; over 175 on Feb 5, 1983, at the Pekelmeer (Voous 1983; Prins et al., 2009). We observed 50 on March 24, 2001, at the Pekelmeer, and 27 at the same location on March 28, 2001, and photographed some of the birds, many approaching full breeding plumage. We observed three nonbreeding birds at the Pekelmeer on July 4, 2001. These same birds were apparently independently sighted by another observer on July 9, 2001, with a single bird sighted in the vicinity through July 15, 2001. It is unclear whether the species remains a regular but overlooked winterer. **Range:** North American subspecies breed in Arctic of Canada and Alaska. Various subspecies winter in different areas, with some wintering in south coastal US and others through Central America and Caribbean to southern South America.

Stilt Sandpiper (*Calidris himantopus*)   **Plate 22**

Papiamento: *Snepi pia largo, Snepi pia largu*
Dutch: *Steltstrandloper*

Description: Medium-sized shorebird, about the size of Lesser Yellowlegs. In nonbreeding plumage (plumage most often seen on the islands), shows flat gray on upperparts and light eye line. Legs are yellow-green. Bill black, rather long, and distinctly down-curved. In breeding plumage, heavily barred underneath, with chestnut cheek patches. Feeds by repeatedly probing with head, causing rear-end to tilt up and down; its feeding style is sometimes compared to the action of a pumping oil well. Shows a white rump in flight, but tail lacks barring as seen in yellowlegs and dowitchers, and instead shows tail feathers with dark edges. Juveniles are like nonbreeding plumaged adults except darker on back, and have wing coverts with light edges to the feathers. **Similar species:** Most easily confused with yellowlegs but note shorter legs and longer, down-curved bill. Also similar to dowitchers but longer legged than dowitchers and note shorter, down-curved (not straight) bill. **Voice:** Makes a low quiet *tew*, similar to some calls of Lesser Yellowlegs. **Status:** Occurs in small numbers on Aruba and Curaçao during migration and in winter, from Aug through Apr. Larger numbers occur on Bonaire, with hundreds sometimes found, especially Nov through Apr, in various shallow lagoons in Pekelmeer, Lac, and at the Gotomeer and other locations in Washington-Slagbaai NP. Like many of the other shorebird species, small numbers of nonbreeders undoubtedly occasionally summer as well. **Range:** Breeds in the Arctic, from northern Alaska east to Nunavut, and with a second disjunct breeding population farther south along the southwestern shores of Hudson Bay. Winters largely in South America but small numbers may occur north to Mexico and southern US.

Sanderling (*Calidris alba*)   **Plate 21**

Papiamento: *Snepi blanco, Snepi blanku*
Dutch: *Drieteenstrandloper*

Description: A small shorebird with short, black bill and black legs, most often seen at the beach running furiously to stay at the edge of the water. In nonbreeding plumage (the plumage type most often seen in the islands), shows a pale, ghostly gray on upperparts; white on underparts extends up onto face; when visible, black shoulder is diagnostic. In flight, shows a bold, white wingstripe. In breeding plumage, reddish coloration on upper breast and head; back checkered with black and reddish coloration. **Similar species:** Nonbreeding Red Knot larger and darker, with lighter legs and flank markings, and lacking bold wingstripe in flight. Nonbreeding Dunlin (a rare species here) is not as pale and has a longer bill (drooped at tip). Other sandpipers smaller or with long wings that extend beyond tail (White-rumped, Baird's) and not as pale. **Voice:** Call is a harsh *chit*. **Status:** Occurs throughout the year as a regular nonbreeding species, in small numbers (typically less than 10 in a flock). Most often seen along hot, sandy beaches or edges of salinas, but sometimes roosts with Ruddy Turnstones on docks and anchored boats. Our largest counts were 30 at the Hadicurari marina, Aruba, Jan 16, 2003, and 27 at the Pekelmeer, Bonaire, March 23, 2001. **Range:** North American population breeds in Arctic Canada. Winters along coasts from US to southern South America.

Dunlin (*Calidris alpina*)   **Plate 22**

Papiamento: *Snepi alpino*
Dutch: *Bonte Strandloper*

Description: In breeding plumage, shows distinctive black belly patch, making the species virtually unmistakable. In nonbreeding and juvenile plumages, more nondescript, with flat gray-brown upperparts and white underparts; note long, black bill, down-curved at tip, and black legs. **Similar species:** Inexperienced observers could mistake long-billed sandpipers like the Western and White-rumped for this species, though the former is much smaller and the latter has a white rump and long wings that extend beyond tail tip at rest. Sanderlings and Red Knots can appear similar—nondescript gray above in nonbreeding plumage—but neither has bills as long and down-curved as in Dunlin. Given that the species normally winters much farther north, any observations of the species should be documented with photos or video if possible. **Voice:** Call is a raspy *cree*. **Status:** Rare nonbreeding vagrant, with perhaps five records (Prins et al. 2009), not all of these records with substantive documentation. No documented records from Aruba. Four reports from Bonaire, but three of those in the same year (1985): one at Gotomeer, March 29, 1985; one at the Pekelmeer, Apr 10, 1985; two in nonbreeding plumage near Kralendijk Harbor, June 4, 1985. Individual reported near Harbor Village Marina, Aug 31, 2003 (Ligon 2006) would be particularly unusual considering the species is generally a late-fall migrant, even in the northern US. One record from

Curaçao of a single bird at Muizenberg, on Oct 18, 2005 (Prins et al. 2009). **Range:** Breeds in Arctic North America and Eurasia. North American populations winter along Pacific and Atlantic coasts of US to Mexico.

Baird's Sandpiper (*Calidris bairdii*)   Plate 22

Papiamento: *Snepi Baird*
Dutch: *Bairds Strandloper*

Description: A small shorebird, only slightly larger than Semipalmated Sandpiper. Has black legs and bill. Wings longer than other "peeps," extending well beyond tail when bird is at rest. Buffy or sandy overall on breast and upperparts. Juveniles show neatly buff-edged feathers to back, scapulars, and wing coverts, giving very scaly effect to upperparts. Nonbreeding adults are fairly nondescript, showing more brown tones (gray on breeding) on upperparts. Adult in breeding plumage has big, black centers to wing coverts that contrast with sandy-gray upperparts. At rest, the long wings give the species a very long, pointy appearance. **Similar species:** Only other species with wingtips extending beyond tail is White-rumped Sandpiper, a feature easy to miss without careful study or experience. In comparison with Semipalmated, Western, or Least sandpipers, will appear larger, slightly taller, and, in the rear, longer and more pointy. Least Sandpiper has yellow legs. More brown-toned on upperparts and upper breast than White-rumped, and lacks white rump; in breeding plumage, lacks red tones and streaks extending down flanks that characterize White-rump. In juvenile plumage, White-rump also shows reddish edges to scapulars (absent in Baird's). **Voice:** Makes a rough *treep*. **Status:** Rare to uncommon migrant visitor, with perhaps five records, though likely easily overlooked. One documented record for Aruba of single bird at Bubali, Sept 8, 1979; one documented record for Curaçao of single bird at Jan Thiel Lagoon, May 28, 1977; at least three documented records for Bonaire: 5 at Lac, Sept 15, 1980 (Voous 1983); 12 at Dos Pos, May 18, 1986; 1 at Lac, Nov 29, 1989 (Prins et al. 2009). Prins et al. (2009) note a record of a larger number of birds recorded by Jerry Ligon in mid-Nov 2000, but Ligon himself makes no mention of such a record in his published checklist (Ligon 2006). **Range:** Breeds in Arctic Alaska and Canada. Winters in southwestern South America from Peru south to Chile and Argentina.

Least Sandpiper (*Calidris minutilla*)   Plate 21

Papiamento: *Snepi chikito*, *Snepi chikí*
Dutch: *Kleinste Strandloper*

Description: A tiny shorebird. Very brownish above, with yellow legs and thin, black bill; streaking on upper breast. **Similar species:** The only small shorebird with yellow legs. Least is smaller than Semipalmated and Western Sandpiper, has yellow (not black) legs, and shows more brown tones on upperparts. **Voice:** Makes a high, thin *kreep*. **Status:** Occurs in small numbers, in wetlands on all three islands. A common migrant and winterer, a few staying through the summer. Ligon

(2006) reported finding 40 near Sorobon, Bonaire, Dec 29, 2001, and about 40 at the Gotomeer, Bonaire, March 2002. Maximum reported on Aruba was 51 at Bubali on Aug 23, 2015 (G. Peterson fide eBird). We generally recorded groups ranging from 1–15. **Range:** Breeds from Alaska across Canada, from northern Boreal forest and Arctic region to Newfoundland. Winters on Pacific Coast of US, from Washington State south; also winters from southern US south to Peru, Bolivia, and southern Brazil.

White-rumped Sandpiper (*Calidris fuscicollis*)   Plate 22

Papiamento: *Snepi Bonaparte*
Dutch: *Bonaparte's Strandloper*

Description: A small shorebird, only slightly larger than Semipalmated Sandpiper. Has black legs and a black bill that droops slightly at the tip. Wings longer than on other "peeps," with wingtips extending well beyond tail when at rest. When it can be seen, the best field mark is the white rump. In breeding plumage, shows rufous on upperparts and dark streaks that extend along flanks. At close range, look for white eye line and light reddish base to lower mandible. In juvenile plumage (not shown), shows crisp light or rufous-edged feathers on scapulars and wing coverts. **Similar species:** Wingtips do not extend beyond tail in Semipalmated and Western sandpipers. Least Sandpiper has yellow legs. All three species are smaller than White-rumped. Baird's Sandpiper also has wingtips that extend beyond tail at rest, but it is buffier overall, with less obvious eye line and more contrast on back. Only White-rumped Sandpiper has a white rump. **Voice:** Makes a distinctive, high, thin *jeeet*, said to be mouse-like and also compared to the sound made by rubbing together two smooth stones. **Status:** Poorly documented but apparently a regular fall visitor, though uncommon to rare, in small numbers (Voous 1983; Prins et al. 2009), with some larger flocks in May. Ligon (2006) reports a flock of 11 on Bonaire, May 12, 2000, and 3 there on Sept 25, 2002. Prins et al. (2009) note a report of 22 from Muizenberg, Curaçao, from Apr 29 to May 20, 2005. **Range:** Breeds in Arctic Alaska and Canada. Winters in southern South America, mostly in Argentina and Chile.

Buff-breasted Sandpiper (*Calidris subruficollis*)   Plate 22

Papiamento: *Snepi blònt boca chikito, Snepi blònt boka chikí*
Dutch: *Blonde Ruiter*

Description: Rather unmistakable with its buff-colored underparts that extend onto face, encircling the dark eye. Legs yellowish. Bill is short, black, and slender. Upperparts scaly, with light edges to dark-centered feathers. Prefers open grassy or grasslike habitats during migration. **Similar species:** No other species shows combination of buff-colored underparts and face and yellowish legs, but sightings should be documented with photos or video as there are few records of the species for the islands. **Voice:** Generally silent, but can give a low *tew*. **Status:** Rare migrant visitor, but detected with increasing frequency on Aruba. Up to four individuals were seen and photographed at the Tierra del Sol golf course on Aruba, Oct 25–27, 2010, by Emile Dirks (photos posted on www.arubabirds.com). An astounding 40 birds were seen (some photographed) near Grapefield Beach, Aruba, on Oct 2, 2011 (G. Peterson fide eBird). Eleven birds were seen near the California Lighthouse, Aruba, on Oct 4, 2015 (M. Oversteegen fide eBird). Sixteen birds were seen (some photographed) near Arashi, Aruba, on Oct 4, 2016 (M. Oversteegen fide eBird). Seventeen birds were seen (some photographed near Grapefield Beach), Aruba, on Oct 7, 2016 (K. Hansen fide eBird). Up to 12 birds were seen (some photographed) near California Lighthouse, Aruba, from Oct 5–16, 2016 (M. Oversteegen fide eBird). On Curaçao, two groups of about 10 birds were reported from Dam Muizenberg, Oct 1, 2000 (Prins et al. 2009). On Bonaire, single birds were found at the Pekelmeer, Oct 23–Nov 1, 1984, and at Malmok, Oct 10–11, 1970 (Prins et al. 2009). **Range:** Breeds in Arctic from northern Alaska east to Nunavut. Winters in grasslands of southern South America.

Pectoral Sandpiper (*Calidris melanotos*)   Plate 22

Papiamento: *Snepi strepi fini*
Dutch: *Gestreepte Strandloper*

Description: Similar in appearance to Least Sandpiper but larger. Warm brown on upperparts; yellow legs; bill with pale base. Note sharp line of demarcation between heavily streaked upper breast and white lower breast and belly. **Similar species:** Most other sandpipers have black legs. Least Sandpiper is tiny in comparison, has all-black bill, and breast streaking does not show sharp line of contrast with white lower breast. Buff-breasted Sandpiper also has yellow legs but is buffy on underside, with no breast streaking. **Voice:** Call is a husky *cherk*. **Status:** Regular migrant visitor in fall (July through Nov), with occasional occurrences from Dec through May (Voous 1983; Prins et al. 2009). Ligon (2006) reports four occurrences on Bonaire, fall 2002, with 10 in Sept and 20 in Nov of that year. Many recent records (2007–2016) at various locations on Aruba, Sept–March, with maximum of 35 reported from Eagle Beach to Tierra del Sol, Aruba, on Oct 27, 2007. Typically seen in numbers of 1–5 at any single location. **Range:** Breeds in Arctic of North America and Siberia. North American breeders winter in southern half of South America.

Semipalmated Sandpiper (*Calidris pusilla*)   Plate 21

Papiamento: *Snepi gris, Snepi pia pretu*
Dutch: *Grijze Strandloper*

Description: A small, rather nondescript shorebird; gray-brown above, white below, with black legs and black, short, thin bill. In nonbreeding plumage, flat gray on back and wings, with white forehead and eyebrow line, gray on upper breast. In breeding plumage, darker brown above, with dark centers to feathers on back and scapulars, and with streaks across upper breast. Juveniles (not shown) show crisp new plumage with broad light edges to feathers on back and scapulars. **Similar species:** Least Sandpiper is smaller and browner, with yellow (not black) legs. Western Sandpiper in nonbreeding plumage not always distinguishable from Semipalmated Sandpiper, even by experts, but note (usually) paler face on Western. In breeding plumage, Western has rufous on head and scapulars; in juvenile, upper scapulars of Western show more rust than on juvenile Semipalmated. Western Sandpipers tend to have longer bills with more droop at the tip, though note that this is not always a reliable key to distinguishing the two. Baird's Sandpiper and White-rumped Sandpiper are larger and have wingtips that extend well beyond tail at rest. **Voice:** Flight call is a short *chert* or *churk*. **Status:** Regular nonbreeding migrant and winterer, with small numbers in summer. Voous (1983) reported that flocks of nearly 300 had been seen in Aruba in Aug of an unspecified year or years. Ligon's (2006) report of hundreds of Western Sandpipers on Bonaire in Sept, 2002, perhaps refers to this species. We counted three flocks totaling about 80 birds at the Pekelmeer, Bonaire, on March 24, 2001, and an additional 10 birds that same day, near Lac Bay. We counted 17 summering birds in nonbreeding plumage on July 4, 2001, at the Pekelmeer, Bonaire. We have not found the species to be a common wintering species on Aruba. **Range:** Breeds from Arctic Alaska east across Canada to Labrador; also breeds south along shores of Hudson Bay to James Bay. Winters from Mexico and Caribbean south to Chile and Brazil.

Western Sandpiper (*Calidris mauri*)   Plate 21

Papiamento: *Snepi mauri*
Dutch: *Alaskasandloper*

Description: A small, rather nondescript shorebird; gray-brown above; white below, with black legs; black, rather long, bill droops at tip. In nonbreeding plumage, appears flat gray on back and wings, with white forehead and eyebrow line, gray on upper breast. In breeding plumage, brightly marked with rufous on head, scapulars, and wing coverts, and with dark arrowhead marks on lower breast and sides. Juveniles (not shown) show crisp, new plumage with broad, light edges to feathers on back and scapulars (note rufous edges on scapulars). **Similar species:** See Semipalmated Sandpiper. **Voice:** Makes a squeaky *cheet* or *jzeet*, higher than calls of Semipalmated Sandpiper. **Status:** In all seasons, less common than Semipalmated Sandpiper, but regular, in small numbers, in salinas and other coastal wetlands. We have encountered the species on the islands only a few times: one at Salina Sint Michiel, Curaçao, Nov 2003; one at the Pekelmeer, Bonaire, March 24, 2001; two near Sorobon, Bonaire, March 28, 2001. Ligon's (2006) report of hundreds of Western Sandpipers on Bonaire, Sept 2002, more likely refers to Semipalmated Sandpiper given the history of the species on the islands. **Range:** Breeds in coastal Alaska. Most of population migrates south along Pacific Coast, but some birds scatter across entire continent, reaching the East Coast. Winters along Pacific Coast, from British Columbia south through Mexico and Central America to Peru, and along East Coast, from Delaware south through Gulf Coast and Caribbean to northern South America.

Short-billed Dowitcher (*Limnodromus griseus*) Plate 23

Papiamento: *Snepi gris chikito, Snepi gris chikí, Snepi*
Dutch: *Kleine Grijze Snip*

Description: Distinctive, long-billed (despite its name), short-legged shorebird that feeds by probing bill up and down into mud, like the needle on a sewing machine. This regular up-and-down feeding pattern can be used to identify dowitchers, even at a great distance. In flight, also distinctive is the wedge of white that extends from the barred tail up the back. In breeding plumage, orangish on head and underparts, with black and gray mottled upperparts. In worn winter plumage, can be gray on upperparts, and with little or no orangish tones to plumage. Juvenile birds in fresh fall plumage show rust on crown, back, and upper breast. Even after juveniles have molted most of their feathers into winter adult-type plumage, they often retain one or more juvenile tertials, which show interior stripes and spots. In contrast, tertials on juvenile Long-billed Dowitchers are rusty-edged, with no internal markings in the feathers. **Similar species:** Shaped like Wilson's Snipe, but that species is all dark above, is always darker on upperparts, and shows dark head stripes. Stilt Sandpiper can be superficially similar and often feeds with regular up-and-down movements like dowitchers but is smaller, with longer legs, longer neck, and shorter, finer bill that has a more down-curved tip. Long-billed Dowitcher is very similar and despite the name cannot be distinguished from Short-billed Dowitcher by bill length. Distinguishing Long-billed Dowitcher from the three variable subspecies of Short-billed Dowitcher can be difficult, even with experience, and may be impossible when looking at birds in worn winter plumage, as is often seen on the islands; but note distinctive call of Short-billed Dowitcher. If one is certain that the bird in question is a juvenile that has retained its juvenile tertials, then identification can be quite straightforward, as described above. We recommend that birders interested in dowitcher identification consult some of the excellent recent North American field guides or shorebird specialty guides now available. **Voice:** Distinctive call is rather mellow *tu-tu-tu* (somewhat like the sound used in science-fiction movies for a ray-gun) and very different from the single, higher-pitched *keek* typically given by the Long-billed Dowitcher. **Status:** In small numbers (typically 1–10), a regular migrant and wintering bird, in mangrove wetlands, estuaries, salinas, and freshwater wetlands on all three ABC islands. (Our largest count was of 50 on March 24, 2001, around the Pekelmeer on Bonaire.) Ligon (2006) reports a high count of 20 on Sept 24, 2002, on Bonaire. Occasional birds summer as well. We found 8–12 birds on Bonaire, July 4–7, 2001. At all seasons, many of the birds we have observed on the islands have been in pale, non-breeding plumage. On Aruba, we have found the species most regularly at Savaneta. Although Voous (1983) notes that the eastern *griseus* is the expected subspecies here, we recorded a single bird in breeding plumage on Bonaire in July 2001 that had extensive reddish coloration on virtually the entire underparts, as is normally typical of the *hendersonii* subspecies. **Range:** Breeds from Alaska east across Boreal forest region of Canada to James Bay; possibly in northern Quebec. Winters on coasts from southern US south through Central America and Caribbean, to northern South America.

Long-billed Dowitcher (*Limnodromus scolopaceus*)   **Plate 23**

Papiamento: *Snepi gris grandi*
Dutch: *Grote Grijze Snip*

Description: To the casual observer, almost identical to Short-billed Dowitcher, and not separable by bill length. **Similar species:** See Short-billed Dowitcher. **Voice:** Call when flushed is a single, high *keek*, unlike the mellow *tu-tu-tu* of Short-billed Dowitcher. **Status:** Although Voous (1983) relied on color photographs (not available for public review as far as we know) from the 1970s to confirm the presence of this species, a lot has since been learned about the identification of the various subspecies and ages of Short-billed Dowitcher as compared to Long-billed Dowitcher; we do not consider the species as having been verified from the islands. No specimen of Long-billed Dowitcher was ever taken on the islands, and the species winters farther north. There are apparently no verifiable records of Long-billed Dowitcher from South America (Prins et al. 2009). We urge observers to carefully photograph and videotape any suspected Long-billed Dowitcher here to clarify the status of the species. **Range:** Breeds in coastal Alaska, Yukon, and Northwest Territories. Winters from US south to Costa Rica, and rarely to Panama.

Wilson's Snipe (*Gallinago delicata*)   **Plate 23**

Papiamento: *Snepi di awa*
Dutch: *Watersnip*

Description: Found along shores of freshwater pools and marshes, this squat, short-legged shorebird has a very long bill. Has dark stripe on head; also has warm-brown upperparts, with prominent light stripes on back and much barring and speckling. Barred on upper breast and sides. Does not show any white on upperparts. Very short tail is orangish. When disturbed, birds crouch low to the ground or flush into the air, sometimes almost from underfoot. **Similar species:** Dowitchers are most similar in appearance and shape, but they always show a prominent white wedge on lower back, while Wilson's Snipe has all-dark upperparts. Note that the South American Snipe (*Gallinago paraguaya*) occurs throughout most of mainland South America, including lowlands of Venezuela and Colombia, and is virtually identical in appearance to Wilson's Snipe (see Hilty 2003 for details). It is possible that this species could occur as a rare vagrant on these islands. Snipes seen May–Aug require further study to determine identity. **Voice:** When flushed, gives a loud, harsh *scraip*. **Status:** Common migrant and winter visitor to shores of temporary freshwater pools and permanent wetlands on all three ABC islands. Although usually encountered in small numbers (perhaps 1–5 encountered in a day of birding), Voous (1983) notes that numbers can be higher Oct–Nov and that hundreds were reported in the wet winter of 1966–67 on Curaçao. Ligon (2006) described counting 20 birds at a single location on Bonaire in Nov 2002. Our highest count was of six birds on Jan 4, 1997, at Bubali, Aruba. Most records are from Sept through Apr, but birds have been reported May–Aug period, though it is unclear if any of these records could pertain to the virtually identical South American Snipe. **Range:** Wilson's Snipe breeds across northern North America and winters from southern half of US south through Central America, to northern South America.

Wilson's Phalarope (*Phalaropus tricolor*)   Plate 23

Papiamento: *Snepi seha grandi*
Dutch: *Grote Franjepoot*

Description: A medium-sized shorebird with long legs and a needle-like, black bill. Adults in nonbreeding plumage show plain gray back and white underparts that extend onto face and into white eye line. Nonbreeders and juveniles show yellow legs; legs become dark in breeding plumage. In flight, shows white rump, gray tail, and no wingstripe. In breeding plumage (unlikely to be seen in the islands), shows chestnut on sides of lower neck; black (female) or gray (male) on sides of head extending through eye; front of neck rusty but note white chin. Juvenile plumage (unlikely to be seen in the islands) resembles that of nonbreeding adult except darker on upperparts, with buffy edges to feathers and buffy on sides of upper breast and neck. All phalaropes typically engage in feeding behavior in which they float on the water and spin in circles as they dip their bill in the water to pick up small aquatic organisms. **Similar species:** Probably most likely confused with faded, nonbreeding Stilt Sandpipers and Lesser Yellowlegs, both of which are roughly similar in size and are grayish on upperparts and white underneath. Stilt Sandpiper also shows yellow legs but has much longer, down-curved bill and usually feeds in a characteristic up-and-down "oil pump" style. Lesser Yellowlegs are often seen feeding on the islands (at least on Bonaire) in a fashion similar to to that of phalaropes—floating and spinning. Wilson's Pharopes are smaller than Lesser Yellowlegs, with a finer bill; upperparts are lighter gray, with no spotting, and breast lacks any streaking or extensive dark gray; also note very white chin and face. Other phalaropes show dark ear patch and rear crown. See Red-necked Phalarope and Red Phalarope for other differences. **Voice:** Makes a nasal *wurf* or *work*. **Status:** Rare migrant with six records, one from Curaçao and five from Bonaire. On Curaçao, two moulting juveniles seen at Jan Thiel, Sept 22, 2001 (Prins et al. 2009). On Bonaire, two seen at Gotomeer, March 5, 2002; one at Playa Lechi, Sept 21, 1984; about 10 birds at Gotomeer, March 8, 1979; one at the Pekelmeer, Jan 1971; and two at the Pekelmeer, Jan 25–27, 1970 (Voous 1983; Ligon 2006; Prins et al. 2009). **Range:** Breeds in prairie region of western Canada and US. Winters in southern South America.

Red-necked Phalarope (*Phalaropus lobatus*)   Plate 23

Papiamento: *Snepi seha boca fina, Snepi seha boka fini*
Dutch: *Grauwe Franjepoot*

Description: A small, short-legged shorebird with a needle-like black bill. Normally spends much of the year far out to sea, feeding by picking small marine organisms from the surface while floating and spinning. In nonbreeding plumage (the plumage most likely seen in the islands), gray on upperparts with streaking on back; dark ear patch, dark rear crown; white chin, face, and forehead. Dark legs. In breeding plumage, shows dark head and throat with red extending up from lower throat to sides of neck, but still retains white chin and has reddish stripes on back. In flight, shows bold, white wingstripe and has gray line down center of rump. **Similar species:** Wilson's Phalarope has longer legs and usually shows yellow (not dark) legs. Wilson's lacks white wingstripe in flight and has a white rump and unstreaked back. Wilson's also does not show dark ear patch and rear crown. In breeding plumage, Wilson's does not show dark on upper breast and lower neck as seen in Red-necked. Red Phalarope has a thicker bill, pale at base. Red also shows a bold wingstripe like Red-necked but has a plain, unstreaked back. In breeding plumage, Red has rufous underparts and is quite unmistakable. **Voice:** Unlikely to be heard, but makes a hard sharp *pik* or *twick*. **Status:** Three reports, two from Bonaire and one from Curaçao: two from the Pekelmeer, Bonaire, Jan 22–28, 1971. and one from same location, Jan 25–27, 1970 (Voous 1983; Prins et al. 2009); one moulting juvenile reported from Jan Thiel, Curaçao, Sept 22, 2001. Any sightings of the species should be documented with photos and video. **Range:** Breeds across Arctic and Subarctic of entire globe. North American populations breed from Alaska east to Labrador. A significant wintering concentration occurs at sea off the Pacific Coast of South America. It is still unknown whether birds breeding in eastern North America migrate to a yet-to-be discovered wintering population perhaps somewhere in the western South Atlantic.

Red Phalarope, Gray Phalarope (*Phalaropus fulicarius*)   Plate 23

Papiamento: *Snepi seha boca diki, Snepi seha boka diki*
Dutch: *Rosse Franjepoot*

Description: A small, fairly short-legged shorebird with a stout, black bill, pale at the base. Normally spends much of the year far out to sea, feeding by picking small marine organisms from the surface while floating and spinning. In nonbreeding plumage (the plumage most likely seen in the islands), gray unstreaked upperparts, dark ear patch, dark rear crown; also note white chin, face, and forehead. Dark legs. In breeding plumage, a stunning bird: rusty-red underparts, yellow bill, black chin and crown, and a white cheek. In flight, shows bold, white wingstripe and has gray line down center of rump. **Similar species:** In breeding plumage, Red Phalarope is unmistakable. Wilson's Phalarope lacks wingstripe and dark ear patch and rear crown and has longer, thinner, all-black bill and white rump. Red-necked Phalarope has a streaked (not plain) back and a needle-like, all-black bill. **Voice:** Unlikely to be heard. Makes a *pit* or *pik*, similar to that of Red-necked Phalarope but slightly higher. **Status:** One report from Playa Frans, Bonaire, Nov 7, 1999 (Prins et al. 2009), but apparently not documented with photos, video, or published/archived written description; validity is uncertain. Any sightings of the species should be documented with photos and video. **Range:** A circumpolar, high-Arctic breeder. In North America, breeds from shores of North Slope of Alaska east to northern tip of Quebec. Winters at sea off Pacific Coast of South America and off Atlantic Coast of Africa.

Spotted Sandpiper (*Actitis macularius*)   **Plate 21**

Papiamento: *Snepi barika pintá*
Dutch: *Oeverloper*

Description: In nonbreeding plumage, a small, rather nondescript shorebird with plain brown back, white underparts, and white eye line. Note brown "comma" mark extending down from neck to sides of breast. Legs pale. Rather short bill usually pale except at tip. Folded wingtips do not extend to end of tail. In breeding plumage, note bold, black spots underneath and bill with orangish base. Behavior is often best key to identification, as the species constantly tips its body back and forth as it walks along, often with a rather hunched, half-crouching posture. When flushed, flies with wings held stiffly and snapped down repeatedly. **Similar species:** No other shorebird shows characteristic "tipping" walk or stiff-winged flight. Solitary Sandpiper is larger and longer-legged, with bold, white eye-ring (no white eye line behind eye), no "comma" mark, and shows wingtips that extend to end of tail. **Voice:** When flushed, makes a sharp *weet-weet*, similar to call of Solitary Sandpiper call but somewhat lower pitched and perhaps more mellow. **Status:** Common nonbreeding species in winter and during migration, with occasional nonbreeding individuals seen during summer. Can be found in a wide variety of marine and freshwater coastline habitats. Usually seen singly, though small flocks may occasionally form during migration periods. **Range:** Breeding range extends over most of North America, north of Mexico. Winters from southern US to southern South America.

Solitary Sandpiper (*Tringa solitaria*)   **Plate 20**

Papiamento: *Snepi solitario*
Dutch: *Amerikaanse Bosruiter*

Description: Medium-sized, brownish shorebird, with long, greenish legs, bold white eye-ring, and long, thin bill. Often appears distinctly spotted on upperparts. In flight, sides of tail are white, with dark barring; central part of tail and rest of upperparts are dark. **Similar species:** Lesser Yellowlegs is slightly larger, without a bold, white eye-ring and with longer, bright yellow legs. In flight, Lesser Yellowlegs shows white rump and upper tail as compared to the dark rump and upper tail in Solitary Sandpiper. **Voice:** When flushed, makes a strident, ringing *peet-weet*. **Status:** Regular but uncommon (probably often overlooked) migrant and winterer, July through May. Typically found in freshwater pools and, as its name suggests, does not form cohesive flocks, although a small number occasionally occur in the same spot. In Jan surveys on Aruba, over several years, we have found single birds at Bubali and Tierra del Sol multiple times, and a single bird at Spanish Lagoon on one instance. Largest number we have recorded was of four birds at Bubali, Jan 4, 1997. On Curaçao, we have observed the species at Klein Hoje, Muizenberg, Schottegat, and Jan Thiel. On Bonaire, Ligon (2006) lists recent reports from Salina Slagbaai and from Washington-Slagbaai NP. **Range:** Breeds in Boreal forest ecoregion of North America, from interior Alaska east to Labrador. Winters from southern Mexico and Caribbean south to northern Argentina.

Greater Yellowlegs (*Tringa melanoleuca*)   **Plate 20**

Papiamento: *Snepi pia hel grandi, Snepi pia hel largu*
Dutch: *Grote Geelpootruiter*

Description: A relatively large, gray-brown shorebird, with bright yellow, long legs and long bill. Bill upturned slightly at tip and noticeably thicker at base; mostly dark but paler, grayish at base. Often shows distinct spots or bars on back, folded wings, and end of tail. In flight, note bold, white upper tail and rump. Most birds seen here in winter are in pale, worn nonbreeding plumage, but in early fall one might see birds retaining features of juvenile plumage. In spring, also note birds in breeding plumage, with darker feathering on back and barring underneath. In migration periods, note adults in various stages of molt. **Similar species:** Most similar in appearance to its smaller look-alike, the Lesser Yellowlegs. Lacking experience, the two can be hard to tell apart. Lesser Yellowlegs is half the size of Greater Yellowlegs, and is shorter and with a slimmer build. Bill of Lesser Yellowlegs usually all black, thinner, and shorter; it is roughly equal in length to the distance between back of head and base of bill; also note that the bill is straight (upturned on Greater). We think of the bill of the Greater Yellowlegs as stout like a pencil, as compared to the more needle-like bill of the Lesser Yellowlegs. Calls of the two species are helpful keys to identification. **Voice:** In flight and when alarmed, gives a high, insistent (but mellow), three- or four-part *tu-tu-tu-tu*. **Status:** On all three islands, a common migrant and winter visitor at coastal wetlands, salinas, and freshwater ponds. Small numbers of nonbreeding birds also occur, even in summer. Usually less numerous than Lesser Yellowlegs, but Voous (1983) notes that Greater Yellowlegs can occasionally outnumber Lesser Yellowlegs during spring (March–Apr) migration, when Greater flocks can number in the hundreds. Largest numbers we have observed were 30 in Jan 1996, at Savaneta, Aruba, 20 there in Jan 1998, and 20 in Washington-Slagbaai NP, Bonaire, Jan 14, 1998. **Range:** Breeds across North American Boreal forest ecoregion, from southern Alaska eastward across Canada to Newfoundland. Winters along Pacific and Atlantic coasts and in southern regions, US, and south, throughout South America.

Willet (*Tringa semipalmata*)   **Plate 20**

Papiamento: *Snepi ala di strepi*
Dutch: *Willet*

Description: Large shorebird with dull, brownish-gray back, long but rather stout and blunt-tipped bill, and thick, gray legs. In flight, drab look transforms, with bold, white wingstripe set against black wing linings and white rump. The two subspecies can be distinguished by skilled observers: western subspecies is noticeably larger than the eastern, and there are other subtle differences. We recommend consulting a shorebird identification guide for more details. **Similar species:** At rest, might be mistaken for a yellowlegs, but note gray (not yellow) legs and much thicker bill. American Oystercatcher and Hudsonian Godwit have striking black-and-white wing pattern, if seen clearly. **Voice:** On breeding grounds, these are noisy birds that make a variety of calls. Most often heard call is a characteristic loud, complaining *pill-will-willet, pill-will-willet*. Birds seen on the ABC islands are usually silent, but flight call is a laughing *kee-dee, kee-dit*, or *kee-dit-dit*. **Status:** Uncommon nonbreeding migrant and winterer on all three islands; has occurred throughout the year. De Boer et al. (2011) mistakenly claimed that the species had not been recorded from Aruba, perhaps based on a similar omission by Prins et al. (2009). In fact, the species had already been reported from Aruba by 1978 and 1982 (Voous 1983). We observed seven Willets at Savaneta, Aruba, on Jan 12, 1996, and one at the same location on Jan 11, 1998. A Willet was photographed by Jan Wierda in a salina near Malmok, Aruba, on Feb 25, 2012 (photograph on www.arubabirds.com) and two were reported by John Galluzo in a salina near the Marriot hotel on Sept 8, 2010 (trip report at www.arubabirds.com). There are now dozens of reports of Willets in recent years from Aruba, many with photographs, with most involving one or two birds (fide eBird). Ligon (2006) reports 8–9 sightings of the species on Bonaire, since 1997. We observed a single bird at Sorobon, Bonaire, on March 24, 2001. A number of specimens of Willet were taken from Curaçao in the 1950s, and we presume the species is still regular but uncommon there. We observed a single Willet at Salina Sint Michiel, Curaçao, on Nov 22, 2003. Maximum counts are: 20 in Feb 1978 and July–Aug 1982, at Bubali, Aruba; and 15 on Sept 22, 1979, on Bonaire (Voous 1983). Most specimens taken were reported to be of the eastern subspecies, except for a single July specimen from Bonaire of the western subspecies (Voous 1983). The 1974 breeding record from an island off the Venezuelan coast was apparently an odd, isolated case, and the species has never been found breeding anywhere within the region since that time. **Range:** Eastern subspecies breeds along Atlantic Coast of North America, from Nova Scotia to Florida; along Gulf Coast south to northeastern Mexico; and in Caribbean to Hispaniola. Unusual isolated breeding documented on an island in the Venezuelan Los Roques archipelago in 1974, but there are no more published breeding records. Eastern subspecies is thought to winter largely in South America. Western subspecies breeds in prairie regions, from southwestern Canada south to California and Nevada, and winters from Pacific and Atlantic coasts of US south, apparently to northern South America. However, the wintering ranges of the two subspecies are poorly documented south of the US.

Lesser Yellowlegs (*Tringa flavipes*)   **Plate 20**

Papiamento: *Snepi pia hel chikito, Snepi pia hel chikitu*
Dutch: *Klein Geelpootruiter*

Description: Medium-sized, gray-brown shorebird, with long, bright yellow legs, and long, thin, dark bill. Often appears distinctly spotted or barred on back, folded wings, and end of tail. In flight, note bold, white upper tail and rump. Most birds seen here in winter are in pale, worn, nonbreeding plumage, but in early fall one may see birds retaining some features of juvenile plumage; in spring, look for birds in breeding plumage, with darker feathering on back. In migration periods, one can see adults in various stages of molt. **Similar species:** Half the size of Greater Yellowlegs, Lesser Yellowlegs is shorter and has a slimmer build. Bill on Lesser Yellowlegs is all black (not gray); thinner and shorter; and straight (not upturned at tip). We think of the bill of the Greater Yellowlegs as stout like a pencil as compared to the more needle-like bill of the Lesser Yellowlegs. Calls of the two species are diagnostic. Solitary Sandpiper is slightly smaller than Lesser Yellowlegs, with a bolder white eye-ring, shorter, more greenish, legs, and, when seen in flight, lacks white rump and upper tail of Lesser Yellowlegs. **Voice:** Makes a *kip* note, singly or in series; flatter and less ringing than call of Greater Yellowlegs. **Status:** On all three ABC islands, a common migrant and winter vistor, at coastal wetlands, salinas, and freshwater ponds. Small numbers of nonbreeding birds also occur, even in summer. Usually more numerous than Greater Yellowlegs. Voous (1983) notes that flocks can occasionally number into the hundreds, especially during spring migration, and he mentions a record of 200 on Apr 1, 1974, at Bubali, Aruba. Largest numbers we have observed were 130 on March 26, 2001, at the Gotomeer, Bonaire; 50 on Nov 22, 2003, at Salina Sint Michiel, Curaçao; 35 on Nov 23, 2003, at the Schottegat, Curaçao; and 30 on Jan 15, 2003, at Spanish Lagoon, Aruba. **Range:** Breeds in Boreal forest ecoregion of North America, from interior Alaska east to southern tip of James Bay. Winters from southern US south, throughout South America.

Family Stercorariidae: Jaegers, Skuas

Great Skua (*Stercorarius skua*)   Plate 24

Papiamento: *Saltadó grandi*
Dutch: *Grote Jager*

Description: A dark brown, stocky bird the size of a large gull, with bright white patches in outer wings. Virtually always seen far out to sea. Tends to show warm, cinnamon-brown tones on plumage and large pale spots and streaks on back and inner wing coverts, very contrasting dark cap, and brown-toned underwing coverts. **Similar species:** Might be easily mistaken initially for a large immature gull but note that normally no gulls show bright white patches in outer wings; most sightings of immature large gulls are made close to the shore, not at sea. Immature and some adult morph jaegers are also all dark, with white patches in outer wing; Parasitic and Pomarine jaegers are gull-sized and could be easily mistaken for Great Skua at a distance. On jaegers, note the generally smaller bills, slimmer bodies, and longer-tailed appearance; immature jaegers show more extensive barring on underwing coverts. Immature Great Skua is difficult to distinguish from South Polar Skua, but South Polar Skua is generally either blacker, with little or no spots or streaking in upperparts, or with dark wings and very blond head and upper breast; all South Polar Skuas show black underwing coverts (brown in Great Skua) and no contrasting, capped appearance. Those seeking more details about skua identification should check one of several in-depth references on the topic. **Status:** There is a single sight record of a large skua species at sea between Curaçao and

Bonaire, on July 21, 1970, made by a seabird expert. While this bird was considered to be a Great Skua, note that the sighting was made before it was known that the South Polar Skua is a regular visitor to the Northern Hemisphere oceans. Several sight records of Great Skua, with credible descriptions, were reported off the western coast of Venezuela in 1996 (Murphy 2000). **Range:** Breeds in Iceland, northern British Isles, and north to Norway and Russia. Winters across Northern Hemisphere seas, though area of greatest concentration is at sea, running from a region to the west of France and moving south to a region west of northern Africa. Apparently generally avoids Caribbean Sea, but at least one specimen of dead bird found on beach in Belize.

South Polar Skua (*Stercorarius maccormicki*)   Plate 24

Papiamento: *Saltadó South Polar*
Dutch: *Zuidpooljager*

Description: This stocky bird is the size of a large gull. Note pale body, dark wings, and bright white patches in outer wings. Virtually always seen far out to sea. Some adults are very dark overall, with very little spotting or streaking on upperparts but usually with a contrasting lighter, yellowish nape. Lighter adults have dark wings with some limited lighter streaking but show pale head and body. All forms usually do not show contrasting capped appearance and have black-toned underwing coverts. **Similar species:** Might easily be mistaken initially for a large immature gull, but note that gulls do not normally show bright white patches in outer wing; most sightings of immature large gulls are made close to shore, not at sea. Immature and some adult morph jaegers are also all dark, with white patches in outer wing; Parasitic and Pomarine jaegers are gull-sized and could be easily mistaken for a South Polar Skua, if seen at a distance. But jaegers generally have smaller bills, slimmer bodies, and a longer-tailed appearance; immature jaegers show more extensive barring on underwing coverts. Immature Great Skua is difficult to distinguish from South Polar Skua but South Polar Skua is generally either blacker, with little or no spots or streaking in upperparts, or with dark wings and very blond head and upper breast; all South Polar Skuas show black underwing coverts (brown in Great Skua) and no contrasting capped appearance (in either dark or light forms). Those seeking more details about skua identification should check one of several in-depth references on the topic. **Status:** A single South Polar Skua was photographed about 0.6 mile (1 km) from the northeast coast of Aruba, on July 15, 2011 (Luksenburg and Sangster 2013). There is a single sight record of a large skua species at sea between Curaçao and Bonaire, on July 21, 1970, made by a seabird expert. While this bird was considered to be a Great Skua, note that the sighting was made before it was known that the South Polar Skua is a regular visitor to the Northern Hemisphere oceans. **Range:** Breeds in sub-Antarctic during austral summer (northern winter) and migrates north to Northern Hemisphere seas during austral winter (northern summer).

Pomarine Jaeger (*Stercorarius pomarinus*)   **Plate 24**

Papiamento: *Saltadó mediano*
Dutch: *Middelste Jager*

Description: This predatory seabird is almost falconlike in flight because of its long, pointed wings and strong flight pattern (but note long tail). Often seen chasing terns, boobies, or gulls in an attempt to steal fish from them. Identification can be difficult, as each of the three jaeger species has a variety of color morphs and plumage changes. Pomarine is the largest of the three jaeger species and has a blocky head and a heavy, pink bill with a black tip. Light morph adults in nonbreeding season are a blackish brown above and show a dark cap extending over eye to base of bill. Juveniles tend to show strong barring in undertail and whitish speckling in underwing that contrasts with darker brown flanks, but there is much variation. Juveniles that show dark heads and light upper tail are Pomarines, as Parasitics never show this combination (Sibley 2000). More extensive details of identification and aging of jaegers are available in a number of other sources. **Similar species:** Juvenile gulls are typically dark overall, but never show a white flash in wings and have different shape and flight style. Also compare with Great Skua (p. 169) and South Polar Skua (p. 170). **Status:** Eleven sightings (some with photographic documentation), 2010–2011, made during cetacean surveys on commercial sport-fishing boats, 0.7–13.9 mi (1.1–22.3 km) from Aruba (Luksenburg and Sangster 2013). Nine records of this species were made kilometers from land by observers experienced in jaeger identification, in the 1970s, Sept–May (Voous 1983). There have been occasional other reports of this species, including one that was photographed, a pale-morph adult, seen over Bubali, on June 18, 1974 (Voous 1983; Vous reviewed the photo). Interestingly, the species has been reported in the hundreds off Venezuela (Olsen and Larsson 1997); it was also reported as the most common jaeger species observed on three cruises between Bonaire and Isla Margarita, Dec–March, with a maximum of 525 birds observed on March 10, 1997 (Murphy 2000). **Range:** Breeds across Arctic regions of the globe. Winters at sea, with some populations ranging through Southern Hemisphere seas. In the Atlantic, it appears that Pomarine Jaegers regularly winter in the eastern and southern Caribbean, off northern South America and Africa, but irregularly off the eastern side of southern South America, unlike Parasitic and Long-tailed Jaegers (Olsen and Larsson 1997).

Parasitic Jaeger (*Stercorarius parasiticus*)   **Plate 24**

Papiamento: *Saltadó chikí*
Dutch: *Klein Jager*

Description: This predatory seabird is almost falconlike in flight because of its long, pointed wings and strong flight pattern (but note long tail). Often seen chasing terns, boobies, or gulls in an attempt to steal fish from them. Seen near shore harassing terns and gulls, more often than Pomarine Jaeger. Identification can be difficult, as each of the three jaeger species has a variety of color morphs and plumage changes. Parasitic sits between the larger Pomarine and much smaller Long-tailed. Parasitic appears slim, with a proportionately small, rounded head, and thin bill, all-dark in adults but light with dark tip in juveniles. Juveniles tend to be more cinnamon or buffy than other jaegers, showing less contrast between flanks and underwing linings than in Pomarine (but there is a lot of variation). Adults and later-stage immatures show a light area at base of bill that is lacking in Pomarine. **Similar species:** Juvenile gulls are typically dark overall but never show white flash in wings and have different shape and flight style. Compare with Great Skua (p. 169) and South Polar Skua (p. 170). **Status:** There are many more records of this species than of Pomarine Jaeger, despite some evidence that Pomarine may be more common than Parasitic Jaeger in South Caribbean waters (Murphy 2000). Offshore surveys in the early 1970s documented at least 17 sightings, Nov–May period, all more than three nautical miles from any of the islands (Voous 1983). On Aruba, there have been at least nine documented sightings from or near shore, most in San Nicolas Bay or Arashi (Prins et al. 2009); interestingly, sightings were made Apr–Aug, when nesting of terns and gulls is at its peak on Aruba. A distant jaeger that we observed harassing terns on Jan 11, 1995, off Arashi, was likely this species. One was photographed by Steve Mlodinow, in Feb 2014, as it flew past the California Lighthouse on Aruba! One record from Curaçao of a dead bird found at Sint Jorisbaai on Jan 3, 1952, and one record from Bonaire of an injured immature bird found and later released on March 7, 1998 (Prins et al. 2009). **Range:** Breeds across the Arctic and sub-Arctic regions of the world (farther south than Pomarine or Long-tailed), wintering at sea, with some populations ranging through Southern Hemisphere seas. In Atlantic, major wintering area is off southern South America, but some winter in eastern and southern Caribbean (Olsen and Larsson 1997; Murphy 2000).

Long-tailed Jaeger (*Stercorarius longicaudus*)   Plate 24

Papiamento: *Saltadó rabo largo, Saltadó rabu largu*
Dutch: *Kleinste Jager*

Description: This predatory seabird is almost falconlike in flight because of its long, pointed wings and strong flight pattern (but note long tail). Long-tailed can also appear ternlike because of its small size. Identification is often difficult, as this species and the other two jaegers on the islands all have a variety of color morphs and plumage changes. Long-tailed is the smallest of the jaegers, with a delicate, almost pigeonlike head, and short, stout bill; in adults bill all dark, in juveniles, pale with dark tip. Adult Long-tailed has less white in outer wing than other two jaegers, and often shows no white flash in underwing, unlike Parasitic and Pomarine jaegers. Some juvenile Long-tailed Jaegers tend to be much lighter, appearing white-headed and white-bellied, but there is a lot of variation. **Similar species:** Juvenile gulls are typically dark overall but never show white flash in wings and have different shape and flight style. Compare with Great Skua (p. 169) and South Polar Skua (p. 170). **Status:** One record of a bird seen and photographed near Baby

Beach, Aruba, on May 18, 1989 (Prins et al. 2009). **Range:** Breeds across the Arctic regions of the world, wintering in Southern Hemisphere seas. In Atlantic, major wintering area is off southern South America and southern Africa, rare in Carribean region (Olsen and Larsson 1997; Murphy 2000).

Family Laridae: Gulls, Terns

Bonaparte's Gull (*Chroicocephalus philadelphia*)   **Plate 25**

Papiamento: *Meuwchi Bonaparte, Kahela Bonaparte*
Dutch: *Kleine Kokmeeuw*

Description: Smaller and slimmer-billed than the common Laughing Gull, with a pale gray back. In adult plumage, back and upperwing pale gray with large, white wedge at wingtip, with black trailing edge. On underwing, note white primaries with narrow black trailing edge. Adult has pink legs and black bill. In breeding plumage, has a black hood, but in nonbreeding plumage this is reduced to a dark ear spot; dusky on hind crown. First-year birds show dusky brown-black mottling in wings and black tail band, pink-toned legs, and largely dark bill. **Similar species:** Adult Black-headed Gull has red bill, more orangish legs, and shows black on underside of outer primaries. First-year Common Black-headed has a pale bill with black tip and more extensive black on wings. Although unrecorded from the islands, the Gray-hooded Gull from South America has occurred in Panama and should not be eliminated as a possibility. Because of widespread vagrancy in gulls and the rarity of any gull species except Laughing Gull in this area, all non-Laughing Gull sightings should be carefully documented with photos and/or video. **Status:** Three records from Curaçao and one from Bonaire. Records from Curaçao may refer to one group but they are as follows: one in winter plumage, Muizenberg, Sept 9, 2000; one at Muizenberg, Sept 16, 2000; three at Muizenberg, Oct 1, 2000 (Prins et al. 2009). Two adult birds in winter plumage were reported from Kralendjik Harbor, Bonaire, Sept 3, 1959 (Voous 1983). **Range:** Breeds in wetlands across Boreal forest region from Alaska east to western Quebec. Winters largely along the Pacific and Atlantic coasts, from southern Canada to northern Mexico, with stragglers to northern Caribbean; rarely to Panama.

Black-headed Gull (*Chroicocephalus ridibundus*)   **Plate 25**

Papiamento: *Meuwchi Oropeo, Kahela Oropeo*
Dutch: *Kokmeeuw*

Description: Smaller and slimmer-billed than the common Laughing Gull, with a pale gray back. In adult plumage, back and upperwing pale gray; note large, white wedge at wingtip, with black trailing edge. On underwing, note black primaries. Adult has reddish legs and bill. In breeding plumage, has a brownish-black hood, but in nonbreeding plumage this is reduced to a dark ear spot with some gray on the rear part of the head. First-year birds show dusky brown-black mottling in wings and black tail band, orangey-toned legs, and pale pinkish bill with dark tip. **Similar species:** Adult Bonaparte's Gull has black bill, more pink-toned legs, and lacks black on underside of outer primaries. First-year Bonaparte's also has a dark bill and less extensive black on wings. Although unrecorded from the islands, the Gray-hooded Gull from South America has occurred in Panama and should not be ruled out as a possibility. Because of widespread vagrancy in gulls and the rarity of any gull species except Laughing Gull in this area, all non-Laughing Gull sightings should be carefully documented with photos and/or video. **Status:** Single record; one immature bird near Kralendjik, Bonaire, Feb 16 to mid-March, 1976 (Voous 1983). **Range:** Breeds in northern Europe and Asia; in the last 50 years has extended breeding range westward to Greenland, Newfoundland, and occasionally Quebec, and south to Canadian Maritimes and New England (Olsen and Larsson 2003).

Laughing Gull (*Leucophaeus atricilla*)   Plate 25

Papiamento: *Meuwchi haridó, Kahela komun*
Dutch: *Lachmeeuw*

Description: The most common gull on the islands. Back and upperwing dark gray, with black wingtips. Legs black in all plumages. Breeding plumaged adult has a black hood, white eye crescents, and reddish bill. In nonbreeding plumage, adult lacks hood and has dark smudge on back of head and black bill. Immature birds show mottled brown and black on upperwing and black tail band that extends to sides of tail. **Similar species:** Herring, Ring-billed, Lesser Black-backed, and Great Black-backed gulls have all-dark bills only as juveniles or through first winter, when their plumages are very different from Laughing Gull. All are larger than Laughing Gull and show pink or yellow, not black, legs. Adults of those four species all have yellow bills as compared to the black (nonbreeding) or reddish (breeding) bill of Laughing Gull. Bonaparte's and Black-headed gulls are smaller than Laughing, with extensive white in wings, lighter gray backs, smaller bills, and reddish or pink, rather than black, legs. Franklin's very similar to Laughing Gull but shows a white band on the inner side of the black in the wingtips, overall giving an impression of a black-and-white wingtip instead of a black wingtip as in Laughing Gull. Franklin's also has a shorter, slimmer bill, and adults in nonbreeding plumage have a more extensive half-hood. First-year Laughing Gull is browner on wings than first-year Franklin's, and tail band on first-year Franklin's Gull does not extend to side of tail as in Laughing Gull; but beware second-year Laughing Gulls, which lack brown in wings and can have an incomplete tail band. **Voice:** As their name suggests, the calls of Laughing Call are reminiscent of high, nasal laughing sounds. **Status:** The South Caribbean race of Laughing Gull is a regular breeder on Aruba and Bonaire, and occasionally on Curaçao, from Apr to Aug (but may arrive as early as late Feb or March). Largest number of nests on Aruba at the San Nicolas reef islands, where numbers increased from 4–5 in 1979 to 400 in the late 2000s (del Nevo 2008; van Halewijn 2009). On Bonaire, 10 pairs were found nesting on an island in Gotomeer and an estimated 50 pairs in the Pekelmeer, in 2002 (Debrot et al. 2009; Prins et al. 2009). On Curaçao, three pairs nested in 2002 and two in 2003, at the Jan Thiel Lagoon (Debrot et al. 2009; Prins et al. 2009). Varying numbers of birds also winter on Aruba, at times numbering in the hundreds. Some or all of these wintering birds are thought to be of the slightly larger North American race of Laughing Gull. **Range:** Breeds on Atlantic Coast of North America, from Maine south to Mexico, along northwestern coast of Mexico, and in Caribbean on south coast of Hispaniola, locally in Lesser Antilles and on Aruba, Bonaire, Curaçao, and Venezualan islands. Winters south to northern South America.

Franklin's Gull (*Leucophaeus pipixcan*)   Plate 25

Papiamento: *Meuwchi Franklin, Kahela Franklin*
Dutch: *Franklin's Meeuw*

Description: Similar to Laughing Gull but in adult plumage note band of white on wingtip separating the black wingtip from the gray back; slightly smaller bill; and, in nonbreeding plumage, the more extensive hood. In winter plumage, adult Laughing Gulls show white tips to outermost primaries although less extensive than in Franklin's Gull. In first winter birds, Franklin's dark tail band does not extend to sides of tail, underwing is white (mottled in Laughing Gull), back of neck is whitish (brownish-gray in Laughing Gull), and upperwing shows less brown coloration. Very rare, so any sighting should be carefully documented. **Similar species:** See Laughing Gull. **Status:** Two sight records from Aruba, and at least two, more recent, photographic records. One adult bird was reported from Bubali, Aruba, Jan 19–20, 1971 (Prins et al. 2009); an adult in winter plumage was photographed by Steve Mlodinow at Malmok, Aruba, March 29, 2004 (fide eBird). Two first cycle birds were seen by multiple observers, Nov 29–30, 2015, at the wetlands behind the Mill Resort near Bubali, Aruba, and photographed by Michiel Oversteegen (fide eBird). A bird was reported without details at Malmok, Aruba, on March 20, 2016 (R. Wauben fide eBird). **Range:** Breeds in prairie wetlands of western US and Canada; winters along Pacific Coast of South America, with some wintering in Andean lakes.

Ring-billed Gull (*Larus delawarensis*)   Plate 25

Papiamento: *Meuwchi pico renchi, Kahela pik renchi*
Dutch: *Ringsnavelmeeuw*

Description: Adult has gray back, black wingtips, yellow legs, and yellow bill with black ring around it. Immature plumages, which are the most likely to be seen in the islands, are mostly white underneath with a gray back and mottled brown and gray on wings; has a narrow black tail band and a pink or pale-yellow bill with a black tip. Pink legs. The identification of gulls of various ages can be a challenge; there are a number of references that can be consulted for more detail. **Similar species:** Slightly larger than common Laughing Gull, with a lighter gray back, yellow (not black) legs, and yellow or pinkish (immature) bill compared to black bill of Laughing Gull. Laughing Gull in breeding plumage is very different, with a black hood and reddish bill. Herring Gull is very similar to Ring-billed Gull but is considerably larger and thicker-billed and always with pink legs. Immature Herring Gulls in first and second years are very brown overall including on the underside, unlike similarly aged Ring-billed Gulls, which are largely white underneath; immature Herring Gulls also show a wider tail band than Ring-billed Gull. Adult Herring Gulls lack the black circle around the bill but in nonbreeding plumage can show a black mark, usually on the lower mandible. Adult Lesser Black-backed Gull and Great Black-backed Gull have much darker backs than Ring-billed Gull, and both are larger. First- and second-year plumages of these species are generally darker and more mottled, with brown or checkered backs, not gray backs as in immature Ring-billed Gull. Because of widespread vagrancy in gulls and the rarity of any gull species except Laughing Gull in this area, all non-Laughing Gull sightings should be carefully documented with photos and/or video. **Status:** Seven records, three from Aruba and four from Bonaire. One sub-adult at Pekelmeer, Bonaire, Jan 23–March 22, 1970; one second-winter bird at Hadicurari, Aruba, Jan 20, 1972; one first-winter bird at Hadicurari, Aruba, Dec 1978; one second-winter bird at Kralendjik, Bonaire, Jan 17–Feb 20, 1982; one (age unspecified) off coast near airport, Bonaire, June 2002 (Prins et al. 2009). We observed a first-winter bird at Hadicurari, Aruba, Jan 7–9 1996, and a second-winter bird (perhaps the same individual?) at Hadicurari, Jan 5, 1997. **Range:** Breeds across southern Canada and northern US. Winters from southern Canada south to Mexico, a few occasionally south to Panama and in Greater Antilles. There are a handful of records from South America, including from Venezuela.

Herring Gull (*Larus argentatus*)   Plate 25

Papiamento: *Meuwchi gris, Kahela gris*
Dutch: *Zilvermeeuw*

Description: A large gull that, in adult plumage, shows gray back and wings (with black wingtips), pink legs, and large, yellow bill; body white, except heavily streaked head and neck in nonbreeding plumage. First- and second-year birds brown overall, with darker brown wingtips and dark brown tail band. Bill all black in juveniles; as birds age, bill becomes paler at base, and by second winter the bill is pink, with black tip; by third year, yellow with black tip. **Similar species:** Ring-billed Gull is smaller; in adult or near-adult plumage, shows yellow legs and yellow bill with black ring. Immature Ring-billed is largely white underneath, rather than brown, and has a gray back. Third-year Herring Gull also shows gray back and brown mottling in wings but note larger size and larger bill. Adult Great Black-backed Gull and Lesser Black-backed Gull have distinctly darker backs, and, in Lesser Black-backed, yellow legs. First- and second-year Great Black-backed Gull show more checkered appearance on back and wing coverts and have whiter head and body than similarly aged Herring Gulls. First- and second-year Lesser Black-backed Gull can be difficult to separate from similarly aged Herring Gulls. Because of widespread vagrancy in gulls and rarity of any gull species except Laughing Gull in this area, all non-Laughing Gull sightings should be carefully documented with photos and/or video. **Status:** Six records, five from Aruba and one from Bonaire. One first-winter bird at Hadicurari, Aruba, Nov 7, 1972; one adult at Bubali, Aruba, March 13–Aug 15, 1975; one immature bird at Bubali, Aruba, Jan 12 and Feb 15, 1980; one immature, Oranjestad reef islands, Aruba, May 22 and June 9, 1986; one immature, San Nicolas Bay, Aruba, May 22, 1987; one (age unspecified) on road east of Rincon, Bonaire, July 21, 2000 (Voous 1983; Prins et al. 2009); one first cycle bird photographed by Ross Wauben at Savaneta, Aruba, on Dec 20, 2014 (fide eBird). **Range:** Extensive worldwide range. Breeds across most of northern North America and winters south regularly, with small numbers to Central America and Greater Antilles. A few records from northern South America.

Lesser Black-backed Gull (*Larus fuscus*) Plate 26

Papiamento: *Meuwchi lomba preto chikito, Kahela lomba pretu chiki*
Dutch: *Kleine Mantelmeeuw*

Description: A large gull that shows, in adult plumage, gray-black back and wings with jet black wingtips, yellow legs and bill; white body overall but heavily streaked on head and neck in nonbreeding plumage. First- and second-year birds have dark brown upperparts with darker brown wingtips, dark brown tail band, and whitish head with more streaked appearance on underparts rather than uniform brown as in Herring Gull of similar age. In flight, upper surface of wing of first- and second-year birds is darker than in very similar Herring Gull. Bill is all black as juvenile and gets progressively paler at base; by second winter has a pink bill with black tip and, by third year, yellow with black tip. Legs pinkish in first year, becoming yellowish by second year. **Similar species:** Distinctly larger than the common Laughing Gull. Adult Ring-billed Gull also has yellow legs but Ring-billed Gull is smaller and has a much lighter gray back. Immature Ring-billed Gulls are smaller and much lighter overall than immature Lesser Black-backed. First- and second-year Lesser Black-backed Gulls can be quite difficult to distinguish from similarly aged Herring Gulls, but note slimmer bill and rounder head of Lesser Black-backed, as well as darker upperwing, whiter head, more streaked underparts and, at rest, wingtips that extend farther beyond tail tip. In adult, plumage differs from Herring in having darker back and wings, yellow legs, slimmer build, and longer legs. Adult Great Black-backed Gull is larger, with whiter head (in nonbreeding plumage), darker back, larger size, and pink rather than yellow legs. Adult Great Black-backed Gull also has two white spots in outer primaries rather than just one in Lesser Black-backed. Note that the southern hemisphere Kelp Gull, which is also a large, dark-backed gull, while still very rare, has been occurring with greater frequency in the Caribbean and North America, so that any dark-backed gull should be carefully studied. Because of widespread vagrancy in gulls and the rarity of any gull species except Laughing Gull in this area, all non-Laughing Gull sightings should be carefully documented with photos and/or video. **Status:** Voous (1983) listed eight records of adults or "near-adults" from Aruba that he considered sufficiently documented, most by photographs. There are at least three more recent records, two from Aruba and one from Bonaire. Oddly enough, there are no documented records from Curaçao. After the first record of a single adult in Nov 1957, at what is referred to as "the Ceru Colorado lagoon," perhaps near Baby Beach, Aruba, the next seven records occurred in the years between 1973 and 1980, in an area between Oranjestad and Palm Beach, Aruba. Some of these records may pertain to the same birds that became resident (at least one bird was known to stay from March through at least Aug one year) or returned to Aruba each winter. Recent records include one photographed by Greg Peterson at Malmok, Aruba, on Apr 21, 2007. A first cycle bird was photographed by Sipke Stapert south of the salt pier at Pekelmeer, Bonaire, on Dec 21, 2015 (first seen there on Dec 17, 2015). A first cycle bird (probably same individual as previous record) was photographed by Peter-Paul Schets at Kralendijk, Bonaire, on Jan 6, 2016 (fide eBird). A first cycle bird was photographed by Ross Wauben at Savaneta, Aruba, on Jan 23, 2016 (fide eBird). A first cycle bird (perhaps the same individual as previous record) was photographed by Michiel Oversteegen at Eagle Beach, Aruba, on Jan 25, 2016 (fide eBird). Given the relatively large numbers that now occur along the east coast of North America, it would not be unlikely to find the species again on the islands. **Range:** Breeds in coastal Europe and Iceland. In Old World, winters south to northern Africa. Increasing numbers found wintering in eastern North America, with occasional stragglers south to Caribbean and Central America and possibly northern South America.

Great Black-backed Gull (*Larus marinus*)   Plate 26

Papiamento: *Meuwchi lomba preto grandi, Kahela lomba pretu*
Dutch: *Grote Mantelmeeuw*

Description: This very large gull would stand out among the common Laughing Gulls. Adult with jet black back and wings, white head and body, pink legs, and very large, yellow bill. First- and second-year birds show checkered brown back and wings, white head, and large black bill, sometimes with a pale base. **Similar species:** Among gulls that have been recorded on the islands, the only other species that in adult (or third-year) plumage would show a dark back is the Lesser Black-backed, which at that age would show yellow legs and a more heavily streaked head (at least in nonbreeding plumage); it is also smaller and has a slighter bill. First- and second-year Lesser Black-backed and Herring gulls are darker underneath and have less of a checkered appearance on the back and wing coverts. Note also that the southern hemisphere Kelp Gull, which is also a large dark-backed gull, while still very rare, has been occurring with greater frequency in the Caribbean and North America, so that any dark-backed gull should be carefully studied. Because of widespread vagrancy in gulls—and the rarity in this area of any gull species except Laughing Gull—sightings of gulls other than Laughing Gull should be carefully documented with photos and/or video. **Status:** Three records: one "in almost fully adult plumage" at Hadicurari, Aruba, Nov 24, 1971, to Feb 1972; one "in similar or more slightly advanced plumage" at Hadicurari, Aruba, Nov 5–Dec 17, 1972 (Voous 1983); one (age unspecified) at the Pekelmeer, Bonaire, Jan 16, 2004 (Ligon 2006). **Range:** In North America, breeds along the Atlantic Coast from northern Quebec south to the Mid-Atlantic states. In winter, birds can range regularly south to Florida.

Brown Noddy (*Anous stolidus*)   **Plate 26**

Papiamento: *Noddy brùin*
Dutch: *Noddy*

Description: A distinctive all-brown tropical tern with whitish forehead that becomes grayer toward back of head. Black bill and legs. Unlike other terns, all dark underneath, including on underwing. Tail is long, with a slight notch rather than a sharp fork as seen in many tern species. **Similar species:** Black Noddy is very similar and can be very difficult to identify at a distance. As its name implies, the Black Noddy is black overall, so appears darker than Brown Noddy, even at a distance. Black Noddy is also slightly smaller, with a slimmer bill and a more extensive, usually more contrasting, white cap. Juvenile Sooty Tern very dark overall but has light underwing and undertail, and speckled with white above. **Status:** Long known to occur throughout the year, in offshore waters around the islands. Recorded historically in roosts and/or in near-shore waters, sometimes in hundreds or even more, off the northern end of Bonaire, southeast side of Aruba, and Westpunt, Curaçao (Voous 1983). First confirmed as a breeding species in 1983, at the San Nicolas reef islands, Aruba, and now typically breeds there in the hundreds of pairs; these birds sometimes accompanied by hundreds of nonbreeding birds (Prins et al. 2009; Van Halewijn 2009). No breeding records from Bonaire or Curaçao, and we are not aware of any recent observations of large numbers of birds away from the Aruba breeding colony and nearby locations. We have observed both Brown and Black noddies by using a telescope set up on the mainland across from the San Nicolas reef islands. The San Nicolas reef islands' nesting sites should not be visited from March through Aug to prevent disturbance and resulting loss of eggs and mortality of young birds in this important South Caribbean tern nesting location. **Range:** Breeds on small islands across the world's tropical regions, including on nearby Venezuelan archipelagos of Los Roques and Las Aves.

Black Noddy (*Anous minutus*)   **Plate 26**

Papiamento: *Noddy preto, Noddy pretu*
Dutch: *Witkapnoddy*

Description: A distinctive all-black tropical tern with white forehead and cap that becomes grayer toward back of head. Black bill and legs. Unlike other terns, this species is all dark underneath, including on underwing. Tail is long, with a slight notch rather than a sharp fork as seen in many tern species. **Similar species:** See Brown Noddy. **Status:** Much less abundant than Brown Noddy, with only three records from Aruba and one from Bonaire prior to 1983 (Voous 1983). The species was confirmed as a breeder on Aruba's San Nicolas reef islands in 1992, where it has continued to breed, with 48 nests confirmed in 2001 (van Halewijn 2009); population reported to be between 78–144 individuals as of a 2008 publication (del Nevo 2008). Since 1983, there have been three additional records on Bonaire. The species still has not been recorded from Curaçao (Prins et al. 2009). We have observed both Brown and Black noddies by using a telescope set up on the mainland across from the San Nicolas reef islands. The San Nicolas reef island nesting sites should not be visited from March through Aug to prevent disturbance and resulting loss of eggs and mortality of young birds in this important South Caribbean tern nesting location. **Range:** Breeds on small islands across tropical Pacific and Atlantic; absent in tropical oceans between Africa and western Australia.

Sooty Tern (*Onychoprion fuscatus*)   **Plate 26**

Papiamento: *Stèrnchi bata preto, Stèrnchi bata pretu, Meuchi bachi pretu*
Dutch: *Bonte Stern*

Description: Same size as Common Tern. Adult has black back, black cap, and white on forehead that extends to top of eye. Black legs and bill. Tail black, with narrow white edges (often difficult to see in flight). Juveniles have black head and underparts, with white flecking and barring on back and wing coverts. **Similar species:** See Bridled Tern. **Voice:** Makes a loud, piercing "wide-a-wake." **Status:** First recorded as a breeding species in 1976, at Aruba's San Nicolas reef islands. Numbers nesting on these islands reached over 6,600 pairs in 2001 (Prins et al. 2009). Birds present around breeding site from Apr through Aug; small numbers occasionally seen around all three islands through Dec. Apparently small flocks sometimes found far offshore throughout the year (Voous 1983), though we are not aware of recent offshore records. We have observed both Sooty and Bridled terns by using a telescope set up on the mainland across from the San Nicolas reef islands. The San Nicolas reef islands' nesting sites should not be visited from March through Aug to prevent disturbance and resulting loss of eggs and mortality of young birds in this important South Caribbean tern nesting location. **Range:** Breeds worldwide, across tropical and subtropical zones.

Bridled Tern (*Onychoprion anaethetus*)   **Plate 26**

Papiamento: *Stèrnchi bril, Meuchi brel*
Dutch: *Brilstern*

Description: Adult has dark-gray back, with black cap and white forehead extending back behind eye. Black legs and bill. White extends back around neck to form partial collar. Tail with broad, white edges. Juveniles show white-edged barring on back and wing coverts and extensive white underparts and head. **Similar species:** Most similar to Sooty Tern and surprisingly difficult to distinguish at a distance. Sooty Tern much darker-backed and lacks contrast between dark cap and light gray back as seen in Bridled. Tail is much darker in Sooty Tern, and white does not extend back behind eye in Sooty. Trailing edge of underwing much darker and more extensive in Sooty Tern. Juvenile Sooty Tern is very distinct, with all-dark head and underparts, as compared to white underparts and head of Bridled. **Voice:** Makes a mellow "hurry-up." **Status:** Breeding documented at Aruba's San Nicolas reef islands since late 1800s, and known to be a regular breeder there in recent decades, sometimes with over 100 nesting pairs (Prins et al. 2009), but in much smaller numbers than Sooty Tern. Birds present around breeding site from Apr through Aug, and occasional small numbers seen around all three islands through Dec. We have observed both Sooty and Bridled terns through use of a telescope set up on the mainland across from the San Nicolas reef islands. The San Nicolas reef islands' nesting sites should not be visited from March through Aug to prevent disturbance and resulting loss of eggs and mortality of young birds in this important South Caribbean tern nesting location. **Range:** Breeds across tropical and subtropical zones, except in central and eastern Pacific.

Least Tern (*Sternula antillarum*)   Plate 27

Papiamento: *Stèrnchi chikito, Meuchi chikitu*
Dutch: *Amerikaanse Dwergstern*

Description: A very tiny tern, with black-tipped, yellow bill, yellow legs, and black cap with white forehead. Back and upperwing gray, with black on leading edge of wing (two outermost primaries). In juvenile and first-winter plumages, shows dark brownish bar across leading edge of inner wing and more extensive dark on outer wing, forming an M-shape. Nonbreeding adults and immatures have an all-black bill. **Similar species:** This is the smallest of the terns that regularly occur in the islands; among the regularly occurring species, only the "Cayenne" Tern morph of the Sandwich Tern has an all-yellow or largely all-yellow bill, and it is two or three times larger than Least Tern, with black legs. The Yellow-billed Tern of South America may occur more regularly than realized; in breeding plumage, the Yellow-billed Tern always shows an all-yellow bill, without a black tip, and the bill is thicker at the base than in Least Tern. Yellow-billed Tern generally shows slightly more black in outer primaries (usually 3–4 outer primaries are black; normally two in Least Tern). Occasionally Least Terns are seen that lack the black tip on bill, and such birds should be studied with carefull attention to bill depth and shape, amount of black in outer wings, and other features, perhaps especially the shape and size of the white forehead and white around base of bill and eye. Least Terns in nonbreeding and immature plumages show an all-dark bill, while Yellow-billed Tern always shows a predominately yellow or pale bill, sometimes with a dusky tip and some duskier area near the nostrils. **Voice:** Calls, often heard around breeding colonies, include a high *kip* and a high, sharp *kid-up, kid-up* and *clee-kid-up, kid-up*. **Status:** A regular and abundant breeding bird on all three ABC islands. Breeding birds typically arrive in Apr and depart by Oct, but there are occasional records of small numbers or single individuals from throughout the year, and several apparently unusual records of large numbers in winter: 400 in first-winter plumage on Bonaire, Jan 22–24, 1971, and flocks at sea into Dec, 1971 (Voous 1983). On Aruba, Least Terns nest on the San Nicolas Bay islands, varying from 20–100+ nests since the 1980s (Prins et al. 2009). They were reported to be nesting on salt flats north of the high rise hotels, presumably near Palm Beach (Tony White, online birding report 1997), and at scattered locations along the northern coastline of the island, including near California Lighthouse and at Boca Prins in Arikok NP (reports received to our website www.arubabirds.com). A more thorough survey of nesting sites should be carried out and mapped. On Curaçao, a survey in 2002 found more than 620 breeding pairs at 16 sites, including more than 250 pairs along the northwestern coast of the island and 140 pairs on Klein Curaçao (Debrot and Wells 2008). On Bonaire, we documented more than 360 adults nesting at 13 sites (Wells and Childs Wells 2006); Debrot et al. (2009) found about 800 pairs nesting at 44 sites. **Range:** Breeds along Atlantic Coast, from Maine south to Hondurus; on Pacific Coast, breeds from California south to Mexico; in interior of US, breeds in Mississippi River drainage and in small numbers through the Caribbean, south to Aruba, Bonaire, and Curaçao, and some nearby offshore Venezuelan islands. Winters south to northern South America.

Yellow-billed Tern (*Sternula superciliaris*)   Plate 27

Dutch: Amazonestern

Description: Very similar to Least Tern, but slightly thicker-based bill and slightly larger, with all-yellow bill in breeding plumage. In nonbreeding plumage, bill is mostly yellow, with dusky tip and dusky area around nostrils. Immatures very similar to immature Least Tern. **Similar species:** See Least Tern. **Status:** Voous notes two possible records of single individuals of the species from Bonaire, one at "Great Salt Lake" (presumably Gotomeer) on Jan 22, 1971, and one off Kralendijk Harbor, on Oct 8, 1979. We photographed a single bird with an all-yellow bill that appears to have a thicker-base on July 4, 2001, inland from Lac Bay, Bonaire. Even more interesting, farther inland from where we documented most of the hundreds of Least Terns that were nesting on the island, we noted several similar birds with all-yellow, thick-based bills sitting on single eggs, each by itself in the middle of a dry riverbed. **Range:** Resident at inland areas, from eastern Colombia south to Brazil. Vagrants have been recorded from Panama and more regularly on Trinidad.

Large-billed Tern (*Phaetusa simplex*)   Plate 27

Papiamento: *Stèrnchi pico grandi, Meuchi pik grandi*
Dutch: *Grootsnavekstern*

Description: A large tern with big, thick yellow bill and black cap in breeding plumage. Distinctive pattern on upperwing, with black outer wing bordered by large, white patch at elbow; also note gray inner wing, back, and tail. Yellowish legs. In nonbreeding plumage, has reduced, more grayish, cap. **Similar species:** Combination of large, yellow bill and sharply contrasting upperwing pattern should be distinctive. On Sandwich "Cayenne" Tern, all-yellow bill is much thinner; it lacks contrasting upperwing pattern of Large-billed Tern. Given that there have not been any records on any of the ABC islands in over 100 years, any sighting should be verified with photographs and/or video. **Status:** One record, a specimen from Aruba collected on May 12, 1908 (Voous 1983). The species does occur regularly along parts of the Venezuelan coast and has occurred as a vagrant in Panama, so it reasonably could occur again on Aruba, Bonaire, or Curaçao. **Range:** Breeds on inland river systems, from northern South America to northern Argentina.

Gull-billed Tern (*Gelochelidon nilotica*)   Plate 27

Papiamento: *Stèrnchi nilótiko, Meuchi nilótiko*
Dutch: *Lachstern*

Description: A relatively large tern, yet smaller than Royal Tern. Has pale gray back, dark legs, and short, stout, black bill. Adult has black cap in breeding season but only a black mark behind eye in nonbreeding season. Tail slightly forked; in flight, shows black on wings, largely restricted to trailing edge of outer primaries. Of the terns that occur in the islands, this is one of the few that frequents freshwater habitats in addition to coastal habitats; sometimes even forages for dragonflies and other large insects over open, grassy areas. **Similar species:** No other tern has stout black bill, black legs, and pale gray upperparts. **Status:** This is a rare to uncommon stop-over migrant, usually between Aug and Nov, with several records in spring (May–early June) and one March and one Jan record. At least 8 documented records from Aruba, 14 from Bonaire, and 8 or more records from Curaçao. Highest numbers recorded were 16 at the Curaçao airport, Sept 11, 1959, and six at Lac, Bonaire, on Aug 28, 1981. These birds were probably migrants from the breeding population in the US, Bahamas, or Greater Antilles, but it is possible that some originated from South American breeding populations. **Range:** Occurs in Europe, Asia, Australia, and North and South America. Small numbers breed in Bahamas and Greater Antilles, and breeding has been considered a possibility along some areas of the Colombian and Venezuelan coast. A disjunct breeding population occurs in southern South America.

Caspian Tern (*Hydroprogne caspia*)   Plate 27

Papiamento: *Stèrnchi gigante, Meuchi gigante*
Dutch: *Reuzenstern*

Description: Large, gull-sized tern that in adult plumage has a gray back and upperwing; black wingtips; large, reddish bill with dusky tip; and black legs and black cap. Extensive black on underside of outer primaries can be seen in flight. Juveniles show brownish scalloping on back and wing coverts. **Similar species:** Unlike the otherwise very similar Royal Tern, Caspian never shows an extensive white forehead, has a larger, brighter red bill with a dusky tip (no dusky tip in Royal), and shows substantiallly more black in underside of outer primaries in flight. **Voice:** Makes a low, heronlike croaking call. **Status:** Rare visitor, with at least six records. On Aruba, two birds at Bubali, Oct 2, 1977; two birds at Bubali, Oct 24, 1978 (Voous 1983). We photographed and videotaped a single individual in the salina behind Palm Beach,

Nov 22–25, 2012. On Bonaire, "a number of birds" were seen at Kralendijk, March 1, 1970; one seen at Sorobon, March 6, 1970 (perhaps same bird as previous record?). On Curaçao, one seen at Santa Martabaai, Nov 27, 2001; and one seen along road to Bullenbaai, Oct 31–Nov 5, 2003 (Voous 1983; Prins et al. 2009). **Range:** A cosmopolitan species with breeding populations in Europe, Asia, Africa, Australia, and North America. North American populations winter from southern US through Central America and Caribbean, to northern South America.

Black Tern (*Chlidonias niger*)   Plate 27

Papiamento: *Stèrnchi preto, Meuchi pretu*
Dutch: *Zwarte Stern*

Description: Quite unmistakable in breeding plumage, with all-black head, back, throat, and belly; gray wings; and white on underparts, from just behind legs to tail. Most sightings, however, have been of birds in nonbreeding or immature plumages. Nonbreeding and juvenile birds show a black "helmet" that extends from behind eye across back of head; a largely white nape; gray back and wings; and white underparts. In flight, note a dark shoulder mark extending down from where the wing meets the body, a very short tail, and all-gray underwing. **Similar species:** In breeding plumage, should be unmistakable. Nonbreeding birds could be confused with Bridled or Sooty terns, but both those species are larger, with longer, deeply forked tails; lack the shoulder bar; and underwing with contrasting tones (rather than all-gray underwing). **Status:** Seven records from Aruba, one from Curaçao, and three from Bonaire. Most records from Aug and Sept, two from May, one from Apr, and one from Jan. Largest numbers 10–15, Sept 9, 2004, along south coast of Bonaire; 10 off east and north coasts of Aruba, on Aug 25, 1972; and 5–8 at Schottegat, Curaçao, Sept 3, 1971 (Prins et al. 2009). **Range:** North American population breeds across southern Canada and northern US. Winters to northern South America.

Roseate Tern (*Sterna dougalli*)   Plate 28

Papiamento: *Stèrnchi Dougall, Meuchi Dougall*
Dutch: *Dougalls Stern*

Description: Smaller than Sandwich Tern but much larger than Least Tern. In adult plumage, a gray-backed tern with black cap and red legs. In flight, upperwing very pale whitish, with limited gray-black on leading edge of outer wing. Has a very long tail that, at rest, projects beyond wingtips. Wings proportionately shorter than wings on Common Tern, so has faster, choppier wingbeats. Bill sometimes all dark, but in Caribbean race the lower two-thirds of the bill is red. All-white tail. In juvenile and first-year plumages, shows dark along front edge of inner wing but lacks any bars on trailing edge of wing. **Similar species:** Common Tern is darker gray on upperwing and back, with a wedge of black on the upperpart of the outer wing and a dark trailing edge to the underside of the primaries (white on Roseate Tern). Common has shorter tail, with thin, dark (not all white) outer edges. In flight, Roseate has much quicker, choppier wingbeats than Common or Sandwich terns. Sandwich Tern larger, with larger bill either all yellow (in "Cayenne" Tern morph) or black with yellow tip; black legs, and more black in outer wing. Juvenile and first-year Common Tern show a dark bar across secondaries and more extensive dark markings in outerwing. Juvenile Roseate Tern has dark, blackish V-shaped marks on back, scapulars, and tertials. On juvenile Common Tern, these are lighter brownish and more like bars rather than V-shaped markings. **Voice:** Makes a high, distinctive *chiv-ee*. **Status:** Uncommon breeding bird on all three ABC islands; usually found from late March and Apr through Aug. Most regular on the San Nicolas Bay reef islands, Aruba, where known as a breeder since 1892, but with numbers fluctuating over the years, reaching a high of 112 nests, 1984–1994; del Nevo (2009) reported that the species had not been confirmed nesting from 2003 to 2009, although small numbers were occasionally observed. Nesting on Bonaire and Curaçao has been more irregular. Nests have been found in the past on Bonaire at the Pekelmeer and Gotomeer, and on Curaçao at the Jan Thiel Lagoon and on Masarigo Island, Spaanse Water, in the 1950s and 1960s (Voous 1983; Prins et al. 2009). On Aruba, terns can be observed from the mainland, across from the San Nicolas reef islands, by using a telescope. The San Nicolas reef islands' nesting sites should not be visited from March through Aug to avoid disturbing the birds and the resulting loss of eggs and mortality of young birds at this important tern nesting site. **Range:** Breeding populations in Europe, Asia, Africa, Australia, and North America and Caribbean Basin.

Common Tern (*Sterna hirundo*)   Plate 28

Papiamento: *Stèrnchi comun, Meuwchi piku kòrá, Meuchi pik kòrá*
Dutch: *Visdief*

Description: Smaller than Sandwich Tern but much larger than Least Tern. In adult plumage, a gray-backed tern with black cap and red legs. In flight, upperwing gray with charcoal on trailing edge of outer wing, often forming a black wedge. At rest, tail shorter than—or even with—wingtips. Proportionately longer wings than those of RoseateTern, so has slower, more fluid wingbeats. During breeding season, bill mostly red with dark tip; in nonbreeding season, bill becomes mostly dark. On the underwing, Common Tern shows a thin, dark trailing edge to the outermost primaries. Tail shows thin, dark edges. In juvenile and first-year plumage, shows dark along front edge of inner wing and on outer primaries, and has a dark bar across trailing edge of inner wing (secondaries). **Similar species:** On Caribbean race of Roseate Tern, lower two-thirds of bill often red (like Common Tern) but usually has a more extensive dark tip. Common Tern is darker gray on upperwing and back, with a wedge of black on the upperpart of the outer wing and a dark trailing edge to the underside of the primaries (in contrast to the very white appearance of Roseate Tern). Common Tern has shorter tail with thin, dark outer edges, not all white as in Roseate. In flight, Roseate has much quicker, choppier wingbeats than Common or Sandwich terns. Sandwich Tern larger, with larger bill either all yellow (in "Cayenne" Tern morph) or black with yellow tip; black legs.

Juvenile and first year Common Tern show a dark bar across secondaries (lacking in Roseate) and more extensive dark markings in outer wing than Roseate. **Voice:** Common call is a high, harsh *kee-urr*. **Status:** An uncommon but regularly occurring breeding bird, from Apr through Nov, but occasionally seen throughout the year. From Dec–March, normally occurs in small numbers, but 144 were reported on Bonaire in Jan, 1971 (Voous 1983). Band returns have demonstrated that migrants from North American breeding areas also occasionally pass through the islands (Voous 1983). On Aruba, Common Terns have regularly nested on the San Nicolas reef islands and the Oranjestad reef islands, with numbers in recent years of about 50–80 nesting pairs (del Nevo 2008). Birds have nested occasionally at other locations on Aruba, including in salt pans and salinas near Bubali, and formerly in dunes behind now built-up beaches (Voous 1983). On Curaçao, birds have nested at the Jan Thiel Lagoon (up to 75 pairs in recent years) and Salina Sint Michiel (up to 15 pairs in recent years), but numbers fluctuate and in some years few birds nest (Debrot and Wells 2008). Surveys in 2002 and 2004 estimated 135 nests at four locations (Prins et al. 2009). Birds have nested on rocky islets in the Schottegat and formally on Isla Macuacu on Curaçao (Voous 1983). On Bonaire, small numbers have nested on various parts of the Pekelmeer, where 30 nests were reported in the saltworks in 2002 (Prins et al. 2009) and at the Gotomeer (5 nests in 2002) and Salina Slagbaai (4 nests in 2002). We counted 18 adults, of which 3–4 were attending nests, and 7 fledglings at the Pekelmeer in July 2002 (Wells and Childs Wells 2006). **Range:** Breeds in North America, Europe, and Asia, with isolated populations in Caribbean and Africa. Winters south to Australia, southern Africa, and southern South America.

Royal Tern (*Thalasseus maximus*)   Plate 28

Papiamento: *Stèrnchi di rey, Bubi chiquitu, Bubi chiki, Meuchi real*
Dutch: *Koningsstern*

Description: A rather large tern, about the same size as the Laughing Gull. Bright, carrot-orange bill somewhat thick and wedge-like (slim in smaller terns). Gray back and wings with dusky, blackish wingtips that have narrower black trailing edge. Underparts white except for narrow black trailing edge to outermost part of underwing. Shows a complete black cap only briefly in spring; in summer months (June–Aug), adults have white forehead and forecrown. White nape separates black crown and gray back. Tail white, with relatively deep fork. Legs black. In winter, head white with black extending from just in front of eye around back of head, to form a thin black border to rear part of crown. Juveniles and immatures similar to winter adult but with darker wingtips and dark bars on inner part of upperwing. **Similar species:** Most similar species is rare Caspian Tern, which has underwing with large area of black on wingtip, in contrast to thin, black trailing edge in Royal Tern. Caspian Tern also has much larger, reddish-orange bill, with darker tip. Very distant birds can be difficult to differentiate, even from Sandwich or other terns. **Voice:** Most often heard in the breeding season; call is high, raspy *kurr-urr* or *kee-kurr-urr*. Juveniles give a high, upslurred whistle. **Status:** A regular year-round resident, though generally in small numbers (5–30) and occurring almost exclusively along leeward coasts, where singles and groups of 2–3 birds may be seen just offshore, flying slowly as they search for fish. Individuals often seen near shore, perched on mooring buoys, breakwaters, pilings, and the like. In the evenings, larger numbers are often seen flying to communal roosting locations. Breeding confirmed on Aruba and Bonaire, but only in very small numbers on Curaçao. Occurs year-round on Aruba, with a breeding population on the San Nicolas reef islands estimated at 800 birds in mid-2000s (del Nevo 2008). From fall through spring, a sprinkling of birds forage up and down the coast from Baby Beach near San Nicolas to Arashi near California Lighthouse, with 5–10 in sight at any one time. In winter, roosting flocks can be found on the small island just offshore from Baby Beach, near San Nicolas, and on the rock breakwater at the marina at Hadicurari. Also occurs year-round on Bonaire; documented breeding at several locations, including an estimated 50 pairs in the Pekelmeer, in 2002 (Prins et al. 2009), where small numbers have sometimes bred along the saltworks dikes (Voous 1983). From fall through spring, counts of 2–10 are typical along the coast from Pekelmeer to Karpata. Larger numbers are sometimes seen in the Pekelmeer area in spring, as birds begin preparing for breeding season; 40 seen on March 24, 2001 (AJW). On windward side, small numbers can occasionally be seen foraging in the bay at Lac, near Marcultura Project (south of Sorobon Beach), or Laguna Washikemba. Occurs year-round on Curaçao, in small numbers; occasionally has bred on small islets such as those at Jan Thiel and on Klein Curaçao (Prins et al. 2009). **Range:** Breeds in western Atlantic, from New Jersey south through Caribbean and Gulf Coast of US to Yucatan, Mexico. From ABCs, breeds east to Trinidad and French Guiana. Caribbean breeding population is highly scattered and of much lower numbers than colonies in the US, where colonies can number in the tens of thousands. Also breeds in Pacific, from California south through Sinaloa, Mexico, and in eastern Atlantic on islands off West Africa. Wintering birds occur from South Carolina south through Caribbean to Argentina; along Pacific Coast, wintering birds occur from California to Peru. Wintering African breeders disperse from Spain to Namibia. Chicks banded in southern Atlantic US states (Maryland, North and South Carolina, Virginia) have been recovered in the ABCs, providing evidence for the origin of many of the wintering and migrant Royal Terns that occur here.

Sandwich Tern (*Thalasseus sandvicensis acuflavida*)   **Plate 28**
Sandwich "Cayenne" Tern (*Thalasseus sandvicensis eurygnathus*)   **Plate 28**

Papiamento: *Stèrnchi grande, Bubi chiquitu, Bubi chiki*
Dutch: *Grote Stern*

Description: A relatively large tern, decidedly larger than Common Tern. Has light gray back, black wingtips, white underside, black cap; sometimes shows a short shaggy crest. Legs usually all black but sometimes all yellow. Bill quite long and slender and is either all yellow (most birds seen on Aruba), black with yellow tip, or black with more substantial yellow at tip. Birds with all-yellow bill are sometimes described as a separate species or subspecies, "Cayenne" Tern (*T. s. eurygnathus*). Black cap extends to the forehead only for a short time; most of the year they have a distinctive white forehead, like the Royal Tern. Juveniles show dark markings on upperparts and dark sides to tail. **Similar species:** Smaller and lighter colored than Royal Tern; long, slender bill is either yellow or mostly black (compare with Royal Tern's stout, carrot-orange bill). Noticeably larger than Common Tern or Roseate Tern, with longer, thicker bill, and, typically, with light-colored rather than dark bill. **Voice:** Makes a harsh *churr-ick*, most commonly heard near breeding colonies. **Status:** Has nested on all three ABC islands, though numbers at breeding colonies highly variable over time. In the late 1950s, up to 1,600 pairs nested at Jan Thiel, Curaçao; up to 2,000 pairs nested on the San Nicolas reef islands, Aruba, in the late 1970s, and between 3,000 and 4,000 pairs were estimated to have nested in the Pekelmeer, Bonaire, in the late 1960s. The species has essentially disappeared as a breeder on Curaçao; today there are much smaller numbers on Bonaire: 180 pairs, Gotomeer, reported in 2002 (Debrot et al. 2009). Largest breeding colonies in recent years on Aruba, at San Nicolas reef islands and Oranjestad reef islands. As many as 3,500 pairs have nested on the San Nicolas reef islands and 1,300 pairs on the Oranjestad reef islands in the last decade (del Nevo 2008). Although nesting is concentrated between May and July, relatively small numbers of birds can be seen feeding and roosting throughout the year at favored spots around Aruba and Bonaire (likely occasionally on Curaçao as well); in Jan, we have recorded numbers as high as 150 on Aruba, but there are usually less than 100 in mid-winter). On Aruba, terns can be observed by using a telescope from the mainland across from the San Nicolas reef islands. The San Nicolas reef islands' nesting sites should not be visited from March through Aug to avoid disturbing the birds and the resulting loss of eggs and mortality of young birds at this important tern nesting site. **Range:** Widespread across coasts of eastern North America, the Caribbean, and South America. Also breeds in Europe and southwestern Asia; winters on coasts of Africa, Middle East, to India.

Black Skimmer (*Rynchops niger*)   **Plate 28**

Papiamento: *Pico di skèr, Bok'i skèr*
Dutch: *Amerikaanse Schaarbek*

Description: Strikingly distinctive, with massive red-and-black bill, black upperparts, white underparts, and reddish legs. Combination of red, white, and black somewhat suggests the colors on a clown's face. The lower mandible of the bill is longer than the upper, and the bill tapers to a point, like a knifeblade; the bird swims with its lower mandible slicing through the water (when it touches a fish, it can snap the lower mandible up quickly to capture it). The South American subspecies shows gray rather than white underwing, and at least one of the South American subspecies shows a thin trailing edge of white on upperwing (broader in North American subspecies) and a darker upperside to tail. **Similar species:** Unmistakable. **Voice:** Makes soft barking notes. **Status:** Birds apparently from South American populations now apparently visit Aruba virtually every year, in numbers ranging from a few to dozens

(33 in Sept 2010). Most records are from March–Oct, when birds from northern South America disperse from flooded areas. Less regular on Curaçao (five records) and Bonaire (two records), with most records of single individuals (one record of four), and all but one record from March–Nov (Prins et al. 2009). Wintering birds from North American breeding areas have not been definitively recorded on the ABC islands, but have been found throughout the Caribbean and, in rare instances, in northern South America. We are aware of only a single record of a Black Skimmer during the mid-winter period, when the North American subspecies might be most expected, a single bird at Santa Martabaai, Curaçao, on Jan 26, 1988 (Prins et al. 2009). In-depth photographic documentation of Black Skimmers, especially during winter months, would help determine their geographic origins. **Range:** At least three subspecies have been described. The North American subspecies is exclusively coastal and breeds from the Mid-Atlantic states south and across Gulf Coast to Mexico; also breeds on the Pacific Coast of Mexico. Another subspecies occurs on large rivers and along coasts of northern South America, and a third in southern South America.

Family Columbidae: Pigeons, Doves

Rock Pigeon (*Columba livia*)   **Plate 29**

Papiamento: *Palomba comun*, *Palomba*
Dutch: *Rotsduif*

Description: The city-dwelling feral pigeon that is a common fixture in many of the world's urban areas. Occurs in a wide variety of color forms. The most common form has a pale gray body and wings, with two black bars across wings, dark head, dark iridescent upper breast, whitish rump, and gray tail with dark terminal band. Dark bill usually with white fleshy area (cere) noticeable at base of bill. **Similar species:** Although Rock Pigeons are almost exclusively found in cities, towns, and, sometimes, farms, other pigeons and doves can occur in these settings as well on the islands. Bare-eyed Pigeon is much lighter and shows bold, white wingbars and a striking, eye-goggle pattern around eyes. Scaly-naped Pigeon is dark gray, much darker than most forms of Rock Pigeon, and is dark overall, without contrasting lighter areas. Scaly-naped also shows a purplish head with scaly, purplish-reddish iridescence on nape and has a bill with reddish base and yellow tip, unlike the all-black bill of Rock Pigeon. Scaly-naped Pigeons are more typically seen in wilder areas, away from towns, though not always. **Voice:** Makes throaty growling *purrs* and *coos*, quite unlike the measured songs of Bare-eyed and Scaly-naped pigeons or the series of single low hoots or cooing sounds of the smaller dove species. **Status:** Commonly found in towns and cities on all three islands; its history on the islands is unclear, as Voous (1983) does not mention the species, although it certainly has been present for decades (Prins et al. 2009). Normally found nesting on ledges of buildings in towns and cities, but apparently has been found recently nesting on rock ledges in natural areas on Curaçao (Prins et al. 2009). **Range:** Native populations are from Europe and Asia, but now introduced to most of the world's urban areas.

Scaly-naped Pigeon (*Patagioenas squamosa*)   Plate 29

Papiamento: *Paloma azul, Blauduif, Palomba di San Kristòf, Palomba pretu, Palomba di baranka*
Dutch: *Roodhalsduif*

Description: Often shy and retiring, this large pigeon is similar in size to the Rock Pigeon. Slate-gray, with purplish tone to head and reddish-purple iridescent lines or scales on nape of neck. Bill reddish at base, with yellowish-white tip. Reddish skin around red eye. **Similar species:** Only a dark form of Rock Pigeon would be all dark and roughly of same size. Note that Scaly-naped Pigeons prefer natural habitat, not urban areas; but in some parts of their Caribbean range, Scaly-naped Pigeons will use backyards in residential areas, and we have seen both Rock Pigeons and Scaly-naped Pigeons flying over at Malpais, Curaçao. In flight, Scaly-naped Pigeons appear longer-tailed and slimmer than Rock Pigeons. Most Rock Pigeons will show some lighter areas or barring or spotting somewhere on the plumage. In all-dark form, a Rock Pigeon shows a dark bill with white fleshy area (cere) around base of bill, in contrast to Scaly-naped Pigeon's red bill with yellowish-white tip, and will not show the more purplish-reddish head and scaly-pattern on nape. **Voice:** Song is similar to commonly heard "Whooo…who-but-you, who-but-you" song of the Bare-eyed Pigeon but Scaly-naped often leaves out the long opening "Whooo" note or, when given, has a rolling, burry quality. Scaly-naped Pigeon song is also slighty lower pitched, and the three-note "WHO-but-YOU?" phrase is delivered more slowly and haltingly, with a pause after the second note. **Status:** Seems to be most widespread on Bonaire, where we have seen small numbers, usually less than 10, at locations across the island. Our maximum on Bonaire was 30 in Washington-Slagbaai NP, Jan 14, 1998. On Curaçao, Voous (1983) noted that the species was largely restricted to the less inhabited western and eastern parts of the island, but Prins et al. (2009) write that it had been seen more frequently outside of those areas. The largest number we have seen on Curaçao was 8–10, at Malpais, on Apr 1, 2000. We also saw a single bird at Klein Hofje, Curaçao, Nov 23, 2003. Although apparently once a breeding species on Aruba, it was last known from the island in 1930, at Rooi Prins, but there have been two reported sightings since then, in 1973 and 1997 (Voous 1983; Prins et al. 2009). Any further sightings on Aruba should be documented with photos or video. **Range:** An endemic species of the Caribbean occurring on Cuba, Hispaniola, Puerto Rico (but not Jamaica), and through the Lesser Antilles. Along with Curaçao and Bonaire, occurs on several small offshore Venezuelan islets.

Bare-eyed Pigeon (*Patagioenas corensis*)   Plate 29

Papiamento: *Ala blanca, barbacoa, paloma di mondi*
Dutch: *Naaktoogduif*

Description: Large, tan-colored pigeon with striking, white wing patches. The bird's beautiful bluish skin around the reddish eye is surrounded by a wider ring of black (sometimes brownish) skin, making this species unmistakable—appearing to wear old-fashioned racing goggles. Bill light yellow, feet dark red. **Similar species:** The only other pigeon or dove on the islands that might have white wing patches is the introduced Rock Pigeon, which is generally found only in cities and urban environments rather than in the natural habitats frequented by Bare-eyed Pigeon. On Curaçao and Bonaire, Scaly-naped Pigeon is similar in size and shape but lacks white wing patches and is a dark slaty-gray (though could be mistaken if seen in silhouette or at a distance.) **Voice:** A rather forceful, low, whistled "Whooo WHO-but-YOU? WHO-but-YOU?" with the first note long and rising then falling. Song often given repeatedly in succession, with several birds vocalizing at the same time. Readily separable from song of Common Ground-Dove, a series of single notes; also separable from song of White-tipped Dove, a single low sound similar to the sound made by blowing into a bottle. Eared Doves have breathy three- or four-parted songs that are much lower, with less variation in pitch. On Bonaire and Curaçao, might be confused most readily with Scaly-naped Pigeon, which typically does not have long opening "Whooo" note or, when given, has a rolling, burry quality. Scaly-naped Pigeon's song is also lower pitched, and the three-note "WHO-but-YOU?" phrase is delivered more slowly and haltingly, with a pause after the second note. **Status:** Common breeder in natural habitats on all three islands, with largest numbers generally seen near water and mangrove habitats, although in recent years it has become increasingly common in residential areas and around grounds of hotels and resorts. Often seen on telephone wires. Our maximum numbers have come from counts of birds entering or exiting night roosts. We counted over 600 leaving a roost near Malpais, Curaçao, on Apr 1, 2000, and an estimated 200 flying into a roost on Nov 24, 2003, near Ruistenberg, Curaçao. Counts of 30–50 are routine on all three islands, in the course of a few hours of birding. **Range:** Caribbean coast of South America, from northeastern Columbia to northeastern Venezuela.

Common Ground-Dove (*Columbina passerina*) Plate 29

Papiamento: *Totolica, Totolika*
Dutch: *Musduif*

Description: The size of a sparrow, with black spots in folded wings and scaly appearance on head and upper breast. Tiny bill is orange with a black tip. In flight, note a startlingly, bright flash of rufous in the outerwing. Often very tame around grounds of hotels and resorts and in residential areas, allowing close approach. **Similar species:** Should be unmistakable. Juvenile Eared Doves, with scaly edges to body and on wing feathers, might be confused with Common Ground-Dove, but it is much larger, with an all-dark bill, and should show two dark marks on each side of the face (marks absent in Common Ground-Dove). **Voice:** Repeats a series of two-parted, rather high-pitched cooing notes, sometimes continuing for minutes without pause: *ooo-oop, ooo-oop, ooo-oop, ooo-oop, ooo-oop.* **Status:** Common throughout the islands, including in residential neighborhoods and on grounds of many hotels and resorts, where it is often possible to find nests easily. Largest numbers we have found have been around water holes, under dry conditions. On July 6, 2001, we counted at least 200 (probably many more) coming and going from the Put Bronswinkel water hole in Washington-Slagbaai NP on Bonaire. On March 25, 2001, we counted over 100 at Pos Mangel water hole in Washington-Slagbaai NP. Interestingly, we estimated that this species was the most common bird we observed on Klein Bonaire on July 7, 2001. **Range:** Breeds from southern US south to northern South America and in Caribbean.

Ruddy Ground-Dove (*Columbina talpacoti*)   Plate 29

Papiamento: *Totolika venesolano*
Dutch: *Steenduif*

Description: A tiny dove. Male shows rusty body with contrasting gray head. Female similar but paler, with reduced rusty tones. Bill black with gray base. **Similar species:** Common Ground-Dove shows scaly appearance on nape and upper breast, and orange-based bill; beware pale Common Ground-Doves on which scaly pattern may be difficult to see. Any possible sighting of Ruddy Ground-Dove should be carefully documented with photos and video. **Status:** Single record of a bird seen and photographed on Aug 8, 1980, in a yard on the outskirts of Kralendijk, Bonaire. **Range:** From northern Mexico south through southern South America; does not generally occur in Caribbean.

White-tipped Dove (*Leptotila verreauxi*)   Plate 29

Papiamento: *Pecho blanch, Ala duru, Buladeifi di hoffi, Jiwiri*
Dutch: *Verreaux' Duif*

Description: A very plain-looking dove. Note pale, pearl-gray head and underparts; slightly darker on back and wings. Lacks any dark markings or white in wings, but, in flight, note white corners to tail and reddish underwing. At close range also note blue skin around eye and all-dark bill. **Similar species:** All other dove species have dark marks on wings or head or have white or reddish in wings. **Voice:** Makes a single long, slow, two-parted note that sounds like someone blowing across the top of a bottle. **Status:** Widespread and relatively common, but generally in lower densities than other dove species; more likely to be seen in areas with relatively pristine habitats, as it seems to avoid disturbed areas. However, it is not uncommon to see them around edges of parking lots and sometimes on hotel grounds. Usually seen singly or in pairs but around water sources may congregate in larger numbers; the highest numbers we have seen have been around the water holes in Washington-Slagbaai NP, where we have observed up to 15 together. **Range:** Occurs from southern Texas south through Central America, to Argentina.

Eared Dove (*Zenaida auriculata*)   **Plate 29**

Papiamento: *Buladeifi di aña*, *Buladeifi*, *Patrushi*
Dutch: *Geoorde Treurduif*

Description: A small dove that might lead birders from North America to think of a Mourning Dove, though it has a shorter, slightly wedge-shaped tail (often appears squared-off), with rufous corners. Also note two black marks on sides of face; black spots on wing evident when wing is folded. Black bill. Pink legs. Juveniles quite different, with white edges on body feathers and wing coverts, giving them a scaly appearance. **Similar species:** Only light-colored dove with two dark marks on face, dark spots on folded wings, and tail with rusty corners. **Voice:** Makes a very low, breathy series of three to four notes, often with second note much shorter: "Whoooo-who-whoooo....whoooo." Much lower and slower than calls of Bare-eyed Pigeon and Scaly-naped Pigeon. Very unlike high, single repetive *ooo-oop* notes of Common Ground-Dove and the single, long note of the White-tipped Dove, which resembles the sound made by someone blowing into a bottle. **Status:** Abundant on all three ABC islands, now often occurring in largest numbers in residential areas and around hotels, although it is also sometimes seen arriving or leaving at roost sites and around water sources. In some areas, one can routinely see 50 or more in a few hours of birding. Our maximum numbers have been from counts of birds flying into or departing night roosts on Aruba. At Divi Divi Resort, Aruba, as we watched birds coming into a roost on the grounds of the hotel, we counted over 1,200 on Jan 4, 1999, and over 440 on Jan 7, 1998. On Nov 25, 2011, we counted an estimated 300 birds departing a roost from somewhere west of Spanish Lagoon (flying over us from west to east). Birds can routinely be seen picking up food scraps around open-air restaurants and nesting on grounds of hotels and resorts. **Range:** Occurs across South America and has recently expanded into southern Lesser Antilles.

Family Cucilidae: Cuckoos

Guira Cuckoo (*Guira guira*)   **Plate 30**

Papiamento: *Cucu guira*, *Kuku guira*
Dutch: *Guirakoekoek*

Description: A prehistoric looking cuckoo, with a shaggy crest, pale bill, and white rump and lower back that contrast with brown streaked wings, creating the appearance of a saddle. **Similar species:** No other commonly occurring bird on the islands is similar. Several South American cuckoos (Pheasant, Pavonine, and Striped cuckoos) are similar but lack a white rump. **Status:** Single record of an emaciated bird found June 12, 1954, at Caracas Bay, Curaçao. Possibility of it being an escaped caged bird cannot be ruled out. Any sightings should be documented with photos and/or video. **Range:** From southern Brazil to Argentina.

Greater Ani (*Crotophaga major*)   Plate 30

Papiamento: *Chuchubi preto mayor, Chuchubi pretu mayó*
Dutch: *Grote ani*

Description: A much larger version of the Groove-billed Ani, with light eyes and larger, more keeled, bill. **Similar species:** Groove-billed Ani is smaller, with dark eyes and smaller, less ridged bill. Male Carib Grackle is all black and has light eyes, but is much smaller and slimmer-billed. Male Great-tailed Grackle is large with light eyes, but has a slim bill that lacks ridge. **Voice:** Most common call is a "a guttural gobbling or bubbling *kro-kro*" (Hilty 2003). **Status:** A photograph has been published online showing one of two birds seen by Eric Newton at Curaçao harbor on Oct 26, 2010 (Neotropical Birds online, accessed Oct 19, 2013). Two birds were reported by Steve Mlodinow at Spanish Lagoon, Aruba, on March 15, 2005 (Mlodinow 2006); a bird was photographed by Ross Wauben at Divi Links, Aruba, on Oct 26, 2015, and resighted on Nov 1, 2015; two birds were seen and one photographed by Michiel Oversteegen at Bubali, Aruba, on Apr 26, 2016 (fide eBird); a bird was photographed at Bubali, Aruba, by Ross Wauben, on Apr 27, 2016, and May 14, 2016 (fide eBird). It is not known whether all Aruba sightings pertain to the same two birds. All sightings should be carefully documented with photos and/ or video. **Range:** Panama to Brazil.

Smooth-billed Ani (*Crotophaga ani*)   Plate 30

Papiamento: *Chuchubi preto pico lisp*
Dutch: *Gladsnavel ani*

Description: A rare ani, with only a single record of two birds from Aruba in 2016. Very similar to resident, and relatively common, Groove-billed Ani. Large, long tail, big bill, and all-black bird, with a blunt, round-headed profile. Dark eyes and bill. Bill is smooth, lacking grooves as in Groove-billed Ani. **Similar species:** Very difficult to separate from Groove-billed Ani, but bill generally higher ridged, with no grooves or with only weak grooves on inner part of bill closest to face. **Voice:** Most definitive way to distinguish Smooth-billed Ani from Groove-billed Ani is by call. Smooth-billed Ani has whining, rising *quweeeik*, while Groove-billed Ani has a more two-syllabled, descending *TEE-who*. **Status:** First record for islands was of two birds seen and photographed on Feb 16, 2016, at Bubali, Aruba (P. Sprockel, R. Wauben fide eBird); sightings of one or two birds (presumably the same ones) at and near Bubali continued until at least Apr 19, 2016. One would assume that the birds might still be present in the area. Any sightings of Smooth-billed Ani should ideally be captured with video or audio (in addition to photographs) to document the distinctive vocal differences between the species. **Range:** From southern Florida through Bahamas, Greater Antilles, Lesser Antilles, and some islands off of Caribbean side of Southern Mexico and northern Central America, then south from Costa Rica through South America to northern Argentina and Uruguay.

Groove-billed Ani (*Crotophaga sulcirostris*)   **Plate 30**

Papiamento: *Chuchubi preto, Chuchubi pretu, Cassilia*
Dutch: *Groefsnavelani*

Description: A distinctive large, long-tailed, big-billed, all-black bird with rather blunt, round-headed profile. Dark eyes and bill. **Similar species:** Smooth-billed Ani has now been documented on Aruba, so all sightings of ani should be checked carefully and their vocalizations studied (see Smooth-billed Ani for more identification details). Greater Ani, which has been photographed once on Curaçao and several times on Aruba, is much larger than Groove-billed Ani, with light eyes and more obvious keel on top of bill. Only commonly occurring all-black species on the islands are the much smaller and slimmer-billed male Carib Grackle, which has a light-colored eye, and the Shiny Cowbird, which has a dark eye and a shorter tail. Great-tailed Grackle has occurred rarely on Curaçao and Aruba; it is similar in size to Groove-billed Ani but has a light-colored eye, slimmer bill, and less blunt-headed head shape. **Voice:** Distinctive two-syllabled descending squeal is unlike vocalizations of any other bird on the islands: *TEE-who*. **Status:** Relatively widespread breeding resident on all three islands, often near water or farms, and usually occurring in groups of 2–4 but occasionally up to 10 birds or so. We counted about 20 around Tierra del Sol golf course on Nov 27, 2013. Commonly seen on Aruba at Bubali and Tierra del Sol, and on Curaçao at Malpais and in Christoffel NP. **Range:** From southern Texas to northern South America.

Yellow-billed Cuckoo (*Coccyzus americanus*)   **Plate 30**

Papiamento: *Cucu pico hel, Kuku pik hel*
Dutch: *Geelsnavelkoekoek*

Description: A distinctively slim, long-tailed, long-billed bird, often rather shy and sluggish in its movements. Clean brown upperparts contrast with stark white underparts. Bill yellow except for dark tip and upperpart of upper mandible. Yellow ring around eye. Bright rufous in flight feathers. Large, white spots in tail visible from below and in flight. **Similar species:** Mangrove Cuckoo also has mainly yellow bill and large, white spots in tail but is very buffy below, lacks rufous in wings, and has a clearly defined black mask through eye. Black-billed Cuckoo has not been recorded from the islands, though it occurs regularly (but rarely) in Colombia and Venezuela, so there is reason to expect that it could occur. Black-billed Cuckoo has a black (not yellow) bill, lacks rufous in wings, has narrow rather than large, white spots in tail, and has reddish rather than yellow ring around eye. The Pearly-breasted Cuckoo, from South America, has never been recorded on the ABCs; it looks like a slightly smaller version of the Yellow-billed Cuckoo but it lacks rufous in the wings and has a slightly contrasting grayish tone to upper breast. **Voice:** The distinctive song given on breeding grounds, a loud *kuk-kuk-kuk-kuk, kowp-kowp-kowp*, is unlikely to be heard on the islands, as the birds are basically silent when migrating and in winter. **Status:** On all three islands, a regular fall migrant (usually Oct–Nov) from North American breeding grounds, sometimes occurring in large numbers, even in the hundreds (Voous 1983), but usually only one or two seen at a time. In Oct 2016, we and many other observers noted multiple birds around Bubali and other locations on Aruba. Observers on an at-sea marine mammal survey cruise many miles north of Aruba in fall 2014 saw a number of Yellow-billed Cuckoos migrating over the open ocean and sometimes perching on the ship as they rested before continuing southward. There is a handful of winter and spring records as well, but it is quite rare after Nov. We observed a single individual in Arikok NP (near what is now the park entrance) on Jan 13, 1999. **Range:** Breeding range extends across eastern and central US, south to Mexico. Winter range extends over much of South America.

Mangrove Cuckoo (*Coccyzus minor*) **Plate 30**

Papiamento: *Cucu mangel, Kuku di mangel*
Dutch: *Mangrovekoekoek*

Description: A distinctively slim, long-tailed, long-billed bird, often rather shy and sluggish in its movements. Clean brown upperparts, buffy underparts. Lower mandible of bill yellow. Black mask through eye. Yellow ring around eye. Large, white spots in tail visible from below and in flight. **Similar species:** Yellow-billed Cuckoo and the possible but not-yet-recorded Black-billed Cuckoo and Pearly-breasted Cuckoo all have starkly white underparts rather than buffy, and lack the obvious dark mask. Yellow-billed Cuckoo also has rufous in wings, which Mangrove Cuckoo lacks. The Dark-billed Cuckoo from South America is very similar to Mangrove Cuckoo but has an all-dark bill. **Voice:** Song is a raspy series of short, low notes that increases in speed, then drops to a lower pitch, and slows down again. Birds observed here have not vocalized and are presumed to be migrants. **Status:** Surprisingly few records of the species on the islands, and all presumed to be migrants. Prins et al. (2009) reported a single record from Aruba, four from Curaçao, and seven from Bonaire. In addition, we observed a single bird on Apr 1, 2000, at Malpais, Curaçao, and a single bird on Jan 15, 2003, at Bubali, Aruba. One was photographed at Spanish Lagoon, Aruba, on Apr 12, 2013 (S. Mlondinow fide eBird). One was photographed at Divi Links, Aruba on Sep 11, 2014 (R. Wauben fide eBird). **Range:** Generally a specialist of lowland coastal habitats, especially mangroves. Its range extends from southern Florida through the Caribbean and coastal Mexico, and Central America south to northern South America. The extent of migratory behavior is unclear in the species.

Gray-capped Cuckoo (*Coccyzus lansbergi*) **Plate 30**

Papiamento: *Cucu gabes gris, Kuku kabes gris*
Dutch: *Grijskopkoekkoek*

Description: Like all cuckoos, slim with a long tail. Rusty upperparts, gray cap, black bill, buffy-orange underside. Tail dark, with white spots. **Similar species:** Darker below than other cuckoos, and darker and more rusty above, with distinctive dark cap. **Status:** A rare, secretive bird. There is only a single record on the ABCs, of a wounded bird that later died at Nikiboko, Bonaire, Oct 14, 1981 (Voous 1983). **Range:** Spottily distributed, from Peru north to Venezuela.

Family Tytonidae: Barn Owls

Barn Owl (*Tyto alba*)   **Plate 31**

Papiamento: *Palabrua*
Dutch: *Kerkuil*

Description: A striking, dark-eyed owl with a white, flat, heart-shaped face, white underparts, long legs, and soft tannish-brown upperparts. **Similar species:** No other owl occurs on Curaçao or Bonaire. **Voice:** A hissing shriek made by most forms, but unclear if it has been recorded on Curaçao or Bonaire. **Status:** Originally known only as a breeding resident of Curaçao (Voous 1983) but confirmed also as a breeding resident on Bonaire, in 1990s (Prins et al. 2009). Breeds and/or roosts mostly in limestone caves (occasionally abandoned buildings) and comes out only at night, so rarely observed. A study published in 2001 estimated the population size on Curaçao at 75 adults (Debrot et al. 2001). Number of birds on Bonaire has been roughly estimated at 20–40 (Prins et al. 2009). Does not occur on Aruba. **Range:** One of few bird species found virtually worldwide, except in colder northern and southern parts of the earth, with dozens of subspecies described. Birds from Curaçao have been described as a unique subspecies, *T. a. bargei*, which is much smaller than mainland South American forms. It is not clear if the form on Bonaire is the same or a different subspecies (Prins et al. 2009).

Family Strigidae: Owls

Burrowing Owl (*Athene cunicularia*)   **Plate 31**

Papiamento: *Choco*, *Shoco*
Dutch: *Holenuil*

Description: Small, long-legged, ground-dwelling owl with yellow eyes. Upperparts brown with white spots, underparts white with brown barring. **Similar species:** The only owl species found on Aruba. **Voice:** Makes a two-note call, with short first note followed quickly by longer and more emphatic second note: *coo, coo-oo*. Local Papiamento name, *choco*, is derived from its call (Voous 1983). When alarmed, gives a variety of chattering calls. **Status:** This resident breeding bird is found in the ABCs only on Aruba, where it nests in burrows in the ground. Population was roughly estimated at 30 breeding pairs in 1977 and 100 breeding pairs in 1992 (Prins et al. 2009), though the methodology and precision of these estimates is difficult to determine. The impact of nest predation by the introduced Boa Constrictor on the Burrowing Owl population on Aruba is of major concern yet not well understood. The Tierra del Sol golf course has a number of nesting pairs on the grounds that are among the most well-known and frequently observed pairs on the island. **Range:** Wide ranging, from western North American south through South America and in South Florida and Greater Antilles. More than a dozen subspecies, including a subspecies found only on Aruba, *A. c. arubensis*, described in 1915 (Voous 1983).

Family Steatornithidae: Oilbird

Oilbird (*Steatornis caripensis*)   **Plate 31**

Papiamento: *Parha zeta, Para zeta*
Dutch: *Vervogel*

Description: A very unusual species. The Oilbird is nocturnal, nests in caves, and flies out at night to eat fruit. It has the elongated body, large eye (shines red when illuminated), and weak feet of most nightjars, but has a larger, hooked beak. Rufous overall, with white spotting in wings and edge of tail, and black barring on tail and inner flight feathers. **Similar species:** Highly unlikely to be found on the islands. Oilbirds are much larger than the resident White-tailed Nightjar, nighthawks, or the rare Chuck-Will's Widow. None of these other species are strongly rufous with a large, hooked beak. Obviously, any possible sighting of the species should be well-documented with photos and/or video. **Status:** Single record of a young female bird found end of Apr, beginning of May, 1976, in a backyard at Brazil, Aruba, where it was harassed by Crested Caracaras and died the same day. It is now preserved as a specimen at the Zoological Museum Amsterdam (Voous 1983). **Range:** Scattered breeding caves are known from Panama south to Bolivia and east to Guyana. Also occurs on Trinidad.

Family Caprimulgidae: Nightjars, Allies

Lesser Nighthawk (*Chordeiles acutipennis*)   **Plate 32**

Papiamento: *Tapa camina menor, Tapa kaminda menor*
Dutch: *Texasnachtzwaluw*

Description: Most likely to be seen in flight, when nighthawks could be mistaken for a falcon because of their long, pointed wings, long tail, and swift flight. Nighthawks fly much more erratically then falcons, though, with sudden changes of directions and quick tips of the body. In flight, note white patches in wings, closer to wingtip than in Common Nighthawk. In female Lesser Nighthawks, white in wing replaced with buff color (female Common and Antillean nighthawks have white wingbar). Male has white stripe near tail tip, and white throat; in female, tail tip and throat are buff-colored. Bases of primaries spotted with buffy colors (usually lacking in Common and Antillean nighthawks), and body is buffier underneath than in Common Nighthawk. At rest, folded wing extends no farther than tail tip. **Similar species:** Nighthawk species are very difficult to distinguish from one another, especially in flight. Smaller than Common Nighthawk, with shorter wings and tail. Wing patch closer to tip of wing than in Common Nighthawk. Wingtips do not extend beyond tail when bird is at rest. Has buffy spotting at base of primaries, more obvious spotting overall on inner flight feathers than do Common and Antillean nighthawks, and more uniformly colored upperwing. At rest, the wingtip of Common Nighthawk extends beyond tail tip. At rest, Antillean Nighthawk wingtip also does not extend beyond tail but note the contrasting, pale tertials. **Voice:** Not heard in the islands but on breeding grounds gives distinctive long trill and stuttering, nasal flight call. **Status:** Two sight records by Voous (1983) of single birds; no photographic or video confirmation or specimen record exist for the islands. One probable female was sighted at Kintjan Hill, Curaçao, on May 26, 1977, and a bird (sex unspecified) was seen over Kralendijk, Bonaire, on Oct 22, 1979 (Voous 1983). Extensive documentation with photos and video of any nighthawks observed would be helpful in clarifying the status of the different species in the islands. **Range:** Southwestern US south through Central America, to Brazil. Northernmost populations are migratory and thought to augment local populations in parts of Central America and South America, but details of migration are poorly known.

Common Nighthawk (*Chordeiles minor*)   Plate 32

Papiamento: *Tapa camina amerikano, Tapa kaminda merikano*
Dutch: *Amerikaanse Nachtzwaluw*

Description: Most likely to be seen in flight, when nighthawks can be mistaken for a falcon because of their long, pointed wings, long tail, and swift flight. Nighthawks fly much more erratically then falcons, with sudden changes of directions and quick tips of the body. In flight, note white patches in wings, which are farther from wingtip than in Lesser Nighthawk and often larger and wider on side and closer to trailing edge of wing. Male has white stripe near tail tip and white throat, but in females there is little contrast. Bases of primaries dark, with no spots; inner flight feathers (secondaries) minimally spotted, with narrow light spots giving much darker appearance to underwing. Body usually whitish, not buffy as in Lesser or Antillean nighthawks. At rest, folded wing extends beyond the tail tip. **Similar species:** Nighthawk species are very difficult to distinguish from each other, especially in flight. Larger than Lesser Nighthawk, with longer wings and tail. Wing patch farther from tip of wing than in Lesser Nighthawk. Compared to Lesser Nighthawk, dark bases to primaries and less obvious, narrow spots or bars on inner flight feathers. Upperwing in Common Nighhawk usually shows more contrasting light bar across inner wing than does Lesser Nighthawk. At rest, wingtip of Lesser Nighthawk does not extend beyond tail tip. At rest, Antillean Nighthawk wingtip also does not extend beyond tail but note contrasting, pale tertials in Antillean, which also show a more contrasting light bar on inner surface of upperwing in flight. **Voice:** Not heard in the islands, but on breeding grounds gives loud, distinctive nasal *BWEEP*. **Status:** On all three islands, uncommon and irregular, mostly as a fall migrant. On Aruba, one record from fall 1977 and another (small flock) from Oct 1978 (Voous 1983). At a distance, we observed one individual that was probably this species, on Aruba (from Bubali), in Nov 2012. On Curaçao, small migrant flocks more regularly noted between Sept and Jan (Prins et al. 2009). On Bonaire, along with several fall records, three records from outside this period: one caught and released on Apr 29, 1985; four observed at Kralendijk on Feb 13, 1986; and one seen near Harbor Village Marina on June 14, 2000 (Prins et al. 2009). **Range:** Breeds across North America and Central America, from northern Canada south to Panama. Winters across much of South America.

Antillean Nighthawk (*Chordeiles gundlachii*)   Plate 32

Papiamento: *Tapa camina antiano, Tapa kaminda antiano*
Dutch: *Antilliaanse Nachtzwaluw*

Description: See Common Nighthawk and Lesser Nighthawk. **Similar species:** Virtually indistinguishable in the field from Common Nighthawk. Also compare with Lesser Nighthawk. **Voice:** Although never heard here, distinctive call on breeding grounds is a rapid, staccato series of short nasal notes, very unlike Common Nighthawk calls. **Status:** Two records, both from Curaçao. One male was collected at Malpais, Apr 19, 1955; and one male was collected at Westpunt, Sept 17, 1955 (Voous 1983). **Range:** Breeds from Florida Keys south through Greater Antilles and Bahamas. Winters in South America, though exact location/s unknown.

White-tailed Nightjar (*Hydropsalis cayennensis*)   Plate 32

Papiamento: *Tapa camina, Batibati, Para karpinté, Tapa kaminda, Palabrua*
Dutch: *Witstaartnachtzwaluw*

Description: Nocturnal, rarely seen during day. Most often seen as silhouette, in headlights on a road or when spotted with flashlights in an open field. Occasionally can be seen flying about in dimly lit areas. When on the ground (which is how it is almost always seen), note long tail, with wingtip extending about halfway down its length. Dark flight feathers and white patch in wing (males only) may be evident; on close inspection, sometimes possible to note reddish nape (reduced in female) and white throat (buffy in female). **Similar species:** Nighthawks' wings are longer than— or almost as long—as tail. Chuck-will's-widow has much shorter tail, brown throat, and barred flight feathers. There are other nightjar species that occur on mainland South America that presumably could occur as rare visitants, so anything that seems out of the ordinary should be documented, if possible, with photographs and video, a challenge at night. **Voice:** Makes an emphatic, single, down-slurred *Swee-eer*, often with three or more seconds between each phrase. At a distance, only beginning part of each note is obvious, sounding more like *Swee*. Mainland forms have a very different song (often rendered as *Pit-sweeer*), with a single short note (missing from all recordings of birds we have from the islands) followed by a second down-slurred note that is two or three times longer than those we have recorded on Bonaire and Curaçao. Oddly, Voous (1983) makes no mention of the difference in song between mainland forms and those of the islands. **Status:** Year-round resident breeding bird on all three islands. Seems relatively common and widespread on Curaçao and Bonaire, though there have been no systematic surveys to assess its status and population trends. We found the species to be relatively abundant in parts of Washington-Slagbaai NP, Bonaire, in March, 2001, and on the road near Klein Hofje, Curaçao, in Nov 2003 (before we were surrounded by armed policemen who thought we were up to no good—take caution!). Status on Aruba is less clear, as we have found it only once there, near Alto Vista, Nov 2012, but J. Wierda discovered birds near Tierra del Sol golf course in Feb 2014 and local birders have documented 3-4 pairs near Bubali in recent years. **Range:** Has an extensive and interesting range, with at least five described subspecies. It occurs from Costa Rica south to Ecuador and east to northeastern Brazil. Also occurs on Trinidad, Tobago, and Martinique. The birds on Aruba, Bonaire, and Curaçao were described in 1902 as the subspecies *Caprimulgis cayennensis insularis* based on smaller size and paler coloration. Similar birds may occur in nearby coastal Venezuela and Colombia, and on Isla Margarita, but there has not been any systematic study of distribution and taxonomy of the various forms of the species. Recordings of birds from Isla Margarita indicate the mainland song rather than the version of the song that we have recorded on Aruba, Bonaire, and Curaçao.

Chuck-will's-widow (*Antrostomus carolinensis*)   Plate 32

Papiamento: *Tapa caminda viuda, Tapa kaminda biuda*
Dutch: *Chuck Will's Widow*

Description: A nocturnal bird, normally virtually impossible to find except when calling on breeding grounds at night. Big-headed, with large, dark eyes (often closed during the day) and small, wide bill with bristles—all adaptations for catching insect prey while flying at night. If you are lucky enough to see one during the day, check for: presence of bristles near bill, which are lacking in nighthawks; absence of bold, white throat patch, as seen in nighthawks; short wings that do not extend to tail; and barred flight feathers. In flight, has all-dark wings with no white patches. Two color morphs; one is very rufous and the other grayer. **Similar species:** Nighthawks slimmer, usually with obvious bold, white patch on throat, dark (not barred) flight feathers, and much longer wings. In flight, nighthawks all show bold, white patches that are lacking in Chuck-will's-widow. Resident White-tailed Nightjar is smaller and slimmer, with long tail that extends well beyond wings. In good light, White-tailed Nightjar shows a white throat, reddish nape, white in wing, and dark (not barred) flight feathers. **Voice:** Not known to call on wintering grounds but gives a distinctive repeated call for which it is named: CHUCK-wills-WIDOW, CHUCK-wills-WIDOW, CHUCK-wills-WIDOW. **Status:** Rare migrant visitor. One specimen record from a bird hit by a car in Oranjestad, Aruba, on Dec 11, 1979. On Bonaire, one specimen from near the Pekelmeer, Nov 26, 1981, and another specimen from Kralendijk, July 27, 1991. On Curaçao, one sight record of a bird flushed during daylight hours, at Santa Maria, Oct 16 and 21, 1966 (Voous 1983; Prins et al. 2009). **Range:** Breeds in southeastern US. Winters in very southern part of breeding range, south through Caribbean and Central America, to northern South America. Southern boundaries of wintering range poorly known.

Family Apodidae: Swifts

Black Swift (*Cypseloides niger*) **Plate 33**

Papiamento: *Veloz preto, Veloz pretu*
Dutch: *Zwarte Gierzwaluw*

Description: Large, all-dark swift with broad-based, long wings, and broad tail. Males have slightly notched tail, but in females, tail is squared (sexes otherwise indistinguishable). The many species of swift are very similar, and beguilingly difficult to identify in the field. **Similar species:** Chimney Swift is much smaller, with shorter, narrow wings, narrow squared-off tail, and gray rather than black body. Spot-fronted and White-chinned swifts, which have occurred in Venezuela, are often impossible to distinguish from this species in the field. Although their speed and rapidly beating wings make them difficult to photograph or videotape, this is the best recourse when one observes any swift species on the islands. **Voice:** Makes chattering chip notes. **Status:** Not confirmed for the islands. A sighting of six birds on May 28, 1991, at Christoffel NP, Curaçao, was thought to be of this species (Prins et al. 2009). **Range:** The species has a very disjunct, poorly known breeding range, extending in pockets from southeastern Alaska and British Columbia, south through western US to Mexico and Central America, with disjunct breeding populations in the Greater Antilles (possibly Lesser Antilles as well). All northern populations and some subset of Greater Antilles populations and other populations migrate south, apparently to northern South America, but there are only a few scattered records, perhaps in part because of the difficulty of distinguishing among northern migrants, residents, and austral migrants.

Chimney Swift (*Chaetura pelagica*) **Plate 33**

Papiamento: *Veloz di chimenea*
Dutch: *Schoorsteengierzwaluw*

Description: A small, sooty-gray swift sometimes described as a "flying cigar" because of its body shape and the fact that the wings are beaten so swiftly that they virtually disappear. Throat slightly paler than rest of body. **Similar species:** Black Swift is much larger, with longer, broader wings, broad tail (slightly forked in males), and black rather than gray body. Chapman's, Ashy-tailed, and Vaux's swifts, which are known from Venezuela, are often impossible to distinguish from this species in the field. Although their speed and rapidly beating wings make them difficult to photograph or videotape, this is the best recourse when one observes any swift species on the islands. **Voice:** Make high-pitched chittering notes. **Status:** Based on expected timing of fall migrant Chimney Swifts from North America, sightings during Oct and Nov consistent in appearance with Chimney Swift are highly likely to be this species. On Aruba, reported to be seen regularly in small numbers in late Oct and Nov (Prins et al. 2009). Two sight records from Curaçao: one bird at Klein Hofje, Oct 20, 1991, and 10 birds at Klein Hofje, Oct 27, 1994 (Prins et al. 2009). Also two sight records from Bonaire: one at Slagbaai Plantation, Oct 28, 1979, and one at Slagbaai Plantation, Oct 24, 1982 (Prins et al. 2009). **Range:** Breeding range extends across eastern North America, from southern Canada to Texas and Florida. Winters in South America, but distribution in winter is poorly known because of difficulties in separating this species from similar-looking species in South America.

Family Trochilidae: Hummingbirds

White-necked Jacobin (*Florisuga mellivora*)   Plate 33

Papiamento: *Blenchi nèk blanco, Blenchi nèk blanku*
Dutch: *Witnekkolibrie*

Description: Larger than either of the two resident hummingbird species on the islands. Males unmistakable: purple head, white belly, mainly white tail, white nape stripe, and green back. Females could be confused with a number of other species. **Similar species:** Any large hummingbird observed on the islands would be an extreme rarity and should be carefully documented with photos and/or video. **Status:** Three records that almost certainly pertain to only two individual birds. One immature male was collected on Aruba, Apr 24, 1908, for the Field Museum of Natural History in Chicago. An adult male was documented (by "colour film" [Voous 1983], a reference perhaps to what we now would now call a video) on June 4, 1958, at Julianadorp, Curaçao (Voous 1983). In Nov 1958, what was almost certainly the same individual was seen at the same location. **Range:** From southern Mexico south through Central America and northern South America, south to Bolivia and Brazil.

Rufous-breasted Hermit (*Glaucis hirsutus*)   Plate 33

Papiamento: *Blenchi pecho còrá, Blenchi pechu kòrá*
Dutch: *Roodborstheremietkolibrie*

Description: Only recorded on the islands once. A very large hummingbird, with conspicuously long, down-curved bill. Tail rufous, except for green central tail feathers and broad black band at end of tail. **Similar species:** Any large hummingbird observed on the islands would be an extreme rarity and should be carefully documented with photos and/or video. There are a number of other species of large hummingbirds with decurved bills that could theoretically occur on the islands. **Status:** One record of an exhausted bird caught by hand, on Oct 26, 1977, at Brother's Convent, Soto, Curaçao (Voous 1983). **Range:** Occurs in most of northern South America, from Panama south to Brazil and Bolivia.

Ruby-topaz Hummingbird (*Chrysolampis mosquitus*)   Plate 33

Papiamento: *Dòrnasol, Blenchi-dornasol* (male), *Blenchi-hudiu* (female)
Dutch: *Muskietkolibrie*

Description: A stunning, beautiful hummingbird, about twice the size of the Blue-tailed Emerald, the other resident hummingbird. In good light, male generally unmistakable, with irridescent orangish throat, reddish head, and rounded rufous tail with black terminal band. In some light conditions, males can appear almost entirely dark brown. Females and immatures with dull green upperparts (sometimes appearing almost light brown) and white underparts, but always with broad reddish-based tailed with white tips to tail feathers. Immature males often get first dark breast feathers along a line in the middle of the breast, as is seen in females of some species of mangos (hummingbirds). Voous (1983) writes that "young males" (perhaps meaning juveniles) have "mainly steel-blue tail feathers," although we have never encountered a Ruby-topaz Hummingbird without the reddish tail coloration. Birds, especially males, often raise feathers on the back of the head, giving them a distinctive tufted or shaggy-headed silhouette. Bill is black and slightly decurved. **Similar species:** The only common hummingbird in the islands is the much smaller Blue-tailed Emerald, which has a finer, straighter bill; shorter, dark, forked tail (all blue but often can appear black). Male Blue-tailed Emerald is entirely dark metallic-green, while female and immature have dark metallic-green upperparts and white underparts. **Voice:** When defending flowering shrubs from other hummingbirds, gives a persistent sizzling buzz, a bit like the sound made by running one's finger over the teeth of a comb. **Status:** Relatively common on all three islands. Often around flowering shrubs, especially in and near natural areas; infrequently observed on grounds of hotels, unlike Blue-tailed Emerald. During dry periods, numbers can diminish greatly; as few shrubs are flowering, it is not known whether birds move regularly between the islands and mainland. On Aruba, we most frequently see this species at Spanish Lagoon, where we recorded 15 birds on Jan 16, 2002. **Range:** Northern South America south to Brazil and Bolivia, but absent from Ecuador and Peru.

Blue-tailed Emerald (*Chlorostilbon mellisugus*)   Plate 33

Papiamento: *Blenchi bèrdè, Blenchi, Blenchi gewoon*
Dutch: *Blauwstaartsmaragdkolibrie*

Description: A tiny hummingbird with a short tail and thin, straight, short bill. Male is entirely dark metallic-green with a dark blue (often appearing black) forked tail. In some light conditions, male can look all dark. Female and immature dark green above and white below, with a light eyebrow line. Bill all black. **Similar species:** This is the only small, dark green hummingbird with short, forked tail that occurs on the islands. See Ruby-topaz Hummingbird. **Voice:** Aggressive calls are a series of rapidly repeated short, squeaky notes, quite unlike buzzing sounds of Ruby-topaz Hummingbird. **Status:** Relatively common across a range of habitats on all ABC islands, but most numerous around flowering shrubs. More likely to be seen in backyard gardens and on grounds of hotels and resorts than Ruby-topaz Hummingbird. **Range:** Across northern South America south to Bolivia and central Brazil.

Family Alcedinidae: Kingfishers

Ringed Kingfisher (*Megaceryle torquata*)   **Plate 34**

Papiamento: *Cabez grandi barica còrá, Kabes grandi barika kòrá*
Dutch: *Amerikaanse Reuzenijsvogel*

Description: A large kingfisher, with blue head and upperparts and white neck collar. No white spotting in wings. Male has entirely rusty underparts. Female has blue breast band above rusty underparts. **Similar species:** In winter, the Belted Kingfisher, a northern migrant, is the common kingfisher in the islands; it also has all-blue upperparts but is smaller, has white spotting in wings, and no or only limited rusty coloration below. Amazon Kingfisher is entirely green above and shows no (or limited) rust colororation below. Any observation of a kingfisher species other than Belted Kingfisher should be documented with photos and/or video. **Voice:** Sibley (2000) describes distinctive "double-note *ktok* call" given in flight. Other rattle sounds are given more slowly and deeply than calls made by Belted Kingfisher. **Status:** Three records: one bird at Bubali, Aruba, Apr 28, 1991; another (perhaps the same bird) at Bubali, Aruba, from July 12 to Aug 29, 1991; and one at Santa Cruz, Curaçao, May 11, 1991 (Prins et al. 2009). **Range:** Southern Texas south through Central America and South America, to Argentina and Chile.

Belted Kingfisher (*Megaceryle alcyon*)   **Plate 34**

Papiamento: *Cabez grandi, Kabes grandi*
Dutch: *Bandijsvogel*

Description: This winter visitor is the most common kingfisher in the islands. Large, with blue head and upperparts and white neck collar. White spotting in wings. Male has single blue breast band, female has blue breast band but with rusty band on belly as well. **Similar species:** Ringed Kingfisher has occurred several times in the islands and also has blue upperparts but lacks white in wings and has extensive rufous underparts. Amazon Kingfisher (has occurred several times) is all green above. Any observation of a kingfisher species other than Belted Kingfisher should be documented with photos and/or video. **Voice:** Makes a distinctive loud rattle that often gives away it presence before it is spotted. **Status:** Regular in small numbers at mangroves, coastline, salinas, and freshwater ponds, where they dive for small prey fish. Typically, several can be seen at scattered wetland locations in a day of winter birding on the islands, perhaps more in rainy periods when the number of wetlands increases. **Range:** Breeds across most of North America; winters south to northern South America.

Amazon Kingfisher (*Chloroceryle amazona*) Plate 34

Papiamento: *Cabez grandi bèrdè, Kabes grandi bèrdè*
Dutch: *Amazoneijsvogel*

Description: A large kingfisher with a dark green head and upperparts; also note white neck collar, no white spotting in wings. Male has rusty chestband. Female has partial green chestband. **Similar species:** In winter, Belted Kingfisher is the common kingfisher in the islands. It has all-blue upperparts and some white spotting in wings. Green Kingfisher has not been recorded from the islands, but is smaller and shows white spotting in wings. Ringed Kingfisher has occurred several times in the islands but has blue upperparts and extensive rufous underparts. Any observation of a kingfisher species other than Belted Kingfisher should be documented with photos and/or video. **Voice:** Howell et al. (2014) say calls include "a low, slightly rasping *krrrik* and a hard, buzzy *zzzrt*," often given in flight. **Status:** A record for a single female bird at Savaneta, Aruba, Aug 1982 (Voous 1983), apparently not documented with photos or video. One female or immature was photographed at Savaneta, Aruba, on June 20 and 21, 2016, by Ross Wauben (fide eBird). A male was photographed at Piscadera Bay, Curaçao, on Aug 13, 2016, and continued to be seen throughout the month (Dirk Hilbers). **Range:** Mexico south through Central America, and South America to Argentina.

American Pygmy Kingfisher (*Chloroceryle aenea*) Plate 34

Description: A tiny kingfisher with dark green head and upperparts and rusty collar, throat, breast and flanks in male. In female and immature, the throat and neck collar are rusty colored and tinged with white; female has green chestband. White belly extends down to tail. **Similar species:** In winter, the Belted Kingfisher is the common kingfisher in the islands. It is much larger and has all-blue upperparts. Rare Amazon Kingfisher is green above but is much larger and has a white throat and collar. Green Kingfisher has not been recorded from the islands. It is relatively small (still larger than American Pygmy Kingfisher) and has a white throat and collar. Green-and rufous Kingfisher has not been recorded from the islands. It is very similar to American Pygmy Kingfisher but is larger, with rusty underside extending to belly and undertail, with a larger bill, and showing white spotting on upperwing. **Voice:** Hilty (2003) describes vocalizations as "an abrupt buzzy *jeeeet* or bi-syllabic *jejeet!*" **Status:** One record. A female or immature bird was photographed on the balcony of an apartment at the Eden Beach Resort in Kralendijk, Bonaire, on Apr 13, 2015 (fide Observado). The bird, which looked sick, flew off when startled. **Range:** Southern Mexico south through Central America and South America, to southern Brazil.

Family Picidae: Woodpeckers

Yellow-bellied Sapsucker (*Sphyrapicus varius*)   Plate 34

Papiamento: *Para carpinté barica hel, Para karpinté barika hel*
Dutch: *Geelbuiksapspecht*

Description: This small migrant is the only species of woodpecker ever recorded in the islands. Note white vertical wingstripe and barring on back, central tail feathers, and rump. Most records (perhaps all) have been of immature birds. Adult male and female show a red cap; adult males also show a red throat. **Similar species:** Should be unmistakable, as no other woodpecker has been observed on the islands. Any sighting of any woodpecker should be carefully documented, especially if lacking the white vertical wingstripe (which would indicate a species other than Yellow-bellied Sapsucker). **Voice:** Although unlikely to be heard on the islands, one of its most characteristics calls is a nasal whining or mewing *queee-ah*. **Status:** Thirteen records from the ABCs, most Oct and Jan, although one occurred from March to June, 1999, at Sero Colorado, Aruba (Prins et al. 2009). Birds have sometimes stayed for weeks. Only four records since 1983, three on Aruba, and one on Bonaire (Prins et al. 2009). **Range:** Breeds throughout Boreal forest region of North America (more than 50% of population is estimated to breed there), from southern Yukon and Northwest Territories east to Quebec and Newfoundland; also in northeastern US and in higher elevation parts of Appalachians, to North Carolina and Tennessee. Highly migratory; in winter, vacates virtually entire breeding range and winters from the southeastern and mid-Atlantic region of the US south to Panama and the Greater Antilles. Handful of winter records from northeastern Colombia, but none at present from Venezuela (Restall et al. 2006).

Family Falconidae: Falcons, Caracaras

Crested Caracara (*Caracara cheriway*)   Plate 14

Papiamento: *Warawara*
Dutch: *Kuifcaracara*

Description: A large hawk with long legs and neck. Large distinctive bill is pale blue, with red (usually) facial skin and flesh over base of bill; also note black cap and white cheek and neck; black belly, back, and upperwing, except for white wingtips. Adult with yellow legs; juvenile with light whitish-gray legs. White tail with fine barring and solid black band at end of tail. Immature shows similar pattern but more faded-looking; juveniles show white edges to back and upperwing feathers. **Similar species:** Only other regularly occurring large hawks on the islands are Osprey and White-tailed Hawk. Ospreys lack the large, brightly colored bill and face and have white (not dark) belly and underwing; show different tail pattern and shape. White-tailed Hawk is much smaller, lacks large, brightly colored bill and face, and has a dark head; has very different plumage coloration and pattern. The rare Yellow-headed

Caracara, similar but smaller, lacks dark cap; red face replaced with yellow. **Status:** In Jan observations on Aruba, from 1993 to 2003, the highest number we saw was four; our most frequent sightings were at Arikok and Sero Colorado, but this bird does wander widely throughout Aruba, though it is not often seen in urban areas. Highest numbers reported for Aruba in Prins et al. (2009) were groups of 9 and 14, though no years given. On Bonaire, highest numbers reported were 10, seen in Jan 1998, but more common to see 1–4 birds at a time (Prins et al. 2009, pers. obs.). On Curaçao, seems more abundant and widespread; we recorded a surprising concentration of 20 at Malpais on Nov 22, 2003. **Range:** Occurs in southern Florida and Cuba, and from Texas south to Central America and throughout South America. Southern South American form may be separate species.

Yellow-headed Caracara (*Milvago chimachima*)   Plate 14

Papiamento: *Warawara cabez hel, Warawara kabes hel*
Dutch: *Geelkopcaracara*

Description: A small, long-legged hawk. Adult with creamy-white head and underparts, dark brown back and wings; dark eyebrow; yellow legs. Tail barred; broad dark band at end of tail. In flight, note white patches at ends of wings contrasting with dark back and upperwing surface. Underwing white, with broad, black trailing edge; white patch in outerwing; and dark wingtips. Immature shows dark streaking on head and underparts. **Similar species:** See Crested Caracara. Immature superficially similar to Merlin but larger, lacking dark cap, and with broader, blunt (not pointed) wings. **Status:** Two sight records: one immature soaring over Julianadorp, Curaçao, Jan 1952 (Voous 1983); one bird near Lac Cai, Bonaire, Dec 1996. Future sightings ideally would be confirmed with photos. **Range:** Occurs from Costa Rica south through Argentina and southern Brazil.

American Kestrel (*Falco sparverius*)   Plate 14

Papiamento: *Kinikini*
Dutch: *Amerikanse Torenvalk*

Description: A petite, brightly colored falcon. The form on Aruba and Curaçao has been described as a distinct subspecies. Adult male with reddish back contrasting with blue wings; blue crown gives helmeted appearance. Black "teardrop" extends from eye downward; note second mark behind eye. Tail reddish, with black terminal band. Female and immature similar to male, but tail, wings, and back reddish, with fine dark barring. Underside rust-colored in both sexes; unstreaked in males, with fine streaks in females and immature. At rest, wings extend only halfway down length of tail. Although not definitively recorded, migrants of subspecies from North America or from the South American mainland possible. Males of North American subspecies would show longer wings, strongly barred back and lower breast, dark sides to tail; would lack strong reddish underside and would have obvious reddish crown patch. **Similar species:** Adult male largely unmistakable, even in flight.

Female and immature most readily confused with migrant Merlins. On Merlin, note dark brown (or blue-gray) back, wings, tail, and crown; lacks distinct facial marks and shows more heavily streaked breast. **Voice:** Makes a high *kee-kee-kee-kee* or *ki-ni-kini-kini*. The Papiamento name—*Kinikini*—is said to derive from the call (Voous 1983). **Status:** Generally common and widespread on Aruba and Curaçao, where it is a resident breeder year-round. On Bonaire, only recorded four times since 1960 (Prins et al. 2009). We estimate 50 to 100 pairs (or more) on Curaçao and dozens of pairs on Aruba, though a thorough survey is needed for more precise census data. **Range:** Occurs throughout most of the Western Hemisphere, with many described subspecies. From Alaska and northern Canada south through Central America and Caribbean; throughout South America except for Amazon Basin.

Merlin (*Falco columbarius*)   Plate 14

Papiamento: *Falki shouru, Kinikini grandi*
Dutch: *Smelleken*

Description: A small, dark falcon, similar in size to more common American Kestrel. Back and upperwing dark blue in male, brown in female and immature. Tail dark, with a narrow white band across tip and several narrow white bands above. Dark cap, narrow light eyebrow, and dark "teardrop" below eye. Underside darkly streaked. In flight, note short, pointed wings and long, narrow tail. Underside of wings darkly marked. Most often seen darting by, low over vegetation or across open areas in search of prey. **Similar species:** In poor light, female and immature American Kestrel might be mistaken for Merlin but look for dark tail with only a few narrow white bands in Merlin. In female or immature American Kestrel, tail is narrowly barred with alternating dark and light bands. Back is solid color in Merlin, while female and immature American Kestrels have a narrowly barred, reddish back. American Kestrel lacks white eyebrow and overall is much lighter in color than Merlin. Peregrine Falcon, especially immature birds, show some plumage similarities but are massively larger, with longer, broader-based wings and tail. **Voice:** Produces a high harsh *ki-ki-ki-ki-ki*, rapidly repeated. **Status:** On all three ABC islands, a regular but somewhat uncommon winter resident, typically arriving in Oct and departing by Apr, during which time it is not uncommon to see 2–4 birds in a day of birding. Three exceptional summer records of single birds on Bonaire and Curaçao (Prins et al. 2009). **Range:** Widespread breeder across northern North America, Europe, and Asia. In Americas, winters from US south to northern South America.

Peregrine Falcon (*Falco peregrinus*)   Plate 14

Papiamento: *Falki peregrino*
Dutch: *Slechtvalk*

Description: Large falcon with broad-based pointed wings and short tail; dark cap and nape; dark "teardrop" extends from eye downward. Adults show slate-blue upperparts, with dark barring on tail, and paler, finely barred, rump. Throat, cheek, and upper breast white in adult, with barring from lower breast to undertail coverts. In first-year birds, slate-blue replaced with brown and crown is light; underside streaked with brown, but still retaining white throat and cheek. Underside of wings show dark barring throughout, slate-blue in adult, brown in immature. **Similar species:** None of the other large hawklike birds that regularly appear on the islands—Crested Caracara, Osprey, and White-tailed Hawk—show pointed wings. Crested Caracara has bold, white marks in wings. Osprey has much longer wings and is very white underneath and on head. White-tailed Hawk is smaller, with shorter tail, blunter wingtips, and different plumage pattern. Merlin and American Kestrel look most similar in shape and plumage pattern but are much smaller. American Kestrel shows reddish hues in plumage and barred back. Merlin lacks strong teardrop or "helmet" pattern on head and has streaked underside extending from upper breast to undertail coverts. **Voice:** Occasionally will give harsh *kek-kek-kek* notes when fighting with another Peregrine Falcon, as sometimes occurs at the relatively few wetland habitats where ducks and shorebirds, their favorite prey, concentrate. **Status:** On all three islands, wintering birds regularly occur, from Sept through May. Not uncommon to see several birds in a day of birding, both immature and adults, usually near wetlands. Sometimes seen soaring over arid hills, as when we observed two calling and chasing each other over Arikok NP, Aruba, on Jan 10, 1998. On Aruba, birds have been regularly noted roosting on ledges of hotels. In approximately 40–50 sightings of Peregrine Falcons in the islands, we have noted about equal numbers of adults and immatures. **Range:** Occurs in many parts of the world. Birds wintering in the islands are expected to be of the subspecies *tundrius*, which breeds across northern North America and Greenland; plumage characteristics of birds observed and photographed are consistent with that designation.

Family Psittaculidae: Old World Parrots

Rose-ringed Parakeet (*Psittacula krameri*)   Plate 35

Papiamento: *Prikichi renchi ros*
Dutch: *Halsbandparkiet*

Description: Slim, green parakeet with a long, thin tail, reddish bill, black flight feathers in flight. Adult male has black collar by second year. Adult females and immatures do not show black collar. **Similar species:** Introduced Blue-crowned Parakeet also possible on Curaçao; shows white patch around eye, pale pinkish upper mandible, dark lower mandible, and, in flight, orangish color on trailing edge of underwing and in tail. Native Brown-throated Parakeet on Curaçao has bright yellow face and dark bill. **Voice:** Makes a variety of loud shrieks and calls including "shrill, rather harsh *kee-ak, kee-ak, kee-ak*" (Juniper and Parr 1998), given in series. **Status:** Not known to currently occur on Aruba or Bonaire, but now an introduced breeding resident on Curaçao. **Range:** An African species that has now been introduced in urban and suburban tropical and subtropical regions of the world.

Family Psittacidae: New World Parrots

Red-lored Parrot (*Amazona autumnalis*)   **Plate 35**

Papiamento: *Lora cara hel, Lora kara hel*
Dutch: *Geelwangamazone*

Description: A large green *Amazona* parrot likely to be seen only on Curaçao. Note short, squared tail, rapid wingbeats in flight, red patch in inner wing, red on forehead. Some forms have yellow cheeks; on forms lacking yellow cheeks, a white eye-ring is sometimes prominent. Bill either all dark or with light upper mandible and dark lower mandible. Extent of head markings can be very difficult to discern in flight. **Similar species:** Parakeets are smaller and show a long tail (short and squared on Red-lored). Red-lored Parrot lacks the yellow crown of Yellow-shouldered, Yellow-crowned, and Orange-winged parrots. **Voice:** Makes a variety of loud shrieks and calls, said to be more metallic than calls of Yellow-crowned Parrot (Juniper and Parr 1998). **Status:** Escaped caged birds seen occasionally in suburban areas on Curaçao; they have bred, so may become resident there. **Range:** From southern Mexico to northern South America (including far western Venezuela). There is a disjunct population in Brazilian Amazon.

Yellow-crowned Parrot (*Amazona ochrocephala*)   **Plate 35**

Papiamento: *Lora skouder còrá, Lora skouder kòrá*
Dutch: *Geelvoorhoofdamazone*

Description: A large green *Amazona* parrot likely to be seen only on Curaçao. Note short, square tail, rapid wingbeats in flight, red patch in inner wing; yellow restricted to forehead. Bill can vary from somewhat dark to light; and in some forms, there is red or pink at base of upper mandible. One subspecies with a very restricted range has a full yellow head, and should show reddish on upper mandible; this subspecies is highly unlikely as an escaped bird. Extent of head markings can be very difficult to discern in flight. **Similar species:** Parakeets are smaller and show a long tail (short and squared on Yellow-crowned). Red-lored Parrot has a red (not yellow) forehead. Yellow-shouldered Parrot and Orange-winged Parrot both have yellow forehead; on Yellow-shouldered Parrot, face usually entirely yellow and thighs prominently yellow (though this can be difficult to see). On Orange-winged Parrot (also with extensive yellow on the face), note greenish-blue line that extends from base of upper mandible through eye to nape. **Voice:** Angehr and Dean (2010) write that "Calls (which include a guttural *cuh-RAO!*) are deeper and hoarser than other *Amazona* parrots." **Status:** Escaped cage birds seen occasionally on Curaçao. Not recorded on Aruba or Bonaire. **Range:** Occurs in much of northern South America, and from Panama south to Bolivia and Brazil.

Yellow-shouldered Parrot (*Amazona barbadensis*)   **Plate 35**

Papiamento: *Lora*
Dutch: *Geelvluegelamazone*

Description: Large, green *Amazona* parrot with extensive yellow face and pale bill. Yellow at bend of wing and small red patch on inner wing. Yellow thighs. Blue flight feathers sometimes look dark in poor light. Compared to parakeets, note the very short, square tail. **Similar species:** Parakeets are smaller and have a long tail (short and square on Yellow-shouldered Parrot). There are no other native *Amazona* species on Bonaire (Yellow-shouldered Parrot is now extinct on Aruba and Curaçao), but exotic, introduced *Amazona* species are possible, especially in suburban areas; in recent decades birds released from captivity have appeared on Aruba and Curaçao. No other introduced *Amazona* species documented on the islands has the full yellow face of the Yellow-shouldered Parrot or the yellow thighs (though this can be hard to see). Orange-winged Parrot has extensive yellow on the face but has a greenish-blue line that extends from base of upper mandible through eye to nape. **Voice:** Like most parrots, makes a variety of loud, screechy calls and soft chortlings. Calls are generally lower-pitched and more full-bodied than the high, thin screeching calls of Brown-throated Parakeets and can be readily distinguished from the calls of that species. Flight call described as "throaty, rolling *cu'r'r'rak-cu'r'r'rak*" (Hilty 2003). **Status:** The Yellow-shouldered Parrot is a rare species currently listed as Vulnerable under the IUCN Red List for birds. It has a very small range and a population size estimated at 1,500–7,000 adults. Historically found on Aruba and Bonaire and, perhaps, Curaçao (some evidence from document published in 1700s), but now only on Bonaire, except for possible occasional birds released from captivity on Curaçao and Aruba that inhabit urban areas. Other *Amazona* species have been released and have possibly bred on Curaçao, and maybe Aruba and Bonaire, making it difficult to sort through the status of various Amazona species. These once captive birds are most likely seen in urban or suburban areas. On Bonaire, recent estimates for Yellow-shouldered put the wild population at 650–800 birds. Although it is now illegal to capture and keep Yellow-shouldered Parrots on Bonaire, there are estimated to be hundreds of birds in captivity on Bonaire and Curaçao. It is possible that many birds are also smuggled overseas for the illegal bird trade. Enforcing the prohibition against illegal trade is tricky, as there is a provision in the law that exempts birds held in captivity prior to the establishment of the law. During dry periods on Bonaire, many birds will come into the suburbs of Kralendjik to feed on fruits in backyard gardens. **Range:** Occurs only on Bonaire, the Venezuelan islands of Margarita and La Blanquilla, and in isolated locations in coastal areas on the Venezuelan mainland.

Orange-winged Parrot (*Amazona amazonica*)   Plate 35

Papiamento: *Lora all oraño*
Dutch: *Oranjevleugelamazone*

Description: A large green *Amazona* parrot that has occurred as an escape on Curaçao and Aruba. Note short, square tail, rapid wingbeats in flight, and red patch in inner wing; yellow on forehead and cheek is divided by a greenish-blue line that extends from base of upper mandible through eye to nape. Extent of head markings can be very difficult to discern in flight. **Similar species:** Parakeets are smaller and show a long tail (short and squared on Orange-winged). Red-lored Parrot has a red (not yellow) foreheard. Yellow-shouldered Parrot and Yellow-crowned Parrot both have yellow forehead; on Yellow-shouldered Parrot, face usually entirely yellow and thighs more prominently yellow (though this can be difficult to see); Yellow-crowned Parrot usually lacks yellow on face and never shows greenish-blue line on face. **Voice:** Hilty (2003) describes the flight calls as "higher, more screeching than in most *Amazona*" and one of the most distinctive flight calls as *cm-quick* or *cm-quick-quick*. **Status:** Escaped cage birds seen occasionally on Curaçao and Aruba. **Range:** Northern South America to southeastern Brazil.

Green-rumped Parrotlet (*Forpus passerinus*)   Plate 36

Papiamento: *Bibitu*
Dutch: *Muspapegaai*

Description: Very small, all-green parrot with an extremely short tail, blue patch in wing, and pale bill. **Similar species:** Except for escaped exotic caged parrot species (and they occur with unfortunate frequency), this species should be readily identifiable. **Voice:** Hilty (2003) describes the vocalizations as "shrill, chattery *chee* and *cheet-it* and *chee-sup* notes when foraging and in flight (like notes of little finches)." **Status:** Small population of released captive birds established itself on Curaçao, but no birds have been seen there for many years. Not found on Aruba or Bonaire. **Range:** Occurs in Northern South America. Introduced on Jamaica and Trinidad.

Brown-throated Parakeet (*Eupsittula pertinax*)   Plate 36

Papiamento: *Prikichi*
Dutch: *Maïsparkiet*

Description: A resident breeding bird; there are three distinct subspecies specific to each island. Green body, wings, and tail. Dark bill. In flight, note long tail and rapid, shallow wingbeats; also note blue on upperwing extending from wingtip to trailing edge of inner wing and in tail tip; yellow in wing lining. At rest the blue wingtip and blue in tail usually visible. Aruba subspecies with very brown face and throat and small, bright yellow patch surrounding eye. Bonaire subspecies with bright yellow extending from crown through face to throat. Curaçao subspecies intermediate, with green crown but extensive yellow face. There have been occasional unintended releases of a subspecies from one island onto another island, so it is possible to occasionally see intermediate birds that are presumed hybrids or a subspecies that should not be on a particular island. **Similar species:** *Amazona* parrots all have very short tails. Introduced parakeets (mostly on Curaçao but they could be seen on any of the islands) lack yellow on head (Curaçao and Bonaire) or around eye (Aruba), lack blue in wings, and will not have a dark bill. **Voice:** Makes a variety of loud, harsh, high-pitched shrieks and other calls. There seems to be a discernible difference in vocalizations between each of the subspecies but more work needs to be done to document the differences. **Status:** Originally quite common on all three islands, but there is a major decline underway, sadly, on Aruba that is likely attributable to the depredations of the introduced Boa Constrictor (*Boa constrictor*), which has proliferated across the island. In Nov 2011 and Nov 2012, on Aruba, we found no Brown-throated Parakeets in most areas where they had been common from 1993–2003, including at Arikok NP (we estimated 25 on a visit to one section of the park in Jan 1998) and Spanish Lagoon (where we regularly counted 10–15 annually between 1998–2003). We found no birds coming into the traditional roost at Alto Vista, which had held 80 or more birds in 2001 (Harms and Eberhard 2003). Small numbers can still be found in some locations on Aruba. Birds remain relatively common and widespread on Curaçao and Bonaire. We counted 165 birds leaving a roost at Malpais, Curaçao, in Nov 2003, and found groups of 2–10 birds at many locations across the island. Similarly, on Bonaire groups of 2–10 (and occasionally up to 30) birds are regularly seen at many locations across the island. **Range:** Northern South America north to Panama, with 14 described subspecies.

Chestnut-fronted Macaw (*Ara severus*)   Plate 36

Papiamento: *Ara enano*
Dutch: *Dwergara*

Description: A large macaw with long tail, white face, and green body with reddish underwing and undertail. **Similar species:** *Amazona* parrots have short tails; Brown-throated Parakeet is much smaller and lacks the white face. **Voice:** Like other macaws, makes loud squawks and screeches. **Status:** In the suburbs of Willemstad, Curaçao, sightings of 2–4 birds have occurred intermittently since 1992 (Prins et al. 2009) and breeding has also been reported (de Boer et al. 2012). We saw 14 birds in the Ruistenberg Park area of Willemstad on Nov 23, 2003, and photographed three birds near there on Nov 24, 2003. On Aruba, three birds were seen near Savaneta in 1996 and two regularly over Tierra del Sol in 1997 (Prins et al. 2009). **Range:** Natural range is from Panama south to Bolivia.

Blue-crowned Parakeet (*Thectocercus acuticaudatus*)   Plate 36

Papiamento: *Prikichi kabez blou, Prikichi kabes blou*
Dutch: *Blauwkoparatinga*

Description: Escaped caged birds seen in suburban areas of Curaçao and may be resident there. Note long tail. All green, with white patch around eye; pale, pinkish upper mandible, dark lower mandible. In flight, orangish on trailing edge of underwing and in tail. Light blue forehead only visible at close range. **Similar species:** *Amazona* parrots have short tails. Chestnut-fronted Macaw is larger, with white face. Introduced Rose-ringed Parakeet also possible on Curaçao; lacks white patch around eye, has red bill, black flight feathers. Native Brown-throated Parakeet, on Curaçao, has bright yellow face and dark bill. **Voice:** Makes a variety of loud shrieks and calls, including "a low, hoarse, rapidly repeated *reedy-reedy-reedy*" (Sibley 2000). Calls said to be readily distinguished from those of Brown-throated Parakeet (Juniper and Parr 1998). **Status:** In the suburbs of Willemstad, Curaçao, a small population of released birds has apparently become resident since at least 2007, though it is unclear if they are breeding. Not known from the other islands. **Range:** Range is highly disjunct, with a population in northern Colombia and Venezuela, another in eastern Brazil, and a large range extending from Brazil south into Argentina. Large numbers of the species have been taken for the pet bird trade from the Argentina population (Juniper and Parr 1998). We are unaware of any information about which population the introduced birds originally came from.

Scarlet-fronted Parakeet (*Psittacara wagleri*)   Plate 36

Papiamento: *Prikichi frente kòrá*
Dutch: *Wagler's parkiet*

Description: A large, green parakeet with red forehead (reduced in young birds), pale bill, white eye-ring (reduced in young birds), and red thighs (lacking in immatures). Likely to be seen only on Curaçao. **Similar species:** *Amazona* parrots have short tails. Brown-throated Parakeet, on Curaçao, has extensive yellow on face (no red) and dark bill. Blue-crowned Parakeet has bluish rather than red forehead, has dark lower mandible, and lacks red thighs. Red on Scarlet-fronted Parakeet can be difficult to discern at a distance and in flight. **Voice:** Flight calls squeaky, "high-pitched and discordant" (Hilty 2003). Juniper and Parr (1998) say some calls are reminiscent of the sounds of domesticated donkeys. **Status:** Apparently feral populations derived from escaped caged birds now occur on Curaçao (de Boer et al. 2011), though the species is not listed in Prins et al. (2009). **Range:** Natural range is in mountains, from Venezuela south to Peru.

Family Tyrannidae: Flycatchers

Caribbean Elaenia (*Elaenia martinica*)   Plate 38

Papiamento: *Elenia caribe*, *Whimpie*, *Chonchorogai*
Dutch: *Witbuikelenia*

Description: A small, plain flycatcher with two white wingbars, and olive-gray upperparts; whitish-gray underparts sometimes washed with very pale yellow on lower belly. On folded wing, white edges to inner flight feathers (secondaries and tertials) are quite conspicuous. Face plain, with prominent dark eye. Small, somewhat broad-based bill with dark upper mandible; lower mandible shows a pink base. When agitated, sometimes raises head feathers, presenting a shaggy crest and white crown patch, but difficult to see. Generally stays hidden in thick shrubs and crowns of trees. **Similar species:** Most similar common species is the Northern Scrub-Flycatcher but note that species' smaller, all-black bill and brighter yellow belly contrasting with pale gray upper breast; back and rump often shows a greener tone than on Caribbean Elaenia. Brown-crested Flycatcher is appreciably larger, with a reddish tail; also note large, broad-based, all-black bill, lemon-yellow lower belly, and reddish in outer wings. Rare austral migrant elaenias of uncertain status are exceedingly difficult to distinguish but look for birds with bolder eye rings, brighter yellow bellies, more obvious white crown stripe, and suggestion of third wingbar above. **Voice:** On Bonaire, commonly makes a descending, whistled *wheer*; our audio recordings are in the collection of the Macaulay Library, at the Cornell Laboratory of Ornithology. Voous (1983) describes the common call as "wee-weew," "wee-wee-weew," or "pee-weet-pee-weet." **Status:** Relatively widespread (but not abundant) on Curaçao and Bonaire. On Curaçao, we have recorded the species at Malpais, Christoffel NP (quite common in Nov 2003), and near Choloma. On Bonaire, we have seen Caribbean Elaenia at Lac Bay, Dos Pos, and throughout Washington-Slagbaai NP (always at Pos Mangel and Put Bronswinkel). Rare on Aruba, with several recent records documented with photographs, at Spanish Lagoon (Prins et al. 2009; eBird 2014; Mlodinow 2014). We have never recorded it with certainty on Aruba but have had brief glimpses of possible elaenia species several times over the years at Spanish Lagoon and on the reef islands off Oranjestad. **Range:** Restricted to southern and eastern Caribbean basin (absent from Cuba, Hispaniola, Bahamas). Occurs on Puerto Rico, Cayman Islands, Virgin Islands, Lesser Antilles, and some other small islands and islets in southern Caribbean.

Small-billed Elaenia (*Elaenia parvirostris*)   Plate 38

Papiamento: *Elenia pico chikito, Elenia pik chiki*
Dutch: *Kortsnavelelaenia*

Description: Elaenias are famous for being incredibly difficult to tell apart. They are all small, rather drab, and unassuming flycatchers with only subtle differences in plumage. Because of the difficulty in identification, there is also much that is unknown about their movements, distribution, geographic variation in plumage characteristics, and species limits. Small-billed Elaenia is very similar to Caribbean Elaenia in appearance except that it typically shows a prominent white eye-ring, often a third wingbar (but Caribbean can show this as well); with little or no crest, it appears more round-headed than Caribbean, but does show a narrow white crown stripe. **Similar species:** Compare with Caribbean Elaenia. **Voice:** Said to be relatively silent on wintering grounds (Hilty 2003). Ridgley and Tudor (2009) describe calls as including a sharp "chu" or "cheeu." Calls are described by Perlo (2009) as "dry *tjip* or chipping *Tjuw* or *ti-sjuw*." **Status:** One specimen record from Aruba, collected May 6, 1908, and now at the Field Museum of Natural History. This species is a relatively common austral migrant from Apr through Sept in Venezuela, arriving from its breeding grounds in southern South America, and seems likely to occur in the islands. **Range:** Breeds from southern Bolivia south to northeastern Argentina, Uruguay, and southwestern Brazil. Winters across much of northern South America.

Lesser Elaenia (*Elaenia chiriquensis*) Plate 38

Papiamento: *Elenia chikito, Elenia chiki*
Dutch: *Kleine Elenia*

Description: One of a group of flycatchers that are exceedingly difficult to distinguish. Hilty (2003) describes this species as "Dull, confusing, and almost devoid of good fieldmarks." Difficulties in identification have led to much confusion about status of this and other species on the ABCs, and throughout their ranges. Lesser Elaenia is very similar to Caribbean Elaenia (as well as Small-billed Elaenia) but is somewhat smaller and less crested. Clearly these are rather subjective and poorly defined characteristics to distinguish the two species, and more ornithological research will be required to better understand the differences between them. **Similar species:** See Caribbean Elaenia. **Voice:** Ridgely and Tudor (2009) write: "Song in most of range a burry, bisyllabic 'chibur' or 'jwebu'; in nw. Ecuador and sw. Colombia (likely not same species) gives a very different burry 'bweer, wheéb, wher'r'r' (P. Coopmans)." Hilty (2003) writes that the most common call is "a clear, whistled *weEEa*" but notes that there are a variety of other soft and plaintive calls. **Status:** Known from two specimens: one collected Oct 27, 1951, at Bullenbaai, Curaçao, and one collected Nov 6, 1951, at Fontein, Bonaire (Voous 1983). There is one unconfirmed sight record from Curaçao, on June 18, 1961, at Groot Sint Joris, and four from Bonaire, the last in 1979 (Voous 1983). **Range:** Occurs from Costa Rica south through South America to Argentina and Brazil. Some populations in Central America and southern South America are migratory, but details of where they winter and timing of migration are virtually unknown because of lack of good field identification characters at this time.

Northern Scrub-Flycatcher (*Sublegatus arenarum*) Plate 38

Papiamento: *Parha bobo, Para Bobo, Chonchoragai*
Dutch: *Noordelijke Struikvliegenpikker*

Description: On all three islands, this is the most common of the small flycatchers. Found regularly in scrubby woodlands and thickets and in mangroves. A small, sprightly bird but rather shy, often staying within the interior of a thick bush or tree. Has a small head, small, all-black bill, and (usually) a pale eye line. Upperparts generally grayish, sometimes with some green tones on back. Note two white wingbars and white edges to inner flight feathers. Grayish-white on throat and upper belly contrasts with yellow lower belly. **Similar species:** Small, all-black bill distinguishes this species from the Caribbean Elaeina (and other, rarer elaenia as well). Also usually shows brighter yellow on belly, has a more small-headed appearance, and lacks any white in crest. **Voice:** Most commonly heard vocalization is a quiet, upslurred *weee* or sometimes a double-noted *wee-doot*, like a child's squeaky toy. **Status:**

Common on all three islands. Note that Caribbean Elaenia is rare on Aruba, with recent documentation only from Spanish Lagoon, so most small flycatchers observed on Aruba are likely to be Northern Scrub-Flycatcher. **Range:** Only recently divided into three similar species (Southern Scrub-Flycatcher and Amazonian Scrub-Flycatcher being the other two species). Northern Scrub-Flycatcher occurs from Costa Rica south across northern Colombia, Venezuela to northern Guianas.

Olive-sided Flycatcher *(Contopus cooperi)*   **Plate 39**

Papiamento: *Pibi canades*
Dutch: *Sparrenpiewie*

Description: The species is rare in the ABCs. A large migrant flycatcher that often sits at top of conspicuous perches. Dark brown upperparts; white underparts with dark along flanks forming dark "vest" when viewed from the front. On perched bird, white tufts can often be seen protruding from behind the folded wings on the back. Often appears tufted or crested toward back of head. Bill large, with a wide base; mostly dark but note extensive orangish base to lower mandible. **Similar species:** No other commonly occurring species would likely be confused. Pewees, also rare migrants in the ABCs, are similar but smaller, with a more obvious fork to tail, smaller bill, and less strikingly marked "vest." Any sightings of this species should be documented with photos and or video. **Voice:** On wintering grounds and during migration, more likely to give distinctive *pip* or *pip-pip-pip* calls. Highly distinctive song, possibly made in migration, is a loudly whistled "Quick-three beers." **Status:** Two records, both from Bonaire: one photographed at Fontein, May 4, 1961 (Voous 1983), and one seen at Dos Pos, May 5, 2001 (Ligon 2006 as reported in Prins et al. 2009). **Range:** Breeds across northern Boreal forest region from Alaska east to Newfoundland and south to northeastern US. In western North America, breeds at high elevations, south to Baja California. Winters from southern Mexico south to Bolivia and Amazonian Brazil.

Eastern Wood-Pewee *(Contopus virens)*   **Plate 39**

Papiamento: *Pibi di este*, *Pibi di ost*
Dutch: *Oostelijke Bospiewie*

Description: Rare in the ABCs. A small migrant flycatcher with dark brown upperparts, white wingbars, and no obvious eye-ring. Light underparts with darker flanks give a "vested" appearance, though not as strongly as in Olive-sided Flycatcher. Tail usually strongly forked. Bill has dark upper mandible with lower mandible often extensively orangish-pink, except for black tip, but this is highly variable. Best identified by vocalizations, if possible. **Similar species:** Western Wood-Pewee is virtually identical in appearance but bill tends to show a more extensively dark lower mandible. Both species winter in South America and could occur as rare migrants in the ABCs. In general, Eastern Wood-Pewee vocalizations are clear and often upslurred, versus the rougher, burry or buzzy, and often down-slurred sounds given by Western

Wood-Pewee. Although there are currently no records of the *Empidonax* flycatchers from the ABCs, Alder, Willow, and Acadian flycatchers occur as migrant winterers in South America and are likely to occur eventually as rare migrants in the ABCs. In fact, an *Empidonax* flycatcher photographed on Aruba, in Oct 2016, may have been a Willow Flycatcher (*Empidonax traillii*), but its identity was undetermined at the time this book was going to press. All *Empidonax* flycatchers are similar to the pewees but show a prominent eye-ring, more obvious wingbars, shorter wings, and a less "vested" appearance. **Voice:** Eastern Wood-Pewee usually gives a sweet, questioning, whistled, upslurred *pee-wee*; sometimes also makes a down-slurred *pee-ur*. Its full song is a strong, clear, whistled *pee-a-wee*. Western Wood-Pewee usually gives a buzzier, emphatic, downslurred *bee-urr*. **Status:** Two photographically documented records from Aruba: one at Bubali on May 5, 2016, and one at Bubali on Oct 8, 2016 (both M. Oversteegen fide eBird). One photographically documented record from Curaçao: one individual at Playa Porto Mari, on Apr 14, 2013 (B. Rodenburg fide Observado). Four sight records, one from Aruba and three from Bonaire. On Aruba, one was reported from Sero Colorado, on Apr 4, 1959. Bonaire sightings are as follows: one near Pekelmeer, Oct 22, 1979; one at Pos Mangel, May 4, 1980; and one at Hato, on Sept 27, 2002 (Voous 1983; Prins et al. 2009). **Range:** Breeds in eastern North America, from southern Canada south to northern Florida and Texas. Winters from Central America through northern South America.

Vermillion Flycatcher (*Pyrocephalus rubinus*)   Plate 39

Papiamento: *Pímpiri còrá*, Pímpiri kòrá
Dutch: *Rode Tiran*

Description: A small flycatcher with only a single record from the ABCs. Male has unmistakable brilliant red underparts and black mask, back, wings, and tail. Females and immatures, the latter probably the most likely to occur in the ABCs, are more easily confused, as they are dark brown on upperparts and pale underneath with brown streaking. Can show a pink or yellow wash on lower belly and vent area, depending on age and geographic origin. The species' habit of sitting on an exposed perch and sallying out for flying insects should also help in identification. **Similar species:** Males unmistakable. Females and immatures identifiable by behavior and plumage. Any sighting of the species should be documented with photos and/or video. **Voice:** Call is a sharp *peek*, sometimes given in rapid sequence. **Status:** A single record of an immature male photographed at Sero Colorado, Aruba, in fall 1957 (Voous 1983). **Range:** Breeding range extends from southwestern edge of US south through Central America, to southern South America.

Pied Water-Tyrant (*Fluvicola pica*)   Plate 40

Description: A small flycatcher with only a single record from the ABCs. Males are strikingly patterned, with white underparts extending onto face and crown. Black nape and back separated from black wings by white stripe. Black tail with white tip and white rump. All-black bill and legs. Females and immatures similar, but black replaced with gray. **Similar species:** Males unmistakable. Females and immature White-headed Marsh Tyrant could be confused with female and immature Pied Water-Tyrant, but White-headed lacks the white stripe between wings and back; base of its bill is pink-orange (all black in Pied Water-Tyrant). **Voice:** Call described by Angehr and Dean (2010) as a "nasal, buzzy *zhreeeoo.*" Hilty (2003) describes another call as "a soft *pick* like bubble bursting." **Status:** A single record of a male photographed (archived via eBird as ML23188771) by Peter-Paul Schets, on Jan 8, 2016, at the sewage treatment plant on Bonaire, and known to be present until at least March 25, 2016 (Herman Sieben reported on Observado.org). **Range:** Breeding range extends from southern Panama across northern South America to northeastern edge of Brazil.

Cattle Tyrant (*Machetornis rixosa*)   Plate 40

Papiamento: *Pímpiri vakero*
Dutch: *Veetiran*

Description: A very distinctive, long-legged flycatcher that spends much of its time foraging on the ground. Bright yellow underneath; brownish upperparts; big, black bill; and red eye. Male and female similar. **Similar species:** Quite distinctive. Tropical Kingbird is also yellow underneath but doesn't forage on the ground or have the long legs of Cattle Tyrant. **Voice:** Makes high-pitched squeaky notes similar to those of Tropical Kingbird but higher. **Status:** First record was of a single bird at Savaneta, Aruba, Apr 2002 through June 2003 (Prins et al. 2009). We photographed and videotaped two birds at Tierra del Sol golf course on Aruba, in Nov 2011; since then, 2–3 birds have continued to be reported from that location through at least May 2016 (fide eBird). Although we are not aware of confirmation of breeding, we suspect that the species had and/or continues to have a small breeding population at Tierra del Sol golf course on Aruba. There are some unconfirmed reports of the species at Divi Links golf course on Aruba and possibly other Aruba locations. We are not aware of any documented sightings of the species from Curaçao or Bonaire. **Range:** This species has a spatially disjunct South American range, with one population occurring south of the Amazon south to Argentina and the other occurring in northern South America, in Venezuela and Colombia. In recent years, it has been expanding into deforested areas and is occurring with greater frequency in Panama.

Brown-crested Flycatcher (*Myiarchus tyrannulus*)   Plate 39

Papiamento: *Tirano grandi, Chonchorogai grandi*
Dutch: *Cayennetiran*

Description: One of the larger flycatchers but shy and retiring, preferring to stay hidden in the middle of thick bushes and trees. Brown upperparts; wings darker brownish-black, with white-edged tertials and two white wingbars, but rusty in outer part of wing (often very hard to see on folded wing). Yellow belly, with gray on upper breast and throat. Tail feathers have rusty inner webs, except for two in center, which are the most visible from above (the tail may not appear rusty at all from above). Even from below, many birds of this pale subspecies, when in worn plumage, can appear to have little or no rufous in tail. Bill is large, hooked at tip, and black, normally showing a small amount of pinkish at the base or in gape, but some birds can appear to show a totally black bill. **Similar species:** Northern Scrub-Flycatcher, which also likes to stay hidden in bushes and trees, also shows a yellow belly but is much smaller, with a small head and bill, and lacks any rufous in wings or tail. Caribbean Elaenia can show a pale yellowish belly but has an obvious pink-based bill; it is much smaller and lacks any rufous in wings or tail. Tropical Kingbird is closer in size but is typically not shy or retiring, has a notched rather than squared tail, has a gray rather than brown head, and whiter throat (versus gray in Brown-crested Flycatcher). **Voice:** Most common call is a sharp *whip*, often given in succession; it is sometimes delivered softly from within a thick shrub or tree, and easily overlooked. When excited, may give a rough *reer* or a series of quick, single *pip* notes followed by *reer*; also makes various combinations of these notes, similar to the calls of other *Myiarchus* flycatchers. Caribbean Elaenia has a somewhat similar *wheer* call but it is more rising and emphatic, and not typically interspersed with *whip* or *pip* notes. **Status:** Common year-round resident on all three ABC islands, in areas of scrub forest. Although Prins et al. (2009) describe it as a scarce breeding resident on Aruba, we have found it there regularly in good numbers over the last 20 years, at Spanish Lagoon, Arikok NP, and in other areas of scrub habitat around the island. The species can easily go undetected if you are not familiar with its often rather quiet and subtle call note. **Range:** Breeding range extends from southwestern edge of US south through Mexico and Central America, and through South America to Argentina; missing in much of Amazonia and western South America. Some northern populations are migratory, and it is likely that some southern South American populations could be austral migrants to areas farther north in South America. (Note: Great Crested Flycatcher [*Myiarchus crinitus*] occurs as an uncommon winterer in Venezuela and, eventually, is likely to be recorded from the ABCs. Great Crested Flycatcher is brighter yellow below, with more reddish in tail and with more extensive pinkish base to lower mandible of bill; it could be a difficult identification unless giving its characteristic loud *wheep* call. There are a number of other *Myiarchus* flycatchers that occur as migrants or residents in northern South America, and some could conceivably occur in the islands; in fact, several suspected rare *Myiarchus* have been photographed on Aruba in recent years. One species that might be a possible rare visitor to the islands is Swainson's Flycatcher [*Myiarchus swainsoni*]. This species has a southern South American austral migrant population that migrates north to Venezuela. The austral migrant subspecies of Swainson's Flycatcher shows a reddish bill and a very pale, yellow-white belly; also note lack of rusty color in tail and wings [though this may be very hard to see]. The Venezuelan resident subspecies of Swainson's Flycatcher shows almost no yellow on belly and has a dark bill. Several other species are virtually identical, so any possible sighting should be extensively documented with photos, video, and sound recordings. Because Brown-crested Flycatcher is a resident breeding species on all three islands, it is possible to see individuals in many different plumages and with varying degrees of feather wear. There is still much to be learned about possible plumage and bill variation in this and similar species.)

Streaked Flycatcher (*Myiodynastes maculatus*) Plate 39

Papiamento: *Pímpiri strepiá*
Dutch: *Gestreepte Tiran*

Description: Rare in the ABCs. A large, boldy streaked flycatcher with a dark mask through eye; rusty rump and tail; large bill with black upper mandible and pink-based lower mandible. **Similar species:** No other regularly occurring species looks similar. Sulphur-bellied Flycatcher, which is very similar, has not been recorded from the ABCs, though it does occur as a winter migrant in northern South America and could occur. Sulphur-bellied Flycatcher has a bolder whisker mark extending from chin down, and framing throat; lower mandible with much less pink at base; underpars are more yellow. Any sighting of either species should be carefully documented with photos and/or video. **Voice:** Calls described by Angehr and Dean (2010) as "a strident whining *kee-YOOO* and a sharp *chik!*" **Status:** Two records, both from Bonaire. One of a single bird photographed near Sorobon Beach Resort, on Sept 25, 1989 (Prins et al. 2009). One bird photographed at at Landhaus Karpata, on Dec 21, 2012 (J. Ligon). **Range:** Breeds from Mexico south to Brazil, Paraguay, and northern Argentina. Southern populations migrate north in austral winter. Some birds from northern populations apparently move south in northern winter.

Tropical Kingbird (*Tyrannus melancholicus*) Plate 40

Papiamento: *Pímpiri hel, Pímpiri*
Dutch: *Tropische Koningstiran*

Description: A large, conspicuous flycatcher, usually seen sitting out in the open on a telephone wire, pole, treetop, shrub, or palm. Bright yellow below, with a darker olive shade to upper breast; gray head; and white throat. Back is greenish-olive; wings and tail dark. Tail is notched. **Similar species:** Cattle Tyrant is superficially similar, but yellow extends to throat (white or gray in Tropical Kingbird). Cattle Tyrant has long legs and spends a lot of time on the ground, even running across the ground, something a Tropical Kingbird never does. Cattle Tyrant does not show the contrasting gray head of Tropical Kingbird and has a ruby-red eye, unlike the dark eye of Tropical Kingbird. Winter migrant Western Kingbirds occur as far south as Panama and could conceivably occur in the islands. Western Kingbird has a square-tipped tail with white edges, and yellow does not extend to upper breast as in Tropical Kingbird. Brown-capped Flycatcher has a yellow belly, but it has reddish tail and red in wings, and is rather shy, tending to stay within the interior of thick shrubs or trees. **Voice:** Similar to the more familiar (in the ABCs) calls of Gray Kingbird, but higher and shriller. Ridgely and Tudor (1994) describe dawn song as a "short series of 'pip' notes followed by a rising twitter 'piririree?'" **Status:** Breeds locally and possibly resident. On Aruba, we have not encountered the species since 1993, although there are historical records of nesting birds from the Yamanota area, 1971–78 (Voous 1983). A bird was photographed in May, 2016 at Bubali, Aruba, by both Peter Sprockel and Michiel Oversteegen (fide eBird). We photographed single Tropical Kingbirds near the Klein Hofje sewage treatment plant on Curaçao, in March 2000 and Nov 2003; others have photographed the species on Curaçao, in 2013 and 2014 (fide eBird). Ligon (2005) has noted only two sightings on Bonaire in recent years, both in July (2000 and 2001). Voous (1983) noted one nesting record on Curaçao, in Oct 1980, and several at Fontein, Bonaire, in 1976 and 1981. Given the irregular and local status of the species, it is worth carefully documenting occurrences of the species on any of the islands. **Range:** Has an extensive breeding range, from southern edge of Arizona south through Central America and South America, to Argentina. Southern birds migrate northward in austral winter, and some northern populations migrate south in northern winter. The species is common on nearby Venezuelan mainland.

Eastern Kingbird (*Tyrannus tyrannus*)   Plate 40

Papiamento: *Tirano cabez preto, Tirano kabes pretu*
Dutch: *Koningstiran*

Description: A distinctive, usually conspicuous flycatcher, all black above, white below with broad white tip to short, square-tipped tail. **Similar species:** Immature Fork-tailed Flycatchers with worn and short tail can look surprisingly similar under some light conditions. Eastern Kingbird has a black back (gray in Fork-tailed) and black half-hood. Fork-tailed Flycatcher also shows some fork to tail and lacks a broad, white tail tip as seen in Eastern Kingbird. In flight, note white underwing of Fork-tailed Flycatcher in contrast to the gray underwing of Eastern Kingbird. Any sightings of Eastern Kingbird should be documented with photos and/or video. **Voice:** Although well-known on breeding grounds for their aggressive behavior and loud, chattery calls, Eastern Kingbirds are largely silent in migration and on wintering grounds. **Status:** Six records: on Aruba, one at Bubali, Apr 5, 1978, and two at the same location on Apr 7, 1978 (Voous 1983); one photographed at Bubali on Apr 28, 2012 (S. Milks fide eBird); on Bonaire, one at Pos Mangel, Oct 15–17, 1977; two at Witte Pan, May 12, 1982; and two at Playa Tam, on Oct 2, 1982 (Prins et al. 2009). No records from Curaçao. **Range:** Breeding range extends from southern Canada through most of US, except the Southwest. Winters in eastern South America south to northern Argentina.

223

Gray Kingbird (*Tyrannus dominicensis*)   **Plate 40**

Papiamento: *Pimpiri gris*, Pímpiri, *Pimpiri*
Dutch: *Grijze Koningstiran*

Description: Best known and most common flycatcher of the islands. Gray above, white below, with dark mask and large, black bill. Tail is dark and notched. Usually quite conspicuous, sitting on tops of shrubs, palms, roof tops, and telephone wires. **Similar species:** Tropical Mockingbird is probably most easily confused but look for that species' longer, tapered (not notched) tail with white corners; lack of dark mask, and yellow (not dark) eye. Tropical Mockingbird has a much smaller and narrower-based bill and has prominent wingbars and rather long legs. An immature Fork-tailed Flycatcher with a worn tail could look similar but note dark half-hood. Eastern Kingbird would be all dark above with squared-off, white-tipped tail. Tropical Kingbird would be yellow underneath. **Voice:** Its loud, exuberant *pee-cherry* or *pit-irree* calls are frequently given and become the defining background noise in areas where common. The local Papiementio name *pimpiri* is presumably derived from its call. **Status:** Generally a common year-round resident on all three islands; historically was much more scarce and irregular on Aruba, for unknown reasons. Quite common and ubiquitous on Curaçao and Bonaire, throughout the year, and clearly breeding. Most abundant in fall and spring, with flocks ranging from 10 to occasionally several hundred birds (especially on Bonaire and Curaçao); these are thought to be migrants moving to and from breeding populations to the North (Voous 1983). On Aruba, from our experience, the species was more irregular and local (occurring at Spanish Lagoon and near the airport, for example), but in recent years the species seems to be more common and widespread. **Range:** Breeds from Gulf Coast of Alabama and Mississippi through Florida, and south through Caribbean, to northern South America. Birds in northern part of range migrate south in winter, occupying an area that includes Panama and northern South America.

Fork-tailed Flycatcher (*Tyrannus savana*)   **Plate 40**

Papiamento: *Tirano rabo fòrki*, *Pimpiri rab'i souchi*
Dutch: *Vorkstaartkoningstiran*

Description: A distinctive flycatcher with a long, forked tail, black half-hood extending down past eye, gray back, black wings, and white underside. In immature birds, black is replaced with dark brown and tail is shorter—sometimes much shorter—but always forked. Underwing is white. **Similar species:** Eastern Kingbird has a black half-hood but has a black (not contrasting gray) back; short, squared-off and white-tipped tail; darker gray underwing, and usually grayer band across upper breast. Although never recorded in the ABCs, Scissor-tailed Flycatcher does occur regularly south to Panama and has occurred in Colombia and Ecuador. Scissor-tailed Flycatcher also has a long, forked tail but note whitish or pearly gray head and pink underwing. **Voice:** Calls include a sharp, dry *tip* and *tip-t-t-t-tip-tr-tr-tr-trrrrrit*. Many calls are dry mechanical twittering or rattling sounding, somewhat like the sound made by running a finger over the teeth of a comb. **Status:** Voous (1983) writes that the species is "of irregular occurrence" and that it is "Most frequently seen in open scrub surrounding fresh water pools, where it congregates in vociferous quarreling flocks of 10-15 birds…" Austral migrants are most common in Venezuela, from March through Oct, and most records from the islands are in this time window, although there are records from throughout the year. Reports of the species were few from 1990 through about 2010, with 10 observations recorded on Bonaire between 2000 and 2010 (Ligon 2006), and perhaps 5 records on Aruba from 1987 to 2010 (2 noted in Prins et al. [2009]). There have been many more records on all three islands since 2010, March through Oct, usually involving 1 or 2 birds but sometimes flocks of up to 7 (fide eBird, personal communication). Sightings of this species in the islands continue to be worth documenting with photos and/or video. **Range:** Breeding range extends from southern Mexico to southern South America. Southern South American subspecies migrates north into northern South America during austral winter.

Family Vireonidae: Vireos

Yellow-throated Vireo (*Vireo flavifrons*)   Plate 45

Papiamento: *Vireo pecho hel, Vireo pechu hel*
Dutch: *Geelborstvireo*

Description: A small songbird, similar in size to Black-whiskered Vireo but larger than a Bananaquit. Has distinctive yellow "spectacles," yellow throat, white belly, and prominent white wingbars. Back and head are greenish. Eye dark. **Similar species:** Unlikely confused with any common resident or migrant species. Perhaps most similar to immature White-eyed Vireo (so far unrecorded in the ABCs), which also shows yellow "spectacles," white wingbars, and dark eye. White-eyed Vireo differs from Yellow-throated Vireo in having a white, not yellow, throat and yellow on flanks. **Voice:** Although song is unlikely to be heard, it is distinctive. Makes a series of slow burry phrases, like a very slow and rough-sounding Black-whiskered Vireo. More likely heard on the ABCs would be its nasal, rasping calls, if agitated: *shiff-shiff-shiff*. **Status:** Only a single documented record of a bird collected at Muizenberg, Curaçao, on March 21, 1957, now a specimen at the Zoological Museum Amsterdam (Prins et al. 2009). One sight record with some supporting written details of a bird at Spanish Lagoon, Aruba, on Nov 25, 2010 (L. Gardella fide eBird). Any sightings of this species should be carefully documented with photos and/or video. **Range:** Breeds across the eastern US, barely reaching southeastern Canada. Wintering range extends from southern Mexico to northern Colombia and Venezuela; also in Greater Antilles.

Philadelphia Vireo (*Vireo philadelphicus*) **Plate 45**

Papiamento: *Vireo Philadelphia*
Dutch: *Philadelphiavireo*

Description: A small songbird, only a bit larger than a Bananaquit, though generally much more sluggish (as are all vireos). Rather nondescript, with greenish upperparts, gray cap, white eye line, with some dark markings between eye and bill. Yellow below can be variable but often extends from throat to underside. Compared to warblers, bill short and thick. **Similar species:** At first sight, could be taken for a wintering warbler, but only Tennessee Warbler (a rare winter visitor to the islands) is similar. Tennessee Warbler shows a thinner, more pointed bill; and undertail coverts are paler than breast. Black-whiskered Vireo and Red-eyed Vireo are larger and lack extensive yellow underparts. Warbling Vireo and Brown-capped Vireo have never occurred in the ABCs but should be considered. Both are paler and lack dark markings between eye and bill, and neither show a yellow throat or contrasting gray cap. **Voice:** Likely to be silent on wintering grounds, but the bird we found on Aruba in 2002 gave scolding, whining, nasal calls in response to pishing. **Status:** Three records of the species, two from Aruba and one from Curaçao. First recorded from Malpais, Curaçao, on Apr 1, 2000 (Wells and Childs Wells 2001). We first obtained photographic documentation when we discovered a bird at Spanish Lagoon, Aruba, on Jan 13, 2002 (Wells and Childs Wells 2004). Another bird was found at Spanish Lagoon, on March 18, 2005 (Mlodinow 2006). **Range:** Breeds largely across Boreal forest region of Canada, to the northeastern edge of US. Winters from southern Mexico to Panama, with a few records from Colombia and a possible sighting in Ecuador.

Red-eyed Vireo (*Vireo olivaceus*) **Plate 45**

Papiamento: *Vireo wowo còrá, Vireo wowo kòrá*
Dutch: *Roodoogvireo*

Description: Well-known to North American birders, the Red-eyed Vireo closely resembles the Black-whiskered Vireo, which is a much more common bird in the ABCs. Appreciably larger than a Bananaquit, Red-eyed Vireo is olive-green above and white below, with variable amount of yellow on undertail coverts. A thin, black line separates gray cap and white eyebrow. As name suggests, eye is red. **Similar species:** Very similar to much more common Black-whiskered Vireo, but lacks the dark "whisker" marks (can be difficult to see) and generally more greenish above (not brown-toned). Black-whiskered often appears to have significantly longer, thicker bill. Yellow-green Vireo has not been recorded in the ABCs; very similar but much more extensively yellow underneath; head markings subdued; typically upperparts more yellow-toned. Given lack of clarity around range of variation in the various populations and subspecies of these vireos, photos and/or videos documenting any sightings are encouraged. **Voice:** Generally silent on wintering grounds, but makes scolding calls similar to those of Black-whiskered Vireo. Song, also unlikely to be heard in the islands, resembles that of Black-whiskered Vireo, but is more liquid, with longer, less snappy phrases. **Status:** Given that the species is a common migrant from North America to South America and that there are also abundant austral migrant and resident South American populations, it is somewhat surprising that there are not more documented occurrences in the islands. There has been an increased number of spring and fall reports from Aruba in recent years, with at least seven reports to eBird from Bubali and Spanish Lagoon, three with photographs showing birds lacking obvious whisker marks. Eight records from Curaçao: specimen taken at Santa Barbara, Oct 7, 1951; specimen taken at Hato, Nov 23, 1978; one seen at Seru Bosman, Christoffel NP, on Feb 28, 1979; two birds observed at Christoffel NP, Nov 23, 1991; two birds observed near the coast at Amphibian Cross, Nov 28, 1991 (all Prins et al. 2009); we observed two Red-eyed Vireos with other migrants at Malpais, on March 31, 2000 (Wells and Childs Wells 2001); a single bird was reported from Christoffel NP, Curaçao, on Nov 8, 2011 (M. Reid and S. Coffey fide eBird); two birds showing no whisker marks were photographed at Parke Publiko Sorsaka, near Jan Thiel on Jan 16, 2015 (J. Gerbracht and B. Van es fide eBird). Five records from Bonaire: two seen at Kralendijk, Oct 9–25, 1977 (Voous 1983); one seen at Kralendijk, Sept 17–18, 1979 (Voous 1983); one seen at Kralendijk, Oct 29, 1979 (Voous 1983); one seen at Put Bronswinkel, Oct 18, 1980 (Voous 1983); and one found dead at Hato, N. Kralendijk, on Sept 13, 2002 (Prins et al. 2009). Additionally, there is a photograph of one individual that was found dead on the deck of a cruise ship that arrived in Aruba on Oct 10, 2005; the ship had traveled overnight from Panama, so it is unclear where the bird met its demise (D. Mudge personal communication, photo on www.arubabirds.com). **Range:** Has a very large range, though defining its range precisely is impossible given that there may be one or more cryptic species. One population breeds across much of North America (absent in Southwest) and winters in South America. South American races occur throughout much of South America, south to Argentina; southern populations are austral migrants, moving to northern South America during the Southern Hemisphere winter.

Black-whiskered Vireo (*Vireo altiloquis*) Plate 45

Papiamento: *Vireo patiya preto, Vireo patia pretu*
Dutch: *Baardvireo*

Description: This is the vireo that you are most likely to see on the islands. Appreciably larger than a Bananaquit, Black-whiskered Vireo is olive-green (though typically duller than Red-eyed Vireo) above and white below, with variable amounts of yellow on undertail coverts. Like Red-eyed Vireo, its gray cap is separated from white eyebrow by a thin, black line; this line is often less distinct than in Red-eyed Vireo. Almost always shows a black "whisker" mark extending down from base of bill, but this can be faint or lacking in worn birds. Sides of face usually grayer than in Red-eyed Vireo, and eye more brown-tinged. Bill usually appears larger than in Red-eyed Vireo. Unlike Red-eyed Vireo, much shyer, often not responding readily to pishing, which can make it difficult to see. **Similar species:** Red-eyed Vireo lacks "whisker" and is usually brighter green above and more distinctly marked on head. Black-whiskered Vireo often has a generally duller appearance and larger bill. Note that the variation among subspecies and populations of these species—as as well as those of Red-eyed and Yellow-green vireos—calls for further study and documentation. **Voice:** Song similar to that of Red-eyed Vireo but quicker and snappier. It is a series of two- and three-note, whistled phrases, sometimes sounding like "sweet-chew, sweet-chew-chew, sweet-chew." In parts of the Caribbean, local names have been ascribed to these two- and three-note phrases, as in "Tom Kelley" and "Whip-Tom-Kelley" (Bond 1980); the same thing has happened in the Dominican Republic, where the song is described as: "julián chí-vi" (Latta et al. 2006). Call also similar to that of Red-eyed Vireo; makes a rather harsh, whining *yeerh*. **Status:** Although it is known to breed (or have bred) on all three islands, its breeding status is rather poorly understood. Whilie it is much more widespread and easy to find on Curaçao and Bonaire, it regularly occurs at Spanish Lagoon, on Aruba. In late Nov 2003, we found singing males at a number of locations on Curaçao, including four birds at Malpais, six at Christoffel NP, one at Westpunt, and two near Sint Joris. In Jan 1998, on Bonaire, we observed 1–2 at Pos Mangel, in Washington-Slagbaai NP. In late March 2001, on Bonaire, we made a number of sightings at Washington-Slagbaai NP: two near Salina Matijs; two at Pos Mangel; two at Put Bronswinkel; and two at Rooi; we also found one or two at Dos Pos and one at Fontein. In July 2001, on Bonaire, we observed two at Dos Pos, three at Pos Mangel in Washington-Slagbaai NP, and three at Put Bronswinkel, also in the park. Ligon (2006) reports that the species is regular on Bonaire at Dos Pos and Fontein; he documented a nesting pair at Peaceful Canyon, in late Nov-early Dec of 2001. On Aruba, at Spanish Lagoon and other locations on the island, we conducted birding visits every year for 10 years, from 1993 to 2003; our first observation of Black-whiskered Vireo was of two birds at Spanish Lagoon, on Jan 15, 2003; they responded to pishing and owl imitations and we suspect they were nesting. Various observers continue to regularly find birds at Spanish Lagoon; we documented three singing males there on Nov 26, 2013, for example. The only other location on Aruba where we have found Black-whiskered Vireo was on the grounds of the Divi Tamarijn Resort, near Oranjestad, on Nov 26, 2001, where we saw a single, apparent migrant bird that quickly moved on. Voous (1983) mentions that the species had occurred at Boca Prins, but we have not found it there nor are we aware of any recent reports from that area. **Range:** Breeding range extends from southern coastal Florida to Greater and Lesser Antilles, the ABCs, and islands off Venezuelan coast. Wintering range extends into northern South America.

Family Corvidae: Crows

House Crow (*Corvus spendens*)   **Plate 57**

Papiamento: *Cap doméstico, Kao doméstiko*
Dutch: *Huiskraai*

Description: An oddity, brought to the islands by ship. Should be readily identifiable, as there is nothing similar on the islands. A large, dark bird with gray nape and black cap and throat. Also note large, black, stout bill and strong, black legs. **Similar species:** Unlikely to be confused with any other species. Any sightings of the species should be documented with photos and/or video. **Voice:** Call is typical of crows: *caw-caw.*

Status: Having apparently arrived by ship, two adults were found on Curaçao near both Salina and Schottegat Harbor, from Apr through Aug 2002 (Prins et al. 2009). Many stories now confirm the ability of this species to survive on large ships (probably often fed by crew) and arrive in ports incredible distances from their original native range in Asia. **Range:** Native to southern Asia, from southern Iran to Thailand, but populations have become established at ports in many parts of the world.

Family Hirundinidae: Swallows, Martins

Southern Rough-winged Swallow (*Stelgidopteryx ruficollis*) Plate 41

Papiamento: *Swalchi di sùit, Souchi di sùit*
Dutch: *Zuid-Amerikaanse Ruwvleugelzwaluw*

Description: A small swallow with all-brown upperparts, pale rump (at least in northern populations), orangish throat, dusky brown upper breast, and white or yellowish-white lower breast and belly. Tail is squared. **Similar species:** Northern Rough-winged Swallow occurs regularly south to Panama; Voous (1983) claimed a possible sighting of three birds on Bonaire, Oct 1979, though Prins et al. (2009) include all sightings of "rough-winged swallows" as Southern Rough-winged Swallow. Northern Rough-winged Swallows should show a more brown-toned throat (though in immatures it can be buffy) and lack a light rump (though austral migrant Southern Rough-winged Swallows can also lack the light rump). Bank Swallow and Brown-chested Martin have brown backs, but both show a clearly demarcated brown breast band. Gray-breasted Martin and immature and female martins of various species can appear brownish backed with dusky throats and upper breasts, but are much larger (though size can be deceiving in flight). **Voice:** Makes a distinctive low, rough *brrrt*. **Status:** Six records listed by Prins et al. (2009) as Southern (rather than Northern) Rough-winged Swallow: three birds near Belnem south of Kralendijk, Bonaire, Oct 23, 1979 (Voous 1983; Voous himself thought these were Northern Rough-winged Swallows); two birds at Kralendijk, Bonaire, from Feb 15 to mid-March, 1986; about 35 birds at other locations on Bonaire, Feb and March, 1986; one to two birds at Bubali, Aruba, on Feb 21, 1993; one on Curaçao, at unspecified location, on Sept 16, 1997; and one bird at Klein Hofje, Curaçao, on Oct 29, 1998 (all Prins et al. 2009). More clarification is needed on the status of the two species in the ABCs. **Range:** Breeding range extends from Costa Rica south to northern Argentina.

Purple Martin (*Progne subis*) Plate 41

Papiamento: *Swalchi azul, Souchi grandi blou*
Dutch: *Purperzwaluw*

Description: A large swallow with a forked tail. Adult male is entirely dark blue, often appearing black. Females and immatures are dark blue above (browner in immature), with gray collar, gray forehead, and sooty-gray underneath. In flight, martins look longer-winged than the smaller swallows. **Similar species:** Although there is apparently a 1955 specimen of Purple Martin from Curaçao (Prins et al. 2009), other records are largely sight records (one bird photographed). The austral migrant Southern Martin is a distinct possibility in the islands, and males are virtually indistinguishable from male Purple Martins. A Southern Martin would be more likely than Purple Martin if seen between May and July, by which time Purple Martins would already be on their northern breeding grounds. Southern Martins have a more deeply forked tail; females and immatures are much darker underneath and lack the gray collar and light gray forehead. Gray-breasted Martin, although not yet recorded in the ABCs, is common throughout much of South America. Both male and female Gray-breasted Martin are lighter underneath than female and immature Purple Martin and lack the gray collar and light gray forehead. Caribbean Martin is known to occur with some regularity in the islands. Males are very similar to Purple Martin but show a sharply demarcated white belly. Female Caribbean Martin lacks the gray collar and light gray forehead of Purple Martin and shows a similar demarcation of the underparts as in males, with a snowy white lower belly but darker breast and throat. Cuban Martin has occurred on Curaçao (specimen records), and females and immatures may be indistinguishable in the field from female and immature Caribbean Martin. Male Cuban Martins appear largely indistinguishable in the field from Purple Martin (though they have more deeply forked tail), but in a preening bird, look for the concealed white belly feathers of male Cuban Martin. Sinaloa Martin could conceivably occur here as well. Female Sinaloa Martins are likely indistinguishable in the field from Caribbean Martin and Cuban Martin. Male Sinaloa Martin is like Caribbean Martin but with more extensive white underparts. Unfortunately most martin sightings in the islands are of flying birds. **Voice:** Makes a sweet, rich, gurgling *churr, churr, churr* and other notes. **Status:** Given difficulty of making a correct identification, it is not surprising that the status of this and other martin species in the ABCs is not entirely clear. An immature bird was photographed on Aruba on Sept 24, 1978 (Voous). There are apparently three records from Curaçao of males with completely dark underparts that were assumed not to be Southern Martin or Cuban Martin: one at Brakke Put, Sept 25, 1951 (Voous 1957); 12 or more over Willemstad, May 2–3, 1961; two (location unspecified) on May 3, 1962 (Voous 1983). There is one sight record of a female at the water purification plant on Curaçao, on Sept 27, 1998 (Prins et al. 2009). There is reportedly a specimen at the Zoological Museum Amsterdam, taken on Curaçao in 1955, though Voous (1983) reports only specimens of Caribbean and Cuban martins from there in 1955. There are two reports from Bonaire, both with unspecified numbers and location: May 5, 1961, and May 4, 1962 (Voous 1983). **Range:** Breeds across much of southern Canada and the US, to northern Mexico, though has more spotty distribution in the western US. Winters in South America.

Cuban Martin (*Progne cryptoleuca*)   Plate 41

Papiamento: *Swalchi grandi cubano, Souchi grandi cubano*
Dutch: *Cubaanse Purperzwaluw*

Description: Males are entirely dark blue, as in Purple Martin, though in the hand they show a narrow, usually hidden, strip of white in center of belly. Tail is somewhat more deeply forked than in Purple Martin, but with current state of knowledge, not reliably separable from Purple Martin in the field. Females and immatures are like Caribbean Martin, but again, current state of knowledge does not allow reliable separation in the field. **Similar species:** See Purple Martin. **Voice:** Similar to Purple Martin. **Status:** Three records from Curaçao, including three specimens now at the Zoological Museum in Amsterdam: an adult female and immature female from Malpais, on Sept 8, 1955; a male from Malpais, on Oct 6, 1955; and a female collected at Malpais, on Sept 30, 1956 (Voous 1983; Prins et al. 2009). **Range:** Breeds in Cuba. Wintering area unknown but thought to be in South America.

Caribbean Martin (*Progne dominicensis*)   Plate 41

Papiamento: *Swalchi grandi caribense, Souchi grandi caribense*
Dutch: *Caribische Purperzwaluw*

Description: A large, dark swallow with long wings and a forked tail. Males are entirely dark blue, with a sharply demarcated area of white extending from center of belly to undertail. Females and immatures are dark above with brownish throat and chest, sharply demarcated from white lower belly and undertail. **Similar species:** See Purple Martin. Male should be readily separable from the all-dark Purple, Cuban, and Southern martins by sharply demarcated white belly and undertail. Females and immatures should be separable from Purple Martin by the lack of gray collar and forehead, and from Southern Martin by dark chin and breast sharply demarcated from white belly and undertail. Male Sinaloa Martin should show a wider area of white in the belly, but females are not reliably separable based on current knowledge. Gray-breasted Martin very similar, but gray-brown throat and breast would not be sharply demarcated from white belly. **Voice:** Makes rich, gurgling notes similar to those of Purple Martin. **Status:** Apparently a regular but uncommon spring (Apr–May) and fall (Sept–Nov) migrant; there is also one winter record. We documented the first record for Aruba, photographing an adult male at the Divi Village Resort near Oranjestad, Jan 15–16, 2002 (Wells and Childs Wells 2005). Another adult male was seen at the Bucuti Beach Resort, Eagle Beach, Aruba, on Apr 3, 2004 (Prins et al. 2009). We saw a group of three rather distant and fast-flying martins along the coast near California Lighthouse on Apr 15, 2000; these birds showed a sharp demarcation between dark throat and chest and white belly, and were likely this species. Voous (1983) writes that there were four records from Bonaire, but lists only the details of the first record: an unspecified number of birds seen May 7, 1966 (Voous 1983). Four individuals were seen over the Plaza Hotel on Bonaire, on Oct 10, 2005 (Prins et al. 2009). The species has apparently been reported more frequently from Curaçao, where Voous (1983) mentioned six occurrences, including six specimens. He also mentions one notable occurrence of a flock of 35–40 that stayed at Malpais, Curaçao, Oct 6–13, 1955 (Voous 1983). Using telescopes, we carefully studied a flock of about 60 circling quite high over Malpais, Curaçao, on Nov 22, 2003. All of the individuals that we were able to clearly see showed a sharp demarcation between dark throat and chest and white belly and undertail. No birds were present when we visited Malpais on Nov 23, 2003. **Range:** Breeds in Greater Antilles (except Cuba) and Lesser Antilles, south to Trinidad. Some have been found wintering in the breeding range (at least in Barbados), but most vacate the breeding range and are thought to winter in South America, though wintering range is still basically unknown.

Brown-chested Martin (*Progne tapera*) Plate 41

Papiamento: *Swalchi pecho brùin, Souchi grandi pechu brùin*
Dutch: *Bruinborstzwaluw*

Description: A large swallow with long wings, a notched tail, and all-brown upperparts. Underparts white, with brown breast band that, in austral migrant subspecies, often extends as a series of brown spots down center of breast. Also note fluffy white undertail coverts that sometimes stick out from sides. **Similar species:** Probably most easily confused with Bank Swallow, which has same basic plumage pattern but is much smaller and does not show fluffy undertail coverts. Southern and Northern Rough-winged swallows are also brown above but lack a breast band. Female and immature martins of a number of species can look brownish above but all lack a breast band. Sightings should be documented with photos and/or video. **Voice:** Makes calls similar to those of Purple Martin and Caribbean Martin. **Status:** First photographically documented records were of up to four birds at Bubali, Aruba, May 6, 2016 (R. Wauben and M. Oversteegen fide eBird). Four sight records, all from Aruba: 25 birds near Oranjestad, Sept 1993; one bird at Frenchmen's Pass, Aug 1994; one bird at Bubali, Aug 19, 1988; and one bird at Pos Chiquito, Aug 29, 1998 (Prins et al. 2009). No records for Curaçao or Bonaire. **Range:** Occurs across much of South America, largely east of Andes Mountains. One population is resident north of Amazonia, while populations south of Amazonia include some austral migrants that move into Amazonia and northern South America during austral winter. Some years, migrants occur to Panama and Costa Rica; this species has also occurred in North America as a very rare vagrant.

White-winged Swallow (*Tachycineta albiventer*) Plate 42

Papiamento: *Swalchi barica blanco, Souchi barika blanku*
Dutch: *Witbuikzwaluw*

Description: Although common in nearby Venezuela, there is only a single record of this striking swallow from the ABCs. Dark green-blue above (immatures browner) with a white rump and white patches in inner wings. Also note white underside and a slightly notched tail. **Similar species:** No other swallow shows combination of white patch in wings and white rump. Mangrove Swallow, from Central America, and Chilean Swallow and White-rumped Swallow, from southern South America, all have white rumps but lack white in wings. Note that the white wing patches can be less obvious in White-winged Swallows in worn plumage. Tree Swallow has occured a number of times in northern South America, but it does not have a white rump or white in wings. **Voice:** Calls described by Hilty (2003) as "Soft, slightly buzzy *tweeĕd* in flight or perched." **Status:** One sight record of a single bird apparently seen multiple times over the freshwater pool at Malpais, Curaçao, in early summer, 1967 (Voous 1983). **Range:** Resident through much of South America, south through Brazil and to the extreme northern edge of Argentina.

Chilean Swallow (*Tachycineta meyeni*)   Plate 42

Papiamento: *Swalchi chileno, Souchi chileno*
Dutch: *Chileense Zwaluw*

Description: A small swallow. Adults dark blue-green above (brown in immature) and white below, with a white rump and slightly notched tail. **Similar species:** Tree Swallow lacks white rump. White-winged Swallow shows white in wings. Greatest difficulty would be separating this species from Mangrove Swallow or White-rumped Swallow. Mangrove Swallow has thin, white line extending from base of bill to top of eye. White-rumped Swallow has an even wider, and more obvious, white line from base of bill to top of eye. **Voice:** Makes gurgling churt calls, as do many other *Tachycineta* swallows; North American birders will recognize the similarity to the calls of Tree Swallows. **Status:** One sight record of two birds at Malpais, Curaçao, on May 15, 1977 (Voous 1983), though there are no published details on how the species was differentiated from Mangrove Swallow or White-rumped Swallow. **Range:** Breeds in southern Chile and Argentina. Southernmost populations migrate north in austral winter, normally as far north as Bolivia and Paraguay and the extreme southern edge of Brazil. There are several extralimital records from Peru, and this one from Curaçao, but no other records from northern South America.

Bank Swallow, Sand Martin (*Riparia riparia*)   Plate 41

Papiamento: *Swalchi ribera, Souchi ribera*
Dutch: *Oeverzwaluw*

Description: A small swallow with a brown back and a single, brown breast band across otherwise all-white underparts. Tail is squared or slightly notched. Rump often appears slightly lighter than back. **Similar species:** Barn Swallow has a long, forked tail and is bluish above, with rusty throat. Cliff Swallow shows a peach-colored rump patch and is dark blue above, with rusty throat. Northern and Southern Rough-winged swallows are also all brown above, and with a squared tail, but both show dusky or orangish throat and upper breast, not a sharply defined single brown breast band as in Bank Swallow. Brown-chested Martin is very similar in pattern but much larger, with breast band less well-defined, and long, white under-tail coverts that sometimes protrude from the sides when seen from above. **Voice:** Very distinctive call is a variety of buzzy rattles, but in migration may only give an occasional, single, rough *brrrt*. **Status:** On all three islands, a regular and sometimes fairly common fall migrant, typically Sept through Nov. Rare in spring, with only 4–5 recent records from Apr to May (Prins et al. 2009). **Range:** North American population breeds across most of Canada and northern half of US. Winters in South America, south to Chile and Argentina.

Barn Swallow (*Hirundo rustica*)   Plate 42

Papiamento: *Swalchi, Swalchi campesino, Souchi kampesino*
Dutch: *Boerenzwaluw*

Description: The most common swallow on the ABC islands. Blue above, with a rusty forehead and throat, buffy underside, and a very long, forked tail. **Similar species:** No other swallow shows the long, deeply forked tail. **Voice:** Not very vocal while on the islands but occasionally gives short, husky one- or two-note calls. **Status:** Most common swallow in the ABCs, especially in fall migration, when occasionally hundreds appear. Large die-offs of exhausted migrants have occurred in fall, one recorded on Aruba in fall 2013, when dead and exhausted birds were found throughout the island. Smaller numbers can be seen in winter and during spring migration, and occasionally a few nonbreeders stay over the summer. **Range:** North American population breeds across much of North America, from northern Canada to Mexico. Winters from Central America south to southern South America.

Cliff Swallow (*Petrochelidon pyrrhonota*)   Plate 42

Papiamento: *Swalchi baranca, Souchi baranka*
Dutch: *Amerikaanse Klifzwaluw*

Description: A stocky, square-tailed swallow, with a pale, peach-colored rump that contrasts with dark back; also note dark upperwing and tail. Adult of widespread eastern subspecies has white forehead, dark crown, and rust-colored sides of face, with rust extending onto sides of black throat. Pale collar often evident. Light underside, with some spotting on undertail coverts. Adults from southwestern US and Mexico have rusty forehead. Immature birds sometimes lack dark throat and have a dark forehead, but usually show dark sides of face. **Similar species:** Other light-rumped swallows lack rust color around head and face. Swallows with rust tones around face or throat do not have highly contrasting pale rumps. Southern Rough-winged Swallow does have a pale rump and is orangish on throat but is all brown above and lacks contrasting pale collar; it does not show a pale forehead. Cave Swallow has occurred and could be hard to distinguish. Adult Cave Swallows normally lack dark throat and always lack whitish forehead. Immature Cave Swallows may be very difficult to differentiate but have a pale throat and forehead, and paler sides of face than immature Cliff Swallow. **Voice:** Makes distinctive creaky calls unlikely to be heard on migration. Other calls rather similar to those of Barn Swallow, but huskier. **Status:** Now known to be a sometimes abundant fall migrant on Aruba (maximums of 50 and 100 birds reported to eBird) from Aug through Nov. Given its current status, there are surprisingly few historical records, with only 14 records listed across all three islands between 1983 and 2006 by Prins et al. (2009). Although likely also a sometimes abundant fall migrant on Bonaire and Curaçao, only five historical records from Bonaire and four from Curaçao. Bonaire: one found dead at Zuidweg, on Oct 22, 1984; three birds seen at Slagbaai, Oct 24, 1997; five birds seen near Sand Dollar Dive Shop, on outskirts of Kralendijk, Nov 18, 1997; five birds seen near Cargill Salt Company, Pekelmeer, Nov 6, 2002; and one bird on pier of Bonaire Dive and Adventure, on Feb 17, 2003 (all from Prins et al. 2009). Curaçao: two birds seen at Klein Hofje, Aug 22 and 24, 1992; one bird seen at Klein Hofje, Apr 17, 1994; one bird seen at Klein Hofje, Apr 20 and 23, 1996; one bird seen at Klein Hofje, May 14, 2000. **Range:** Breeds across much of North America, from Canada to Mexico. Winters in South America, south to northern Chile and Argentina.

Cave Swallow (*Petrochelidon fulva*)   **Plate 42**

Papiamento: *Swalchi cueba, Souchi kueba*
Dutch: *Holenswaluw*

Description: A stocky, square-tailed swallow. Shows rust-colored or peach forehead, usually lacks dark tones on throat. Sometimes shows rust on flanks. Immatures very similar to immature Cliff Swallow but usually show lighter sides of face. **Similar species:** See Cliff Swallow. **Status:** Very rare. One specimen of immature bird said to be of Mexican *pallida* subspecies taken on Oct 6, 1952 (Voous 1983). Specimen is in the Zoological Museum Amsterdam (Voous 1983; Prins et al. 2009). Three studied and one photographed on Bonaire (exact location not indicated) on Apr 26, 2014 (J. Ligon and E. Lenting fide eBird). **Range:** Breeds from Texas and New Mexico south to southern Texas (subspecies *pallida*); another subspecies (*fulva*) breeds from southern Florida through Greater Antilles. Northern populations in US, Cuba, and northern Mexico migrate south, but it is unclear if some birds from other populations also migrate south, and unclear where they winter. The limits of the wintering range are poorly described, but there are a number of records from Costa Rica and Panama; none from the South American continent.

Family Muscicapidae: Old World Flycatchers

Northern Wheatear (*Oenanthe oenanthe*)   **Plate 43**

Papiamento: *Chuchubi ala preto, Chuchubi ala pretu*
Dutch: *Tapuit*

Description: A small, active, ground-loving bird with long, black legs and short tail; it frequently bobs up and down. Although males in breeding plumage are a striking black and gray, the rare visitor to the ABCs would likely be in immature, female, or nonbreeding male plumage, which is brownish on upperparts and cinnamon on throat and belly. Key fieldmark in any plumage is white rump and upside down black T at tail tip. **Similar species:** Tail pattern is distinct from other bird species likely to occur in the islands. This is a very rare species, so any sightings should be carefully documented with photos and/or video. **Voice:** Calls include "a hard 'chak' and a sharp, whistling 'wheet'" (Beaman and Madge 1998). **Status:** Two records. One seen at Malpais, Curaçao, on Nov 4, 1962; it "repeatedly perched on top of mist net, but carefully avoided being caught and banded" (Voous 1983). One photographed at Amboina Plantation, Bonaire, on Dec 18, 1975 (Voous 1983). **Range:** Breeds across northern Europe, Asia, and Greenland. North American populations breed in Alaska and Yukon, and in northern Quebec and Labrador (one breeding record for insular Newfoundland). Western North American breeders apparently winter in eastern Africa, while those from eastern North America winter in western Africa (Bairlein et al. 2012). Occurs as a rarity in southern Canada and US, especially along Atlantic and Pacific coasts, but with scattered records across the interior.

Family Turdidae: Thrushes

Veery (*Catharus fuscescens*) **Plate 43**

Papiamento: *Chuchubi garganta hel*
Dutch: *Veery*

Description: A medium-sized thrush, slightly smaller than Tropical Mockingbird, and with a shorter tail. Upperparts entirely rust colored; limited spotting on upper breast; no obvious eye-ring; white belly and grayish flanks. Typically seen on or near the ground, and very shy. **Similar species:** Wood Thrush is larger, with large black spots underneath and more intense reddish-orange on head, contrasting with duller rusty-brown on rest of upperparts. Swainson's and Gray-cheeked thrushes usually show duller brown upperparts (not rusty) and more intense spotting on upper breast. Swainson's Thrush shows obvious yellowish "spectacles," while Gray-cheeked Thrush has plain gray face. **Voice:** Very distinctive song is a descending series of resonating *veer* notes, often described as a descending spiral of notes. Typical call note is a sharp, plaintive *veeer*. **Status:** Fall (Oct–Nov) and spring (Apr–May) migrant. Eleven records, including two from Aruba, four from Curaçao, and six from Bonaire. Aruba: one seen at Savaneta pools, Apr 8, 2013 (S. Mlodinow fide eBird); one heard singing at Spanish Lagoon, Apr 30, 2014 (K. Hansen and R. Merrill fide eBird). Bonaire: one at Put Bronswinkel, in Washington-Slagbaai NP, Oct 22, 1975; one at Fontein, Nov 1, 1979; one at Pos Mangel, in Washington-Slagbaai NP, Oct 24, 1997; one at Hato, Oct 1978; and one in Washington-Slagbaai NP (exact location not specified), Oct 12, 2005 (all Prins et al. 2009). Curaçao: one found dead from crashing into a window, at Rio Canario, Oct 14, 1954; individual birds on Apr 25–May 1, 1960, at Julianadorp (one found dead from crashing into a window); one banded on May 6, 1963, at an unspecified location (all Voous 1983). **Range:** Breeds across southern Canada and northern US. Winters in South America, south to Bolivia and southwestern Brazil.

Gray-cheeked Thrush (*Catharus minimus*)   **Plate 43**

Papiamento: *Chuchubi garganta pintá*
Dutch: *Grijswangdwerglijster*

Description: A medium-sized thrush, slightly smaller than Tropical Mockingbird, and with a shorter tail. Upperparts dull brownish-gray; extensive spotting on upper breast; sides of face gray, with indistinct "spectacles" around eye. Typically seen on or near the ground, and very shy. **Similar species:** Veery and Wood Thrush show strong rust or rufous tones to upperside. Swainson's Thrush is very similar but has obvious yellowish "spectacles" and more buffy sides of face. **Voice:** Typical call is a nasal, rising, then falling *wheere*, higher and more nasal than calls of Veery. **Status:** Fall (Oct–Nov) and spring (Apr–May) migrant. Eight records, including two from Curaçao (both specimens) and six from Bonaire; none currently from Aruba. <u>Bonaire</u>: one at Fontein, Oct 16–17 and 22, 1976 (Prins et al. 2009); one at Fontein, Nov 1, 1979, seen with a Veery and Swainson's Thrush "after a night of heavy passage of thrushes in which Swainson's Thrush and Veery had taken part" (Voous 1983); one at Pos Mangel in Washington-Slagbaai NP, Apr 3, 1985; one at Pos Mangel, Washington-Slagbaai NP, Oct 24, 1997; three at Put Bronswinkel, Washington-Slagbaai NP, on Oct 14, 2000; and one at Put Bronswinkel, Washington-Slagbaai NP, Oct 12, 2005 (all Prins et al. 2009). <u>Curaçao</u>: one specimen from Malpais, May 11, 1957; one found dead after hitting a window, at Julianadorp, Apr 25, 1960 (Voous 1983). **Range:** Breeds across northern Canada and Alaska; winters in northern South America, south to northern Peru and northern Brazil.

Swainson's Thrush (*Catharus ustulatus*)   **Plate 43**

Papiamento: *Chuchubi wowo rant blanco, Chuchubi wowo rant blanku*
Dutch: *Dwerglijster*

Description: A medium-sized thrush, slightly smaller than Tropical Mockingbird, and with a shorter tail. Upperparts brownish-gray; extensive spotting on upper breast; bold yellow "spectacles" and sides of face buffy. Typically seen on or near the ground, and very shy. **Similar species:** Veery and Wood Thrush show strong rust or rufous tones to upperside. Gray-cheeked Thrush is very similar but lacks the obvious yellowish "spectacles" of Swainson's Thrush and gray sides of face. **Voice:** Typical call includes liquid *whit* or *pip* notes, very different from calls of Gray-cheeked Thrush or Veery. Swainson's Thrush flight call, given during migration, is a distinctive *heep*. **Status:** Fall (Oct–Nov) and spring (Apr–May) migrant. Nine records, including two from Curaçao (both specimens) and six from Bonaire; one currently from Aruba. <u>Aruba</u>: one photographed and initially identified as a Hermit Thrush on Oct 15, 2015, at Divi Links (R. Wauben). <u>Bonaire</u>: one at Fontein, Apr 10–19, 1977; one at Dos Pos, Apr 13, 1977; one at Slagbaai, Oct 28, 1979; one at Fontein, Nov 1, 1979; one at Put Bronswinkel, in Washington-Slagbaai NP, Oct 18, 1980; and one in Kralendijk, Oct 21, 1982 (all Voous 1983; Prins et al.

2009). <u>Curaçao</u>: one specimen on Oct 23, 1951, from Grote Knip, where bird was feeding in "dense manchioneel thicket" (Voous 1983); one found dead in Willemstad, Apr 23, 1960, "during strong thrush-migration" (Voous 1983). **Range:** Breeds across Boreal forest region of Canada and Alaska, and to northeastern US; also in mountains in western US, New Mexico, and California. Winters in eastern South America, along the Andes south to Bolivia and northern Argentina.

Wood Thrush (*Hylocichla mustelina*)   Plate 43

Papiamento: *Chuchubi barica pintá, Chuchubi barika pintá*
Dutch: *Amerikaanse Boslijster*

Description: A medium-sized thrush, slightly smaller than Tropical Mockingbird, and with a shorter tail; plumper and a bit larger than the *Catharus* thrushes. Reddish-brown above, but head and nape brighter and orangish, contrasting with duller back and wings. Big black spots on throat and breast extend down to belly. Prominent white eye-ring. Typically seen on or near the ground, and very shy. **Similar species:** Both Veery and Wood Thrush show strong rust or rufous tones to upperside, but on Veery head and back show uniform coloration, while on Wood Thrush color of head contrasts with color of back; also, spotting on Veery is confined to upper breast. Swainson's Thrush and Gray-cheeked Thrush are brown above, lacking any rusty tones; and spots below are smaller and do not extend as far down. **Voice:** Distinctive call has been likened to the sound of a machine gun: *pit-pit-pit.* **Status:** One record of a specimen (now at the Zoological Museum Amsterdam) from Klein Sint Joris, Curaçao, on Oct 25, 1951 (Voous 1983; Prins et al. 2009). **Range:** Breeds across eastern US; winters from southern Mexico to Panama.

Family Mimidae: Mockingbirds, Thrashers

Pearly-eyed Thrasher (*Margarops fuscatus*)   Plate 44

Papiamento: *Palabrua Boko Duru, Chuchubi spañó, Chuchubi Wowo Blanku*
Dutch: *Witoogspotlijster*

Description: Pearly-eyed Thrasher is a bit larger than a Tropical Mockingbird. Brownish upperparts, with no wingbars; streaked underparts, white-tipped outer tail feathers; stout, pale bill; white eye; and pink legs. Usually a shy and retiring bird. **Similar species:** Tropical Mockingbird is lighter gray, with wingbars, un-streaked underside, and black bill and legs. Immature Tropical Mockingbird is streaked underneath but has dark eyes, black bill, black legs, and two white wing-bars. **Voice:** Ligon (2006) compares the song to a violinist tuning up, with a series of short phrases of one to three notes with long pauses in-between. Also make harsh, often double-noted, calls. **Status:** Resident on Bonaire only, where it seems to be regular at a number of places, including water holes in Washington-Slagbaai NP, Dos Pos, and Fontein, with usually only one or two birds seen at any one time.

Voous (1983) lists a sight record from Rio Canario, Curaçao, but without details. An individual was photographed near Westpunt, Curaçao (Kura Hulanda resort), on July 20, 2015, and was reported intermittently through at least June 2016 (Marvin Thodé, Rob Wellens). **Range:** Has a very limited range, occurring as a resident in the Bahamas, Puerto Rico, Virgin Islands, Lesser Antilles, and Bonaire.

Brown Thrasher (*Toxostoma rufum*)   Plate 44

Papiamento: *Chuchubi barica marcá, Chuchubi barika marká*
Dutch: Rosse *Spotlijster*

Description: A distinctive bird, rust-brown above, with two white wingbars. Heavily streaked below. Similar in size to a Tropical Mockingbird. **Similar species:** There are other thrashers that occur in the southwestern US and Mexico but none are likely to occur in the ABCs. **Status:** One specimen taken on Oct 2, 1957, at the oil refinery on Curaçao (Voous 1983). **Range:** Breeds from southern Canada across eastern US, south to Gulf States. Winters in southeastern US.

Tropical Mockingbird (*Mimus gilvus*)   Plate 44

Papiamento: *Chuchubi*
Dutch: *Tropische Spotlijster*

Description: One of the most common, well-known, and conspicuous birds of the ABCs, found on hotel grounds, backyard gardens, in scrubby undeveloped areas, and other places. Bird enthusiasts from many places in the US will recognize it as a close cousin to the familiar Northern Mockingbird. A medium-sized bird with gray upperparts, two white wingbars, and white eyebrow; stout, black bill; long tail with large patches of white in corners; and pale underside. Adults have a pale yellow eye (dark in immatures). Some individuals (perhaps immature birds that have not fully moulted into adult plumage) show streaking along lower flanks. Often sits at top of buildings, trees, or other exposed perches, from which it sings loudly, especially at first light of dawn. At other times may be seen hopping around in open areas, looking for food; will come readily to fruit and juices left out for it. **Similar species:** Gray Kingbird is perhaps most similar in size, appearance, and habit of sitting on exposed perches, but it lacks white wingbars, white in tail, and white eyebrow. It also has a broader base to its bill; a shorter tail; and usually shows a dark mask (not line) through eye. On Bonaire, Pearly-eyed Thrasher could be confused with Tropical Mockingbird but note the thrasher's pale (not all-black) bill and streaked underparts. **Voice:** Its song is a series of repeated phrases, of varying length, with a loud musical quality. Apparently the Papiamento name *chuchubi* is derived from its song (Voous 1983). Unlike Northern Mockingbird, does not obviously mimic. Calls include a rather harsh-sounding *tsheck*, similar to that of Northern Mockingbird. **Status:** Common on all three islands. **Range:** One resident population ranges from southern Mexico to Honduras. Introduced into Panama, where now resident and perhaps the origin of increasing numbers in Costa Rica (Garrigues and Dean 2007; Angehr and Dean 2010). Also resident in Lesser Antilles and northern South America, including Trinidad and other islands of northern coast of South America. Another population has a disjunct range extending along coastal Brazil, from near mouth of Amazon south to Rio de Janiero (Brewer 2001).

Family Sturnidae: Starlings

European Starling (*Sturnus vulgaris*)   **Plate 44**

Papiamento: *Chuchubi oropeo*
Dutch: *Spreeuw*

Description: A dark, chunky bird with a short tail. It will be familiar to most North American and European birders. Adults in nonbreeding plumage have a black bill and show white freckling on black body. In breeding plumage, adults have a yellow bill and lose white freckling. Juveniles are drab gray and have a black bill. Immature birds sometimes have a gray head but adultlike plumage on remainder of body. **Similar species:** Caribbean Grackle and Shiny Cowbird could be mistaken for a European Starling, but note that those species have longer tails and lack any white spotting. Given the rarity of European Starling in the ABCs, any sighting should be documented with photos and/or video. **Status:** Three records, all from winter period. One sight record, presumably of a single bird, from Aruba, Nov 3, 1960 (Voous 1983). Two birds in nonbreeding plumage were photographed at Bubali, Aruba, from Nov 18, 1977, to Jan 2, 1978 (Voous 1983). A single bird was photographed near a Ruddy Turnstone along the rocky shore of Bonaire, between Playa Funchi and Wayaca, Nov 10, 1980 (Voous 1983). **Range:** Originally a species of the Old World, but now introduced in many parts of the world and resident across North America, Greater Antilles, and Puerto Rico. Only established South American population is in Argentina (Ridgely and Tudor 2009).

Family Bombycillidae: Waxwings

Cedar Waxwing (*Bombycilla cedrorum*)   **Plate 44**

Papiamento: *Parha di cedro, Para di seder*
Dutch: *Cederpestvogel*

Description: A small, tan-and-gray songbird with distinctive crest, black mask, and yellow tail tip. In normal range, travels in flocks, but the few sightings in South America have been of single individuals. **Similar species:** Unlikely to be confused with any regularly occurring species in the ABCs. **Voice:** Makes a high-pitched, hissing *seeeeeee*. **Status:** One record of a male found dead (specimen at Zoological Museum Amsterdam) in Oranjestad, Aruba, on Feb 22, 1979 (Voous 1983; Prins et al. 2009). **Range:** Breeds across most of southern Canada and northern US.

Winters from northern US (sometimes extreme southern Canada) to Cuba (rarely Hispaniola and other Caribbean islands) and Central America; and rarely to Colombia. One record from Venezuela.

Family Parulidae: New World Warblers

Ovenbird (*Seiurus aurocapilla*)   Plate 49

Papiamento: *Chipe di fòrno, Chipe di fòrnu*
Dutch: *Ovenvogel*

Description: Similar in size to Bananaquit. This secretive warbler prefers to stay low in thick mangroves, so it is often overlooked. Unmarked brownish-olive upperparts; bold, white eye-ring, and orange cap with black borders. Underside with bold streaking. When foraging, prefers to walk on or near the ground. **Similar species:** Most similar in appearance and habits to Northern Waterthrush and Louisiana Waterthrush, but both of those species show a conspicuous white eyebrow stripe rather than a bold, white eye-ring, and neither has an orange cap. Without reference to size, some observers may confuse Ovenbird with a thrush, but note the smaller size of Ovenbird, orange cap, and lack of warm rusty tones anywhere in the plumage. **Voice:** Dunn and Garrett (1997) describe call as "a loud and rather sharp *tsick, chut,* or *tsuck*." They point out that "the call is not quite as sharp and metallic as the *chink* call of the waterthrushes or of the Hooded Warbler," species that share similar wintering habitat in the ABCs. The song, very unlikely on wintering grounds, is an explosive, loud *TEACH-er-TEACH-er-TEACH-er,* very familiar to birders from eastern North America. **Status:** Regular but uncommon migrant and winter resident, Sept–May, usually in mangroves. Prins et al. (2009) list nine records for Aruba. Additionally, we have recorded the species nine times on Aruba between 1995 and 2011, Nov–Apr, with all records from Bubali and Spanish Lagoon. On four occasions, we have encountered at least two individuals together, though most of our records pertain to single birds. There are 17 records for Bonaire (Prins et al. 2009, J. Ligon fide eBird) and 10 records for Curaçao, including our Apr 2000 record from Malpais (Prins et al. 2009). **Range:** Breeds from southern Boreal forest region of Canada south to southeastern US; breeds patchily in interior of the US, south to Colorado. Winters from southern Florida south through Greater Antilles and Mexico, to Panama. Only seven records from Venezuela noted in Hilty (2003).

Worm-eating Warbler (*Helmitheros vermivorum*)   Plate 50

Papiamento: *Chipe comedor di bichi, Chipe komedó di bichi*
Dutch: *Streepkopzanger*

Description: A relatively large warbler with a large, pale bill. When seen clearly, very distinctive, with four prominent black stripes on head—one through each eye and two on the crown. Upperparts are an unmarked olive-green, underparts tan or buffy. **Similar species:** No other species shows combination of four black head stripes, unmarked tan underside, and plain, unmarked olive-green back, wings, and tail. **Voice:** Often gives its double, buzzy *zeep-zeep* flight call on the ground; also makes a flat chip call (Dunn and Garrett 1997; Sibley 2000; Stephenson and Whittle 2013). Song, unlikely to be heard on the islands, is a dry trill. **Status:** Two records from Aruba, but probably of same bird. We videotaped one at Spanish Lagoon, Nov 23, 2012, and Jan Wierda saw one at the same location on Feb 12, 2013 (details reported to us at www.arubabirds.com). We had a brief glimpse of what was almost certainly a Worm-eating Warbler at Spanish Lagoon, on Jan 13, 2003, but we could not refind the bird for photographic or video confirmation. Three records from Bonaire: one near Punt Vierkunt, Oct 23, 1979; one at Fontein, Nov 2, 1989 (Prins et al. 2009); one at Pos Mangel, Washington-Slagbaai NP, Jan 10, 2011 (J. Heller fide eBird). Two Bonaire records described in Prins et al. (2009) as being adults, although this would be very difficult to ascertain except by specimen or banding records. No records from Curaçao. **Range:** Breeds in eastern US north to southern Wisconsin, Indiana, New York, and Massachusetts. Winters in Greater Antilles and from southern Mexico south to Panama. Two records from Venezuela (Hilty 2003), one from mainland Colombia (McMullan et al. 2010).

Louisiana Waterthrush (*Parkesia motacilla*)   Plate 50

Papiamento: *Chipe di suela surenō*
Dutch: *Louisianwaterlijster*

Description: A shy, secretive species that occurs near water, most commonly in mangroves or thick brush. Habitually bobs its tail up and down. All brown above; light eye line widens behind eye, often buffier in front of eye and whiter behind. Throat usually unstreaked. Legs, at least in spring, usually appear brighter pink than those of Northern Waterthrush. **Similar species:** Northern Waterthrush eye line narrows behind eye and is usually buffy (not white) throughout. Throat is usually unstreaked in Louisiana and streaked in Northern. Northern Waterthrush has a bill that is thinner and shorter than that of Louisiana; also shows more streaking below and less contrast between flanks and breast. Louisiana bobs its tail up and down more vigorously than does Northern. Note that there is just a single specimen record of Louisiana Waterthrush from the islands (Bonaire), and only one published photograph or video that we are aware of. **Voice:** Call similar to that of Northern but not as sharp or metallic. Song, unlikely to be heard in the islands, is very different from Northern. It consists of three or four clear, slurred notes followed by a jumble of twittering notes. **Status:** A bit unclear given difficulties in identification and secretive nature of the species. There is a single specimen record of a bird found dead in Kralendijk, Bonaire, Oct 3, 1975. As far as we can determine, most other reports have been sight records without supporting photographic or video documentation. One photographed at Divi Links, Aruba, on Oct 17, 2015 (R. Wauben fide eBird), seems to show some characteristics consistent with Louisiana Waterthrush. Prins et al. (2009) report three sight records from Aruba, two from Curaçao, and four from Bonaire. <u>Aruba</u>: one at Sero Colorado, Oct 7,

1956; one at Bubali, March 29, 2004; and one at Bubali, March 12, 2005. Bonaire: one at Dos Pos, Oct 4, 1976; one at Fontein, Dec 28, 1976; one at an unspecified location, Nov 3, 1979; and at an unspecified location, March 1, 2000. Curaçao: one at Willemstad, Feb 8, 1970; and one at Klein Hofje, Dec 29, 1991. **Range:** Breeds across eastern US. Winters in Greater Antilles and from Mexico south to Panama, rarely to northern Colombia.

Northern Waterthrush (*Parkesia noveboracensis*) Plate 50

Papiamento: *Chipe di suela nortenõ*
Dutch: *Noordse Waterlijster*

Description: This shy and secretive warbler occurs on the islands almost exclusively near water, usually in mangroves or buttonbush. It is well-known for its habit of constantly bobbing its tail. All brown above, with an obvious light eye stripe (yellowish or white but uniformly colored throughout) that tapers to a point behind eye. Underparts fairly uniformly yellowish-white or white, with dense blackish-brown streaking. Throat usually streaked or spotted. **Similar species:** Louisiana Waterthrush is very similar in appearance and can be difficult to distinguish without experience. It has an eye stripe that widens behind the eye, a slightly longer and thicker bill, usually an unstreaked throat, and typically shows a white breast that contrasts with yellowish-buffy flanks. Eye stripe is often buffy in front of eye and white behind eye, but it can be difficult to see. Ovenbirds are found in same habitats and show similar behavior but have an eye-ring, not an eye stripe, as in the waterthrushes. Northern Waterthrush, the most abundant overwintering migrant warbler in the islands, is much more likely than Louisiana Waterthrush. There is only a single specimen record of Louisiana Waterthrush from the islands (Bonaire) and only one published photograph or video that we are aware of. **Voice:** Call, a harsh *chink*, is often heard emanating from mangroves, especially if one pishes or the birds are startled. Song, unlikely to be heard in the islands is a loud, ringing *chip-chip-chip-sweet-sweet-sweet-chew-chew-chew*, with many variations. **Status:** Common wintering migrant, Aug–May, in mangroves and other thick, brushy areas, almost always near water. Voous (1983) notes without detail that occasional birds have been found in the ABCs in the northern summer (June–July) as well. **Range:** Breeds across Boreal forest region of Alaska and Canada, south into northern US. Winters from Mexico and Greater Antilles south into northern South America.

Golden-winged Warbler (*Vermivora chrysoptera*) Plate 50

Papiamento: *Chipe ala di oro*
Dutch: *Geelvleugelzanger*

Description: A small songbird (similar in size to Bananaquit) with gray upperparts and gold wing patch; has whitish underparts. Note striking head pattern: dark mask and bib and gold crown contrast with white eyebrow and white moustache stripe. **Similar species:** Not easily confused with any other species. **Voice:**

Call is a quick, dry *schik* or *chip*. While not likely to be heard on the islands, song is a high-pitched, buzzy *zee-bee-bee-bee*. **Status:** Bonaire has a single record of a sub-adult, possibly a first winter male, seen Oct 19, 1983, at Trans World Radio Building. No records from Aruba or Curaçao (Prins et al. 2009). **Range:** Winters from southern Mexico to central Colombia and northern Venezuela. Breeds in northern and eastern US and southern Canada.

Blue-Winged Warbler (*Vermivora cyanoptera*) Plate 50

Papiamento: *Chipe ala blou*
Dutch: *Blauwvleugelzanger*

Description: A small (similar in size to Bananaquit), brightly colored songbird with yellow head and underparts; black line through eye; short, bluish-gray tail and wings, and two white wingbars. **Similar species:** Although there are several yellow-colored birds on the islands, black line through the eye makes Blue-winged Warbler distinctive. **Voice:** Call is a quick, dry *schik* or *chip*, indistinguishable from call of Golden-winged Warbler. Song, unlikely to be heard on the islands, is a two-part buzzy *bee–buzzzz*. **Status:** One record from Aruba of a single bird seen at Palm Beach, on Nov 15, 1991 (Prins et al. 2009). No records from Bonaire or Curaçao (Prins et al. 2009). **Range:** Winter range is southern Mexico to Panama. Single records from Colombia and Venezuela (Ridgely and Tudor 2009). Breeds in eastern US north to southernmost tip of Ontario, Canada.

Black-and-white Warbler (*Mniotilta varia*) Plate 49

Papiamento: *Chipe trepador, Chipe kabes abou*
Dutch: *Bonte Zanger*

Description: A small songbird, similar in size to Bananaquit. Striped black-and-white "zebra" pattern of this warbler is distinctive. White eyebrow; white central crown stripe; black streaking on breast, sides, and undertail coverts; and black legs. Adult males have black throat. Creeps up and down tree trunks and limbs, unlike any other warbler. **Similar species:** Very distinctive. Breeding adult male Blackpoll Warbler lacks white eyebrow, white central crown stripe, streaking on undertail coverts, and all-black legs. In Blackpoll Warbler, bill is shorter. **Voice:** Call described by Dunn and Garrett (1997) as "a dull *chip* or *tik*." Song, unlikely to be heard on the islands, is a repeated series of high, two-syllable notes: *weesa-weesa-weesa-weesa*. **Status:** An uncommon visitor, Sept through Apr; often seen in mangrove habitats, usually a single bird but occasionally two to three. When Voous published his important work on the birds of the ABCs (1983), he noted 3 records from Aruba, 14 from Bonaire, and 9 from Curaçao. Prins et al. (2009) list three additional records from Aruba and one each from Bonaire and Curaçao. We found the species six times on Aruba over a 13 year period (twice at Bubali, three times at Spanish Lagoon, once in mangroves near Parkientenbos), Nov through Apr. Additionally, there are at least five records for Aruba that have been reported to our website (www.arubabirds.com), some with photos, since 2003. Finally, there are at least 17 additional records reported to eBird since 2009. This totals to a minimum of 34 records that we are aware of from Aruba, 31 of them since 1983. We observed the species once on Curaçao, Christoffel NP, on Nov 24, 2003. **Range:** Breeds across Boreal forest region of Canada, from Northwest Territories and northeastern British Columbia east to Newfoundland, and south across eastern US. Winters from US Gulf Coast south through Central America, to northwestern South America.

Prothonotary Warbler (*Protonotaria citrea*)   Plate 49

Papiamento: *Chipe protonotario*
Dutch: *Citroezanger*

Description: A very distinctive, large warbler, with a relatively thick, long bill; in spring, bill on adult male is entirely black. No other warbler has combination of bright yellow head and underparts with striking dark eye, greenish back, gray wings with no wingbars, and white undertail coverts. Females and immatures can show some duskiness on cap and nape. **Similar species:** Migrant male Yellow "Northern" Warbler (and females of resident Yellow "Golden" Warbler) also have yellow head and body but yellow extends through undertail coverts (white in Prothonotary) and tail. Yellow Warbler is also much smaller and does not show contrasting darker bluish-gray wings and tail. Immature Hooded Warbler might be confused with Prothonotary, but note that Hooded has yellow undertail coverts, darker olive neck and cap, longer tail, and lacks bluish-gray wings and tail. Rare Blue-winged Warbler shows two white wingbars and dark line through eye. **Voice:** Call note described by Dunn and Garrett (1997) as a "dry, loud *chip*" and by Sibley (2000) as a "clear, metallic squeak *tsiip*." Song, unlikely to be heard on the islands, is a loud, ringing *sweet, sweet, sweet,* given on a single pitch. **Status:** Regular but uncommon winter resident, Aug–May, especially in mangroves but occasionally seen in other habitats. Prins et al. (2009) tallied 12 records for Aruba. We documented 8 additional records from Aruba, between 1996 and 2011, 7 from Jan and 1 from Nov. Of our 8 Aruba records, 6 were of single individuals and 2 were of two birds together. Five of our 8 records were from Bubali, 2 from Spanish Lagoon and 1 from mangroves near Parkientenbos. In addition, the species has been documented at Spanish Lagoon, Feb 2012 and Feb 2013, by Jan Wierda (photo of one can be seen on our website www.arubabirds.com) and there are 5–6 additional records for Aruba reported to eBird since 2007. Prins et al. (2009) tallied 6 records from Curaçao. We have 1 additional record of a single bird at Muizenberg, Nov 23, 2003, and 2 additional records have been reported to eBird in 2011 and 2016. Prins et al. (2009) tallied 20 records for Bonaire. We have 1 additional record from Bonaire of a single bird seen at Lac Bay, March 24, 2001, and 6 additional records have been reported to eBird since 2009. **Range:** Breeds across southeastern US north to extreme southern Ontario, southern Minnesota, Wisconsin, Michigan, and New York. Winters from southern Mexico south to northern South America.

Tennessee Warbler (*Oreothlypis peregrina*) Plate 46

Papiamento: *Chipe tennessee*
Dutch: *Tennesseezanger*

Description: Small, drab songbird (similar in size to Bananaquit). Dull olive-green back. Adult male in spring has white underparts, gray head contrasting with brighter green back and wings, and white eyebrow. Adult females have yellowish throat and breast, yellowish eyebrow; may show very faint, thin wingbars and slightly contrasting grayish crown. Immature birds (probably most likely seen on the islands, especially in fall) similar to adult females but more extensively yellow underneath (though undertail coverts always lighter than rest of underparts); usually show more distinct yellowish wingbars and lack any gray tones on head. Bill is thin and pointy. Tail appears short. **Similar species:** Care should be given to distinguish this species from immature and molting Yellow "Golden" Warblers, which can appear in an astonishing variety of plumages; but even in worn, immature, or molting individuals, note that they are larger than Tennessee, do not show obvious pale eyebrow line, and have a thicker, blunter bill. Philadelphia Vireo is slightly larger, with stronger eye lines and thicker, less pointed bill. **Voice:** Call is a crisp *tsip* or *tit* (Dunn and Garrett 1997). Song, unlikely to be heard on the islands, is a loud, mechanical, three-part series, with each part progressively faster and louder: *sit-sit-sit, see-see-see, se-se-se-se-se-se.* **Status:** Rare. One sight record of a male at Spanish Lagoon, Aruba, on March 25, 2003 (Mlodinow 2004). A male specimen died after landing on an oil tanker traveling between Curaçao and Venezuela, on Nov 17, 1954 (Voous 1983). Four records from Bonaire: two at Fontein, Apr 15–19 and Nov 12, 1977; one at Gotomeer, Nov 3, 1982, and May 15, 2006 (Prins et al. 2009). Given the species is a regular winterer in northern South America, including Venezuela, it is somewhat surprising that there are not more records of this species in the ABCs. **Range:** Winters in southern Mexico to northern South America. Not uncommon in West Indies during nonbreeding season. Breeds in northern North America, largely in the Boreal forest region of Canada.

Connecticut Warbler (*Oporornis agilis*) Plate 51

Papiamento: *Chipe di Connecticut*
Dutch: *Connecticutzanger*

Description: Similar in size to Bananaquit. This rather secretive and skulking warbler is often seen walking on the ground, with a long-legged and long-toed appearance. Bulky with thick bill. Olive-green above with full hood that is gray in adults, brownish in immature birds. Full white (buffy in immatures) eye-ring is distinctive (fall birds occasionally show small break in rear part of the eye-ring). Yellow underside, including undertail coverts that are very long and extend almost to tip of tail. **Similar species:** The Mourning Warbler has a full hood but usually lacks eye-ring; in rare cases in which it does have an eye-ring, it is thin and partial. Immature Mourning Warbler usually shows yellowish tones extending from eye to above base of bill, while in immature Connecticut this area is usually not very contrasting. Mourning usually shows more contrasting yellow throat compared to browner throat of immature Connecticut. MacGillivray's Warber has not been recorded from the islands; it also has a full hood and yellow underparts but usually has bold white eye arcs above and below eye. **Voice:** According to Dunn and Garrett (1997), "The rarely heard call note is a loud, nasal *chimp* or *poitch*, lacking the raspy quality of the call of the Mourning Warbler." Flight call is a short, buzzy, *zeep* like that given by Yellow Warbler and Blackpoll Warbler. Although unlikely to be heard in migration, the song is a distinctive, loud series of four-syllable phrases that speed up at the end. Described by Sibley (2000) as *tup-a-teepo tup-a-teepo tupateepotupateepo*. **Status:** Connecticut Warbler is one of the few warblers that migrates both spring and fall across the Caribbean, to and from its South American wintering grounds. There are a large number of records from the islands (36), though 23 are from Bonaire (Prins et al. 2009). All records to date recorded Sept–Nov, with the overwhelming majority from Oct (Voous 1983; Prins et al. 2009). Five records for Aruba but only one since 1983 (Prins et al. 2009), that of a bird photographed at Bubali on Sep 27, 2016 (M. Oversteegen fide eBird). Eight records from Curaçao but only two since 1983, the most recent from Nov 1998 (Prins et al. 2009). Five of the eight records pertain to specimens and three of the records are from Malpais (Prins et al. 2009); several of these may pertain to birds captured in mist nets. Five records from Bonaire since 1983, all in Oct, most recent from 1997. Given the lack of recent records, it is worth carefully documenting with photos and/or video any sightings of the species. **Range:** Breeding range largely confined to southern portion of Boreal forest region of Canada, from British Columbia east to central Ontario and, sparingly, east to Quebec and south to northern Minnesota, Wisconsin, and Michigan. Wintering range poorly delineated but apparently Venezuela and eastern Colombia south to portion of northern Brazilian Amazon.

Mourning Warbler (*Geothlypis philadelphia*)   **Plate 51**

Papiamento: *Chipe di luto*
Dutch: *Grijskopzanger*

Description: A small warbler, similar in size to a Bananaquit. Usually secretive and skulking; stays low to ground, hopping rather than walking. Olive-green above; adults with a full gray hood; no eye-ring, except in very rare individuals, and even then never has a complete eye-ring. Adult male has black bib and dark face; adult male and female usually show noticeably darker lores than immature birds. Immature birds have a yellowish throat and yellow area that extends from eye to above base of bill; may show a very narrow partial eye-ring, broken in back and front of eye. **Similar species:** See Connecticut Warbler. Since there are no confirmed records on the islands for Mourning Warbler, any possible sightings should be documented with photos and/or video. **Voice:** Sibley (2000) describes the call as "a dry, flat husky *pwich*" and the flight call as "a clear, high *svit*."

Although highly unlikely to be heard on the islands, song is a rich, low, *cheery-cheery-chorry-chorry*. **Status:** Four possible records (two each from Aruba and Curaçao), but none documented with photos or video or confirmed in any other way (Voous 1993; Prins et al. 2009). Aruba possible records: two at Gran Tonel, Oct 24, 1974; one at Brazil, Oct 30–Nov 4, 1978 (Voous 1983). Curaçao possible records: one at Parera, Nov 1, 1951 (seen by Voous himself, who noted no obvious eye-ring); one at Wechi, Nov 25, 1984 (Voous 1983; Prins et al. 2009). **Range:** Breeding range extends across southern Boreal forest region of Canada, from extreme southeastern corner of Yukon across to Newfoundland, and then south into northeastern US and down through Appalachian Mountains, south, spottily, to Virginia and West Virginia. Winter range extends from Nicuaraga south to northern South America, from western Venezuela through Colombia to northern Ecuador.

Kentucky Warbler (*Geothlypis formosa*) Plate 51

Papiamento: *Chipe di Kentucky*
Dutch: *Kentuckyzanger*

Description: A small, secretive warbler. All yellow below, with no streaking; olive-green above, with no wingbars. Distinctive yellow "spectacles" around eyes framed in black (most striking in adult males) or dark olive (in immature birds). **Similar species:** Canada Warbler also has "spectacled" look but has necklace of dark streaks across breast, bluish upperparts (versus olive-green), and white undertail coverts (all-yellow underside in Kentucky). Common Yellowthroat lacks yellow spectacles in all plumages; yellow on underside is confined to throat, upper breast, and undertail coverts (not on belly). Hooded Warbler has a large area of yellow on face and shows white in tail. **Voice:** Call described by Dunn and Garrett (1997) as "a low, distinctive *chup* or *chuck*." Has been likened to the *chup* note of Hermit Thrush or the *chip* portion of the *chip-burr* call of the Scarlet Tanager (for birders from eastern North America). Sibley (2000) describes the flight call as "a short, rough buzz *drrt*." Although unlikely to be heard on the islands, the song of the Kentucky Warbler is a loud, rollicking series of two-parted phrases often described as *tory-tory-tory-tory-tory*. **Status:** Nine records, seven from Bonaire and one each from Aruba and Curaçao, with eight in Oct and one in Sept (Prins et al. 2009). Aruba sight record is of a bird that lingered Sept 8–25, 1977, at Paraquana (Voous 1983—note that Prins et al. [2009] spell the location as Paraguaná). Bonaire records from Prins et al. (2009) are as follows: one at Fontein, Oct 6–10, 1974; one at Fontein, Oct 16–22, 1976; one at Fontein, Oct 13, 1977; one at Kralendijk, Oct 28, 1978; one at Kralendijk, Oct 20 and 24, 1981; and one at Put Bronswinkel, Oct 9, 1983. Single Curaçao record is for a bird seen at Wechi on Oct 10, 1984 (Prins et al. 2009). **Range:** Breeding range extends across eastern US, and from southern Wisconsin, Michigan, and New York south to northern Florida and the Gulf States, to eastern Texas. Wintering range extends from southern Mexico to Panama, and rarely to northern Colombia and northwestern Venezuela.

Common Yellowthroat (*Geothlypis trichas*)   **Plate 51**

Papiamento: *Mascarita comin, Maskarita komun*
Dutch: *Gewone maskerzanger*

Description: A small songbird, similar in size to Bananaquit. Prefers to stay in vegetation, close to the ground and usually close to water. Presence often given away by its loud, distinctive call. Adult males are unmistakable with their black "robber" mask bordered by white above; bright yellow throat; and olive-brown upperparts. Female and immature have similar unmarked olive-brown upperparts but lack black mask (immature males have a gray mask, with some black that increases as spring approaches); they typically have a yellow throat (some birds lack yellow throat); and show brownish band across breast and on flanks, with a whitish (or at least not yellow) belly contrasting with yellow undertail coverts. **Similar species:** South America is home to yellowthroat species (not yet documented in the ABCs) that are very similar to the Common Yellowthroat; they are entirely yellow below and males lack white border along top edge of mask. Female Common Yellowthroat might be confused with Connecticut or Mourning warblers, but both of those species have all-yellow underparts and are larger, with larger-bills. Kentucky Warbler also stays low to the ground but always shows yellow "spectacles" and has all-yellow underside. Canada Warbler has an eye-ring, black "necklace," and white undertail coverts. **Voice:** Call is a distinctive low, husky *check* or *tchep*. Although unlikely to be heard on the islands, song is a loud, boisterous *witchety-witchety-witchety-witchety*. **Status:** When Voous (1983) published his authoritative work on the birds of the Netherlands Antilles, there was only a single record of Common Yellowthroat from the islands. Prins et al. (2009) list nine additional records. We recorded the species nine times between 1995 and 2012. Of these 18 records, 13 have come from Bubali, Aruba. Except for an exceptional June record, all records come from Nov through Apr (latest Apr 14). <u>Aruba</u> records, in chronological order: one male at Bubali, Jan 26, 1979 (Voous et al. 1983); one female or immature at Bubali, Jan 9, 1995 (our record); one adult male at Bubali, on Jan 4, 1997 (our record); two at Bubali, Dec 7, 1997 (Prins et al. 2009); nine at Bubali, Jan 12, 1998, and one there on Jan 16, 1998 (our record); two or three males and one female at Bubali, Apr 12, 2000, and one male and one female there on Apr 14, 2000; one male at Bubali, Dec 6, 2001 (Prins et al. 2009); one female or immature at Bubali, Jan 12, 2002 (our record); five birds (including two adult males) at Bubali, March 27, 2004 (Prins et al. 2009); 10 at Bubali, one at Bucuti Beach Resort, one at Spanish Lagoon, and one at Tierra del Sol golf course, March 12–18, 2005 (Mlodinow 2006; Prins et al. 2009); one or two at Bubali, Nov 23, 2011 (our record); two at Bubali, Nov 21–22, 2012, and one there on Nov 27, 2012 (our record); one at Spanish Lagoon, Nov 23, 2012. Many of these records have been documented with photos and/or video. Five additional records have been reported to eBird since 2011. <u>Bonaire</u>: There is only a single record from Bonaire, a male specimen at or near the airport, June 11, 1995, late for this migrant species to be so far south of its breeding grounds. <u>Curaçao</u>: Prins et al. (2009) list only a single record for Curaçao, to which we can add one of our own; one male was seen at Wechi, Nov 18 and Dec 2, 1984 (Prins et al. 2009); we observed an adult male at Malpais, March 31 to Apr 1, 2000. **Range:** Breeds in most of North America, except for northern extremes, and south into Mexico. Winter range extends from southern US to northern Colombia, with only a single record from Venezuela (Hilty 2003).

Hooded Warbler (*Setophaga citrina*)   Plate 51

Papiamento: *Chipe velo preto, Chipe belo pretu*
Dutch: *Monnikszanger*

Description: Hooded Warbler is one of the larger warbler species. It is usually seen in the islands in mangroves or tall, brushy habitats near water. Adult males are distinctive, with black hoods framing a bright yellow face and forehead that encircle a large, dark eye. Some distinctly marked females show a partial black hood framing a bright yellow face. Dully marked females and immature birds have an olive cap that still shows striking contrast with yellow face. In all plumages, upperparts are unmarked olive above, with no wingbars, and all show strong, white flashes in the tail as the bird flits about, occasionally spreading tail feathers. From below, Hooded Warbler of any age or sex shows unstreaked yellow underparts, including yellow undertail coverts, and an all-white tail framed along the edges with dark. **Similar species:** On adult male, combination of black hood, yellow face, unstreaked yellow underparts, and white in tail is definitive. Canada Warbler lacks yellow face, is blue-gray above, has white undertail coverts, and a dark tail. Kentucky Warbler has yellow "spectacles" but not a full yellow face and lacks white in tail. Immature or female Yellow "Golden" Warblers in various stages of molt could potentially be confused but have all-yellow tails (or at least lack white in tail), are much smaller, and do not show obvious contrast between cap and face. In all plumages, Common Yellowthroat lacks full yellow face, is not entirely yellow below, and does not have white in tail. Wilson's Warbler has not been recorded from the islands; it is much smaller and lacks white in tail; adult male has black cap. **Voice:** Call, quite distinctive, is described by Dunn and Garrett (1997) as "a sharp, loud *chink* or *chip*." Song, unlikely to be heard on the islands, is a loud series of whistled notes. Dunn and Garrett (1997) render one version of the song as *taw-ee-tawee-tawee-TEE-to*. **Status:** A migrant visitor and apparent wintering bird, in small numbers. Voous (1983) noted a surprising number of occurrences on Aruba and Bonaire (but none on Curaçao) considering the rarity of the species in Venezuela and Colombia. We estimate a minimum of 32 Hooded Warbler records for Aruba and Bonaire based on Voous (1983), Prins et al. (2009), and our own observations. All 13 Aruba records, in chronological order: one at Bubali, Apr 22, 1977 (Voous 1983); one at Sero Colorado, Nov 1978 (Voous 1983); one female or immature at Bubali, Jan 10, 1995 (our record); one male and one female or immature at Bubali, Jan 4, 1997 (our record); one male at Spanish Lagoon, Jan 7 and 11, 1997 (our record); two females or immatures at Arikok NP, near the entrance, Jan 14, 1998 (our record); one male at Spanish Lagoon, Jan 16, 2002 (our record); one male at Spanish Lagoon, Jan 13, 2003 (our record); one female at Spanish Lagoon,

March 23, 2003 (Prins et al. 2009); one male at Bubali, March 24 and 30, 2003 (Prins et al. 2009); one female at Spanish Lagoon, March 28, 2004 (Prins et al. 2009); and one bird at Spanish Lagoon, March 18, 2005 (Prins et al. 2009). Voous (1983) had listed 13 records for Bonaire, to which Prins et al. (2009) added at least five (note that the number of records is hard to enumerate because of a set of sightings from Dos Pos, March to Apr 2002, that could refer to the same birds). We have documented the species twice on Bonaire, although one record is likely a sighting of the same adult male mentioned by Prins et al. (2009) at Dos Pos, Apr 1 and 4, 2001 (our sighting of an adult male at Dos Pos was on March 26, 2001). The records since Voous (1983) for Bonaire are listed here in chronological order: one at Dos Pos, March 1, 1997 (Prins et al. 2009); one male at Put Bronswinkel, Washington-Slagbaai NP, Jan 14, 1998 (our record); one male at Dos Pos, March 26, 2001 (our record); one at Dos Pos, Apr 1 and 4, 2001 (Prins et al. 2009); five sightings of a male at Dos Pos, March 3–Apr 13, 2002, might have been the same bird (Prins et al. 2009); one at Fontein, Oct 28, 2003 (Prins et al. 2009); and one male near Dos Pos, Dec 28, 2004 (Prins et al. 2009). There are no well-documented records of Hooded Warbler from Curaçao. **Range:** Breeds from extreme southern Ontario, southern Wisconsin, Michigan, and New York south to Florida and Gulf States, to eastern Texas. Winters from Mexico south to Panama (where rare). Rare winter visitor along Caribbean Coast of Colombia and Venezuela, where there are only five records (Hilty 2003).

American Redstart (*Setophaga ruticilla*)   Plate 49

Papiamento: *Rabo còrá americana, Rabu kòrá*
Dutch: *Amerikaanse Roodstart*

Description: A small songbird, similar in size to Bananaquit. Adult male has jet black upperside, head, and throat, with bright orange patches in wing, sides of breast, and tail. Females and first-year males have yellow instead of orange and show greenish-gray back and wings, and gray head. Regularly fans tail out as it hops around. **Similar species:** Adult male is unmistakable. Females and immature males always show the distinctive tail pattern and tail-fanning behavior. **Voice:** Call described by Dunn and Garrett (1997) as "a thin, slurred *chip* or *tsip*, recalling the chip of Yellow "Golden" Warbler but thinner and more sibilant, often with an almost hissing quality." Although unlikely to be heard on the islands, song is typically a high *see-see-see-seeo*. **Status:** Regular but uncommon winter resident on all three islands, Aug–May, especially in mangroves but occasionally seen in other habitats. We observed one in shrubs, on the grounds of the Divi Tamarijn Resort, Aruba, in Nov 2013, and three together in a small park across the street from the resort in Oct 2016. The species is one of the more regularly observed wintering warbler species in the islands and during migration can be occasionally abundant, as in Sept 2016, when Michiel Oversteegen found 19 at Bubali, Aruba. We have recorded the species at least 16 times on Aruba, usually as a single bird, but on six occasions we found two or more together. Although most of our sightings have been of birds in female-type plumage, we have observed four first-year males and three adult males over the years. Most of our American Redstart sightings on Aruba have been at Spanish Lagoon and Bubali, though we have seen the species in mangroves near Parktienbos (as well as the above mentioned sightings at or near the Divi Tamarijn Resort). Ligon (2006) lists six records of American Redstart on Bonaire since 1997 and there at least six additional records reported to eBird for Bonaire since 2006. **Range:** Breeds across eastern US and throughout Canada. Winters from Mexico and West Indies south, through Central America to northern South America, and south to Peru.

Cape May Warbler *(Setophaga tigrina)*   **Plate 47**

Papiamento: *Chipe Cabo May, Chipe Kabo May*
Dutch: *Tijgerzanger*

Description: A small songbird, with fine, pointed, black bill. Adult males in breeding plumage distinctive, with chestnut cheeks framed in yellow, yellow underside with extensive dark streaking, white patch in wing, and contrasting yellowish rump. In Jan, we have seen adult males on Aruba that were already largely in this spring plumage. Adult males that have just molted into nonbreeding plumage are duller and sometimes show very little chestnut in cheeks, but quite variable. Drabness of immatures and females can make identification difficult for observers not experienced with the species. Females have brownish-green upperparts with an unstreaked back, gray-brown ear patch framed in drab yellow, contrasting yellowish rump, and two wingbars (upperpart often whiter and more prominent); underneath, shows yellow from throat to belly, but has whitish undertail coverts; shows extensive fine streaking. Immature birds similar to female but can be almost entirely gray on upperparts (though rump is still lighter and often yellowish), with obscure wingbars. Underparts on immatures are streaked and often lack any yellow. In all plumages, shows greenish-yellow edges to flight feathers in folded wing. **Similar species:** North American birders often find it surprising to see the range of variation in molting Yellow "Golden" Warblers in the ABCs, many of which show a patchwork of gray and yellow that approximates the grays and yellows of immature or female Cape May Warblers. Yellow "Golden" Warblers have thicker, stubbier bill, lack contrasting facial pattern shown by even pale immature Cape May Warblers, and never show contrasting yellowish rump of Cape May. Immature and nonbreeding adult Yellow-rumped Warbler could be readily confused but larger; shows thicker, stubbier bill, browner upperparts, bright lemon-yellow rump, usually some yellow on flanks, and with coarser brown streaking below. Palm Warbler always shows yellow undertail coverts and almost constantly pumps tail. **Voice:** Call note is a "distinctive high, thin, sharp *seet* or *tsip*" (Dunn and Garrett 1997). Song, not likely to be heard on the islands, is a series of 5–6 thin, high-pitched *seet* notes; sometimes sings fewer notes or more than the 5 to 6. **Status:** Most common on Aruba. Seven records listed in Prins et al. (2009) for Aruba, one Apr 2015 record listed in eBird, plus five additional records from our observations, making a total of 13 records from Aruba, from six different locales: adult male at Frenchman's Pass, Apr 2, 1961 (Voous 1983); one at Rooi Lamoenchi, Jamanota, Jan 26, 1979 (Voous 1983); one female or immature seen at Spanish Lagoon, Jan 10, 1995 (our observation); one adult male at Bubali, Jan 4, 1997 (our observation); six, including 1–2 adult males, at Bubali, Jan 13, 1998 (our observation); one female type, at Spanish Lagoon, Apr 14, 2000 (our observation); two at Spanish Lagoon, Jan 13, 2002 (our observation); one at Bubali, March 24, 2003 (Prins et al 2009); two at Bubali, March 12, 2005; two at Savaneta, March 13, 2005; four at Spanish Lagoon, March 13–18, 2005; two at Eagle Beach, March 14, 2005 (all Mlodinow 2006); up to three individuals, some photographed, at Spanish Lagoon, Apr 2015 (R. Wauben, G. Peterson, and B. Van es fide eBird). Note that of the 13 records for Aruba, 10 were made by observers very familiar with the identification of northern warblers, thus suggesting that the species is likely more regular than the historical record suggests. Curaçao has a single record, from Piscadera Bay, Apr 18, 1964. Four records from Bonaire: one from Lima-Sorobon Road, Apr 27, 1976; two from Seru Largu, Jan 8, 1977, and Apr 28, 1977; one bird at Barbara Crown Heights, Jan 24, 2012 (J. Ligon fide eBird). **Range:** Winter range extends from Florida to West Indies; it occurs as a rare visitor to Central America, south to Panama, and in coastal Colombia and northeastern Venezuela. In breeding season, largely restricted to foreal forest region of Canada and northern parts of eastern US.

Cerulean Warbler (*Setophaga cerulea*)   Plate 49

Papiamento: *Chipe blou garganta blanco, Chipe blou garganta blanku*
Dutch: *Azuurzanger*

Description: A small songbird, similar in size to Bananaquit. One of the shortest-tailed warblers, but with a fairly thick-based and blunt bill. Adult males are a striking sky-blue (cerulean) above, with two white wingbars; white throat bordered at top of breast by blue breast band; note dark blue streaking on sides. Female or immature plumaged birds probably more likely; individuals in this plumage easily misidentified or overlooked. Female and immature greenish-gray above with two white wingsbars, pale line over eye, and whitish (adult female) or yellowish (immature) underneath, with some streaking on sides. **Similar species:** Some immature Blackburnian Warblers can be very similar in appearance to immature Cerulean but are larger, longer-tailed, and yellowish color below does not extend beyond upper breast. Immature Blackburnian has light stripes on sides of back (no stripes on back of immature or female Cerulean) and some show yellowish stripe or patch on front part of cap (always lacking in Cerulean). Nonbreeding adult and immature Blackpoll Warblers can look surprisingly similar to female or immature Cerulean Warbler. Blackpoll Warbler is larger and longer-tailed than Cerulean, usually shows strongly streaked back, and lacks obvious pale eyebrow line or, if present, eyebrow line does not widen behind the eye (as it does in immature Cerulean and Blackburnian). Blackpoll Warbler also almost always has noticeable yellow-orange feet, with black legs versus all-black legs and feet in Cerulean. Nonbreeding male and female Blackpoll Warblers can show dark stripes along edges of throat that are always lacking in Cerulean. **Voice:** Call described by Dunn and Garrett (1997) as a "full, slurred *chip.*" Birds are known to sing during migration and on South American wintering grounds. Song is a quite distinctive, usually three-part series, with 2–5 low, buzzy or trilled, notes given in first two parts, and ending with a buzzy, non-rising trill: *zee-zee-zee-zizizizi-zreee*. **Status:** Three records from Bonaire: one found dead, Oct 1975; one at Slagbaai plantation, Oct 28, 1979; one male with summer plummage, at Kralendijk, Apr 4, 1987 (Prins et al. 2009). No records from Aruba or Curaçao. Despite the major decline that this species has undergone in the past 50 years, it is nonetheless surprising that there are so few records, as the species winters in northern South America and is regular in Venezuela. We would guess that the species may occur more frequently than the few records would indicate, and that birds in female and immature plumage are easily overlooked. **Range:** South American winterer, from Colombia and northwestern Venezuela south along Andes, to Peru and Bolivia. Breeds in eastern mid-Atlantic region to mid-western states, and north to extreme southern Ontario. Major declines over the last 40–50 years are a concern (Wells 2007).

Northern Parula (*Setophaga americana*)   Plate 46

Papiamento: *Chipe parula*
Dutch: *Brilparulazanger*

Description: A small songbird, about the same size as the Bananaquit. Shows distinctive bluish upperparts, greenish-yellow patch on back, two prominent, white wingbars, white arcs above and below eye; yellow throat and upper breast contrast with white on remainder of underparts. Breeding male has bands of burnt orange and black across upper breast, more subdued in female. Immature birds similar but lack breast band. **Similar species:** Although never recorded in the ABCs, Tropical Parula is common in northern South America and should be considered. Tropical Parula lacks white eye arcs and is more extensively yellow underneath. Magnolia Warbler can show greenish on back in some plumages, but note that that species has streaking on flanks, yellowish rump, and striking tail pattern. Other similar warblers lack wingbars. **Voice:** Call is sharp *tsip* (Dunn and Garrett 1997). Song, unlikely to be heard on the islands, is a buzzy, rising trill ending with a sharp note, often described as *zeeeeeeee-up*. **Status:** Occurs in small numbers as a winter or migrant visitor, on all three islands. Voous (1983) noted a total of 10 records from Curaçao and Bonaire, though Prins et al. (2009) appears to list only four for each island for that same time period. In total, Prins et al. (2009) listed 6 occurrences on Curaçao, 7 on Bonaire, and 11 on Aruba. In addition, we have documented the species at least 15 times on Aruba, making a grand total of at least 39 occurrences, with 31 of these since 1983. Pooling the records from Prins et al. (2009) and our own, we found that a single bird was recorded in 60% of the observations; two birds in 23% of the observations; and 3–6 birds in 16% of the observations. Six or more individuals have been documented on Aruba on two occasions: 11 at Bubali, Jan 2003 (our observation); 12–18 at Spanish Lagoon, March 2005 (Mlodinow 2006). Observations of the species in the ABCs have spanned from Oct through Apr. Although most of the occurrences have been documented in Jan (11) and March (9), this likely results more from the timing of surveys than from when the birds occur. In fact, we suspect that the species is likely much more regular on all three islands than is reflected in the number of documented occurrences, as 87% of occurrences since 1983 can be accounted for by only three sets of observers. Oddly, while Northern Parula occurs quite regularly in the ABCs, it has never been documented in mainland Colombia and has only been documented three times in Venezuela, including two records from islands off the Venezuelan coast (Hilty 2003). On Aruba, there were no records up to 1983 (Voous 1983). Since that time, we have documented at least 15 occurrences on Aruba; others have documented at least 11 occurrences (Prins et al. 2009). The first record for Aruba was our sighting of two individuals at Bubali, Jan 9, 1995. Our other records are as follows (female/immature plumaged birds unless noted): 1–2 birds at Bubali, Jan 9, 1996; an adult male at Bubali, Jan 4, 1997; 1–2 birds at Bubali, Jan 5, 1997; an adult male at Spanish Lagoon, Jan 11, 1997; one at Arikok NP (near today's visitor's center), Jan 10, 1998; one at Bubali, Jan 6, 1999; one adult male at Spanish Lagoon, Jan 13, 1999; two at Spanish Lagoon, Jan 14, 1999; one at Bubali, Apr 12, 2000; one at Spanish Lagoon, Apr 14, 2000; one at Bubali, Jan 12, 2002; two adult males and one or two female-immatures at Spanish Lagoon, Jan 13, 2002 (two of the same birds also appeared on Jan 15 and 16; at Bubali, two adult males and four female-immatures, on Jan 11, 2003, and two female-immatures on Jan 14, 2003; at Spanish Lagoon, one on Jan 13, 2003, and two on Jan 15, 2003; one at Spanish Lagoon, Nov 23, 2012; and one at Bubali, Nov 26, 2012. This species has also been documented on Aruba at: Casibari, Ayo rock formations, Pos Chiquito, and Savaneta (Prins et al. 2009); note, however, that the repeated records at the same locations certainly reflect the fact that observers have tended to visit these locations because they are known to be excellent birding hot spots. On Curaçao, the species has been documented from Mount Sint Christoffel (Sint Christoffelberg), Malpais, and near Bullenbaai (Prins et al. 2009). On Bonaire, the species is known to have occurred at Put Bronswinkel, in Washington-Slagbaai NP, and at Dos Pos (Prins et al. 2009). With more birding, the species may be found to occur more widely at locations across all three islands. **Range:** Winters in southern US south through Mexico, Costa Rica, and the Greater Antilles. Very rare in northern South America, so the number of records in the ABCs is somewhat unexpected. Breeds extensively throughout eastern US and Canada.

Magnolia Warbler (*Setophaga magnolia*)   Plate 47

Papiamento: *Chipe barica castaño, Chipe barika kastaño*
Dutch: *Kastanjezanger*

Description: A small songbird, similar in size to Bananaquit. Most likely to see birds in immature or winter plumage on the islands. From throat to mid-belly, note bright yellow; white from lower belly to undertail coverts; also note faint streaks on sides (dark streaks in adults) and gray line across base of throat. Upperparts bluish-gray, with two white wingbars; greenish back with contrasting yellow rump. Grayish-blue face with narrow white eye-ring. In all plumages, note unique tail pattern when seen from below: wide black tail tip and white base. From above: when bird fans tail note blocks of white on either side of tail. Breeding plumage shows stronger colors: bright yellow underparts with black streaking and black "necklace" across breast; upperparts blue-gray, with large, white wing patch in male. **Similar species:** Tail pattern is unique among warblers. Observers are most likely to confuse this species with winter-plumaged Northern Parula, as both have wingbars, yellowish throat and breast, and greenish patch on the back. Magnolia is larger and longer-tailed, with streaking usually evident on sides, more extensive yellow on underside, and a yellow rump and tail pattern very different from that of Northern Parula. Winter-plumaged Prairie Warbler similar but lacks white wingbars; yellow extends to undertail coverts; lacks contrasting yellow rump; has a different facial pattern; and does not show the distinctive tail pattern. **Voice:** Dunn and Garrett (1997) describe the call note as "unique among our warblers, being a rather nasal, dry chip represented variously as *nieff, schlep, tlep, tzep.*" One of our good friends described the call as sounding like a constipated chip note. Song, not likely to be heard on the islands, is a whistled *wheata, wheata, wheata* or *wheata, wheata, wheata,WEE-teeo.* **Status:** Rare but found with increasing frequency in Oct–Apr period. Seven records from Aruba, including three of our observations: one at Bubali, Jan 12, 1998 (our record); one at Bubali, Jan 6, 1999 (our record); one or two at Bubali, Jan 11 and 14, 2003 (our record); one at Spanish Lagoon, March 28, 2004 (Prins et al. 2009); and one at Bubali, Oct 27, 2007 (Prins et al. 2009); one photographed at Bubali, Apr 10, 2010 (S. Mlodinow fide eBird); two (one photographed) at Spanish Lagoon, Oct 15–16, 2016 (M. Oversteegen, G. Peterson fide eBird). One record from Curaçao: one bird at Christoffel NP, Nov 21, 2003 (our record cited in Prins et al. 2009). Four records from Bonaire summarized in Prins et al. (2009), plus six additional reported to eBird: one immature at Put Bronswinkel, Washington-Slagbaai NP, Feb 27, 1993; one female at Put Bronswinkel, Washington-Slagbaai NP, Feb 27, 2001; one male at Dos Pos, March 12 and 23, 2004; and one in "winter plumage" at Dos Pos, Dec 28, 2004; one at Pos Mangel, Washington-Slagbaai NP, Jan 20, 2010 (J. Heller fide eBird); one at Dos Pos, Dec 19, 2013 (J. Ligon fide eBird); one at Fontein, Oct 20, 2014 (S. Williams); one at Pos Mangel, Washington-Slagbaai NP, Dec 7, 2014 (J. Chard, M. Timpf fide eBird); one at Dos Pos, Dec 23, 2014 (J. Ligon, E. Bevins fide eBird); one at Dos Pos, Mar 21, 2015 (J. Ligon, E. Bevins fide eBird). **Range:** Winters in Mexico and Central America, into Panama; also winters in Greater Antilles. Single record from Venezuela (Hilty 2003); only four occurrences from mainland Colombia noted in Hilty and Brown (1986). Breeds across most of Boreal forest region of Canada, south into northeastern US.

Bay-breasted Warbler (*Setophaga castanea*)　　　　　　　　　　**Plate 48**

Papiamento: *Chipe barica castaño, Chipe barika kastaño*
Dutch: *Kastanjezanger*

Description: A small songbird, similar in size to Bananaquit. Birds in immature and adult-winter plumage (what you are most likely to see on the islands) are greenish-yellow above, often (but not always) with dark streaks on back, two strongly contrasting, white wingbars, and show yellowish wash on throat and breast with little or no darker streaking on breast. Undertail coverts usually buffy. Often have rust colored flanks, though sometimes shows as dull pinkish color or just as a yellowish-buff. Feet and legs dark. Sides of neck usually a contrasting brighter yellow. In spring, adult females similar to immature and adult-winter plumage, but always show rusty flanks and fairly dark back, wings, crown, and face. Breeding plumaged male is quite dramatic and generally unmistakable, with his chestnut cap, throat, and flanks, black face, and light nape. **Similar species:** In immature and adult-winter plumage, very similar to the much more common Blackpoll Warbler and easily confused. Individuals with rust or pinkish wash on the flanks are generally easily separable from immature and adult-winter Blackpoll Warblers. Bay-breasted Warblers lacking the rust or pinkish on flanks will show buffy flanks (pale or whitish in Blackpoll), dark legs and feet (yellow, on feet or soles of feet in Blackpoll), buffy undertail coverts (versus contrasting white in Blackpoll), little or no streaking on breast (obvious streaking in Blackpoll), contrasting yellow-green sides of neck (grayer with little or no contrast in Blackpoll), and more contrasting white wingbars (area between wing-bars not as dark and contrasting in Blackpoll). Some Chestnut-sided Warblers, in female plumage or in immature plumage moulting into spring plumage, can have some rust on flanks while showing greenish-yellow upperparts, but note that Chestnut-sided Warbler always has prominent, white eye-ring, is boldly white below, and is a bright lime-green above. **Voice:** Call note is a rather rich *chip*; also makes a short, buzzy *zeep* flight call; both similar to those of Yellow "Golden" Warbler. Although unlikely to be heard on the islands, song is a series of doubly repeated notes, so high that many people are not able to hear them. **Status:** Because this bird can be difficult to identify, it may be less rare on the islands than it seems. Prins et al. (2009) list only six records, including two specimens. We have identified the species only once in the islands, when we photographed and videotaped an individual with rust-colored sides at Bubali, Aruba, Nov 22–23, 2011, making a total of seven published records. For Aruba, two published records: one at Spanish Lagoon, Nov 2, 2007 (S. Mlodinow cited in Prins et al. 2009) and our record noted above. Bonaire has three records: one unconfirmed sighting during a mass passage of warblers, in Oct 1959 (Voous 1983); one male specimen found between airport and Punt Vierkant, Apr 14, 1978 (Voous 1983; Prins et al. 2009); and one specimen at same location, Oct 31, 1981 (Voous 1983; Prins et al. 2009). Curaçao has two records: one seen Oct 28, 1951 (no location specified); and one see at Suffisant, Oct 14, 1962 (Voous 1983; Prins et al. 2009). **Range:** Breeding range is almost completely con-fined to Boreal forest region of Canada, extending from the southern Northwest Territories eastward to Newfoundland and south into New England and New York. Winter range extends from Costa Rica south to northeastern South America.

Blackburnian Warbler (*Setophaga fusca*)   Plate 48

Papiamento: *Chipe Blackburn*
Dutch: *Sparrenzanger*

Description: A small songbird, similar in size to Bananaquit. Birds in immature and adult-winter plumage (what you are most likely to see on the islands) are rather nondescript gray-brown on upperparts, with two white wingbars and two pale, contrasting lines on back; yellow or yellow-orange on upper breast and throat extends onto face to frame a dark, arrow-shaped cheek patch. Yellow forehead patch, when present, is unique. In spring, adult male has fiery orange throat and large, white wing patches and is generally quite unmistakable. **Similar species:** Drab fall birds could be mistaken for Blackpoll or Bay-breasted warblers, but those species lack the two pale lines on the back and show less contrast on face. Blackpoll Warbler almost always has noticeable yellow-orange feet with black legs versus all-black legs and feet in Blackburnian. Some immature Blackburnian Warblers are very similar in appearance to immature Cerulean but are larger, longer-tailed, and yellowish color underneath does not extend beyond the upper breast. Immature Blackburnian also has light stripes on sides of back (no stripes on back of immature or female Cerulean) and some birds show a yellowish stripe or patch on front part of cap (always lacking in Cerulean). **Voice:** Dunn and Garrett (1997) describe call note as "a rich *chip* or *tsip*" and the flight call "a buzzy *zzee*." Although unlikely to be heard on the islands, songs of Blackburnian Warbler are all very high pitched. One common version is a series of high-pitched doubled notes followed by a loose trill and then a rising note that seems to go so high that you can no longer hear it, sometimes likened to *seebit-seebit-seebit-ti-ti-ti-ti-teeeeeeeeeee*. **Status:** Paucity of records from the ABCs is somewhat surprising, as the species is a common winterer in Venezuela and Colombia. Ten records listed by Prins et al. (2009), eight from Bonaire, and singles from Aruba and Curaçao. Two additional documented records from Aruba in 2016. Two of the records are specimens now housed in the Amsterdam Zoological Museum. One of two birds seen at Bubali, Aruba, on Apr 17, 2016, was photographed by Ross Wauben to provide Aruba's first documented record of the species (fide eBird). A single birds was photographed by Michiel Oversteegen at Bubali, Aruba, on Sept 18, 2016 (fide eBird). Aruba sight record was on Nov 21, 1971, at an unspecified location (Voous 1983). Bonaire records (as listed in Prins et al. 2009) are as follows: one specimen at Kralendijk, Apr 30, 1958; one at unspecified location on Apr 22, 1975; one photographed at Kralendijk, Nov 3, 1979; one at unspecified location on Oct 15, 1981; one at Pos Mangel, Oct 15, 1981; one at Pos Mangel, Oct 14, 1983; one male at Pos Mangel, May 2, 1991; a male at Dos Pos, Dec 23, 2000; and one described as an immature male at Dos Pos, May 26, 2001. Curaçao record is of a specimen taken at Malpais, Sept 4, 1955 (Prins et al. 2009). **Range:** Breeds across southern Boreal forest region of Canada, from Alberta east to Quebec and south; breeds in eastern US in Appalachian Mountains to northern Georgia. Winter range extends from Costa Rica south through Venezuela and Colombia, south in Andes to northern Bolivia.

Yellow "Golden" Warbler (*Setophaga petechia rufopileata*)   **Plate 46**
Yellow "Northern" Warbler (*Setophaga petechia aestivia* Group)   **Plate 46**

Papiamento: *Parha di misa, Chibichibi hel Cu, Para di misa*
Dutch: *Gele Zanger Ar*

Description: A tiny, bright yellow bird common on all three islands. Note dark eye and dark bill. Males show distinctive red cap and red streaks on breast. Females lack cap, are paler yellow, and have reduced red streaks on breast. Immature, molting, and faded birds can be washed out and pale yellow-green or with mottled plumage reminiscent of various, much rarer, vagrant warblers. The North American breeding subspecies of Yellow "Golden" Warbler, which lacks a red cap, winters regularly in Venezuela, but no photograph or specimen has been procured for the islands (Prins et al. 2009). However, Voous (1983) reports two possible records of the northern form, and we observed a breeding plumaged male of the northern form (i.e., bright yellow with bright red breast streaks but lacking a red cap) on March 29, 2000, near the Klein Hofje sewage treatment plant on Curaçao. **Similar species:** Saffron Finch, which now is a common breeding resident on Curaçao and a localized resident on Aruba and Bonaire, is the only other commonly occurring small, all-yellow bird on the islands. Larger than Yellow "Golden" Warbler, with a thicker bill, and lacks red breast streaks. Several other warbler species with extensive yellow, especially on the underparts, can occur as rare or uncommon migrants or winter visitors, and pale or moulting Yellow "Golden" Warblers can bear a surprising initial similarity to rare species like Tennessee Warbler, Cape May Warbler, and others. No other warbler besides Yellow will show yellow tail and yellow undertail coverts. **Voice:** Song quite varied but always makes a series of sweet musical notes given in series. Sometimes "sweet-sweet-sweet-too-I'm a sweet or too-sweet-a-weet," and many other variations, quite unlike any other common singing bird to be heard in the islands; call is a rich *chup*. **Status:** Abundant in mangroves and shrubs on all three islands. On Aruba it is fairly common to see 30–40 birds in mangroves and shrubs surrounding Bubali wetlands and 20–30 around Spanish Lagoon; also possible to see many dozens along trails in Christoffel NP, Curaçao, and at Washington Slagbaai NP, on Bonaire. In dry years, large numbers sometimes occur in the vicinity of freshwater sources, as in March, 2001, when we observed more than 100 birds at the freshwater pond near entrance to Washington-Slagbaai NP, Bonaire. In many locations, birds become very tame, even feeding from the hand and searching for food scraps in and around cars and garbage facilities. Breeding can occur throughout the year but is said to be concentrated in Nov–Jan (Voous 1983). On Aruba, in Jan, we have regularly recorded adults feeding young, and many singing birds. **Range:** Yellow "Golden" Warbler complex, which contains dozens of forms, is divided by some authors into several species. If considered as a single species, the range extends from northern North America through Central America and the Caribbean, to northern South America.

Chestnut-sided Warbler *(Setophaga pensylvanica)* **Plate 46**

Papiamento: *Chipe flanco castanō, Chipe banda koló kastanō*
Dutch: *Roestflankzanger*

Description: A small songbird, similar in size to Bananaquit. Birds in immature or winter plumage (what you are likely to see on the islands) have lime-green upperparts with unstreaked back; bold, white eye-ring, and two prominent white or yellowish wingbars. Underparts white. Adult males in winter plumage can show chestnut stripe of varying length along sides of breast and flanks. In breeding plumage, readily identified by yellow crown and chestnut streaks along sides of breast and flanks, which contrast strikingly with white underparts. **Similar species:** Generally a quite unmistakable warbler. Winter plumaged male

Blackpoll Warbler may have a greenish tinge to upperparts, though never as green as Chestnut-sided; it also shows some streaking on breast, typically with yellow wash on throat and breast, compared to white, unmarked breast in Chestnut-sided Warbler. Blackpoll also lacks bold, white eye-ring. Winter-plumaged Bay-breasted Warbler typically has some buffy on underparts, often near throat and undertail coverts, and diffuse rust-colored area on flanks (not dark chestnut color as in Chestnut-sided). Bay-breasted is always brighter green above and clean white below, with a prominent white eye-ring. **Voice:** Call described by Dunn and Garrett (1997) as a "husky slurred *chip*, very similar to some of the lower pitched chips given by Yellow Warblers." Song, unlikely to be heard on the islands, is an emphatic whistled "pleased, pleased, pleased-to-meet-cha." **Status:** Based on records from Sept through Apr, a rare and irregular winter migrant visitor. At least 10 records for Aruba, including four of our own observations: one at unspecified location, Oct 19, 1971 (Voous 1983); one in immature plumage, Jan 7–10, 1994, on grounds of Divi-Divi Resort, near Oranjestad (our record); one in immature plumage, Jan 13–14, 1999, at Spanish Lagoon (our record); one, Jan 7, 2005, at Spanish Lagoon (J. Valimont fide eBird); three birds at Bubali, March 12, 2005 (S. Mlodinow in Prins et al. 2009); one in immature plumage, Nov 25, 2011, at Spanish Lagoon (our record); one, Feb 3, 2014, at Bubali (S. Mlodinow fide eBird); up to 4 birds, Apr 26–29, 2016, at Bubali (M. Oversteegen, R. Wauben fide eBird); up to 5 birds, Sept 27–29, 2016 at Bubali (M. Oversteegen fide eBird); one in immature plumage, Oct 2–8, 2016, at Bubali (R. Wauben, J. Wells fide eBird). Eleven records for Bonaire are summarized in Prins et al. (2009); our record included in that publication is of a bird with some chestnut on flanks, seen near Salina Matijs, in Washington-Slagbaai NP, on March 25 and 27, 2001. We have documented two occurrences of the species on Curaçao: an adult male almost into full breeding plumage, at Klein Hofje, March 30, 2000; and one in immature plumage, at Malpais, March 31, 2000. **Range:** Winters in Central America and Greater Antilles, occasionally as far south as Venezuela. Breeds from southern Canada south to northeastern US.

Blackpoll Warbler (*Setophaga striata*)   **Plate 48**

Papiamento: *Chipe pèchi preto, Chipe pèchi pretu*
Dutch: *Zwartkopzanger*

Description: A small songbird, similar in size to Bananaquit. Birds in immature or adult-winter plumage (what you are most likely to see on the islands) are drab greenish-olive above, with dark steaks on back, two white wingbars, yellowish wash on throat and breast, with some darker streaking on breast. White on long undertail coverts contrasts with rest of tail. Feet (sometimes only soles of feet) and often legs (especially in spring) are bright yellow. In fall, when most birds have darker legs, contrasting yellow feet can be quite striking. Some birds seen on the islands beginning in Jan and continuing through the early spring begin to molt into breeding plumage; it is not uncommon to see individual males showing partial black caps and distinct black streaking along sides of throat and on breast. Females in breeding plumage similar to immature and adult-winter birds but lack yellow suffusion on underparts and are grayer on upperparts. Breeding male is striking, with black cap and white cheek. **Similar species:** In immature or adult-winter plumage, most similar to much rarer Bay-breasted Warbler, which usually shows some rust or buffy coloration on flanks; with no (or very little) streaking on breast; usually has buffy (not white) under-tail coverts; dark legs and feet (not bright yellow as in Blackpoll), and usually shows a more contrasting yellow-green patch on side of neck. See also Cerulean Warbler. **Voice:** Call note is a rather rich *chip*, quite similar to call of Yellow "Golden" Warbler; flight call—a short, buzzy *zeep*—is also similar. Although unlikely to be heard on the islands, song is a distinctive series of very high notes, like the sound of a squeaking belt or wheel on a car—*seet-seet-seet-seet-seet*—rising in volume before becoming softer again. **Status:** During fall migration, Blackpoll Warblers are sometimes the most abundant migrant warbler on the islands. Blackpoll Warblers are famous for their epic fall migrations, in which they set off at night from Maritime Canada and northeastern US for a non-stop, 3–5 day flight that relies on prevailing westerly winds from Africa to deposit them in northern South America. Storms can blow large numbers off course, such that they will seek any place to land and find water, food, and rest. Voous (1983) writes of fall migrations in the 1950s and 1970s in which loose flocks of 30–50 Blackpoll Warblers could be found for a few weeks throughout the islands, after a major grounding event that very likely involved many thousands, at least for a short time. Voous (1983) also notes that many were ex-hausted, and specimens were extremely emaciated. Certainly many individuals die under these circumstances. In recent years, the only documentation of ex-tremely large numbers of Blackpoll Warblers we are aware of is that of Ligon (2007), who noted more than 100 on Bonaire, Oct 1998, 40 tallied between Eagle Beach and Tierra del Sol, Aruba, on Oct 27, 2007 (S. Mlodinow fide eBird), and multiple daily tallies of 40–50 birds at Bubali and Spanish Lagoon, Aruba, from late Sept through Oct, 2016 (M. Oversteegen, G. Peterson, R. Wauben fide eBird). We tallied a minimum of 46 at seven locations on Curaçao, Nov 21–24, 2003, including 20 at Christoffel NP, Nov 24, 2003. Small numbers apparently winter on the islands, at least occasionally, as we have recorded the species in Jan, on Aruba, in six years between 1995 and 2002, including three at Bubali, Jan 5, 1997, and four, also at Bubali, Jan 6, 1999. The species is decidedly less common in spring migration than during fall migration, though there is one May record noted by Prins et al. (2009) and a number of records from Aruba in May, 2016 (fide eBird). From March 29 to Apr 1, 2000, we tallied at least seven on Curaçao, including one male in full breeding plumage. **Range:** Breeding range extends across Boreal forest region of North America, from Alaska across Canada to Newfoundland, and south in northeastern US, to high-est peaks of Adirondack and Catskill Mountains, New York. Wintering range generally extends over northeastern South America, east of Andes and south to Amazon basin of northern Peru and northern Brazil. Small numbers do winter on Aruba, Bonaire, and Curaçao.

Black-throated Blue Warbler (*Setophaga caerulescens*) Plate 47

Papiamento: *Chipe blou garganta preto, Chipe blou garganta pretu*
Dutch: *Blauwe Zwartkeelzanger*

Description: A small songbird, similar in size to Bananaquit. Male very distinctive; blue upperparts; black, on throat and cheek and along sides, contrasts with white underside. White "handkerchief in the pocket" patch on wing is a unique characteristic seen in all plumages. Female and immature grayish-green above, buffy-yellowish and unstreaked below, with light stripe above eye, and thin, white arc below eye. **Similar species:** Males should be unmistakable. Females could be confused with a number of other species, but most have wingbars and streaked underparts, and all lack white "handkerchief" in wing. Immature Tennessee Warbler could look particularly similar, but note Black-throated Blue's strong white patch on wing and darker cheek. **Voice:** Very distinctive call is a dry *tup*, like the sound made by pulling the tongue back rapidly from the top of the mouth. Song, unlikely to be heard on the islands, is a series of short, accelerating buzzy notes often described as "zur-zur-zur-zree" or "I'm so laz-eee." **Status:** Six records from Aruba: one of unspecified sex at Sero Colorado, Sept 1973 (Voous 1983); one female at Bubali, Jan 12, 1998 (our record); one male at Spanish Lagoon, March 28 and Apr 2, 2004 (Prins et al. 2009); one female at Spanish Lagoon, March 15, 2005 (Mlodinow 2006); one male at Spanish Lagoon, Feb 5, 2013 (J. Wierda pers. comm.); one female at Spanish Lagoon, Apr 12, 2013 (S. Mlodinow fide eBird). Six records from Bonaire: one male at Fontein, Nov 13–21, 1975; one male at Kralendijk, Oct 25, 1977; one male found dead at Nikiboko, Oct 26, 1977; at Dos Pos, one female on Apr 10, 2001, and two females on Apr 24, 2001 (all from Ligon 2006); one adult male at Dos Pos, May 18, 2008 (J. Ligon fide eBird). No records from Curaçao. **Range:** Winter range primarily restricted to Bahamas and Greater Antilles; a rare winter visitor to Central America, Colombia, and Venezuela. Breeds from southeastern Canada south into higher elevation mountains of southeastern US.

Palm Warbler (*Setophaga palmarum*) Plate 48

Papiamento: *Chipe di palma*
Dutch: *Palmzanger*

Description: A small songbird, similar in size to Bananaquit. In all plumages, drab brown above, with pale eyebrow line, contrasting yellowish rump, and yellow undertail coverts. The subspecies that breeds in the eastern part of the range shows entirely yellow underside, while those from the rest of the breeding range have a whitish belly with blurry gray streaking. Adults in breeding plumage have rufous cap; breeding plumaged adults from the eastern subspecies have reddish breast streaking. Typically likes to stay low, even feeding from the ground. One of most distinctive identification features is that Palm Warblers almost constantly bob their tails up and down. **Similar species:** Prairie Warbler also bobs tail, but has a longer tail, lacks contrasting yellowish rump, lacks streaking across breast (confined to sides), has distinctive facial pattern, and lacks brown tones seen in upperparts of Palm Warbler. Cape May and

Yellow-rumped Warblers both have yellow rump but have white undertail coverts and do not bob tail. **Voice:** Call is a low, often loud, emphatic, hollow sounding *chik*. Described by Dunn and Garrett (1997) as "a sharp, slurred *tsik* or *tsup*, quite distinctive once learned (but does recall note of Prairie)." Song, unlikely to be heard in the islands, is a loose, buzzy trill. **Status:** Five records from Aruba including two listed in Prins et al. and three others. Aruba records are as follows: one at Sero Colorado, Nov 4, 1956 (Prins et al. 2009); one at Bubali, Dec 7, 1997 (Prins et al. 2009); two at Bubali, Jan 12, 1998 (our record); a possible intergrade between the two subspecies seen at Tierra del Sol Golf Course, Apr 7, 2013 (S. Mlodinow fide eBird); three of western subspecies, at Spanish Lagoon, Apr 12, 2013 (S. Mlodinow fide eBird). Two records from Bonaire: one at Dos Pos, March 29, 2003; one in Hato region, Apr 30, 2004 (Prins et al. 2009). Three records from Curaçao: one collected at Muizenberg, Feb 23, 1957; one collected on at Muizenberg, March 9, 1957; one seen (tail bobbing behavior noted) near Santa Martha Baai, Mar 28, 1977 (K. Overman, J. Gee fide eBird). **Range:** Winters from Atlantic Coast of US through Florida, to Gulf Coast and Greater Antilles; and in Central America, south to Honduras and rarely south to Costa Rica and Panama. One record each from Colombia and Venezuela (Ridgely and Tudor 2009). Breeding range largely confined to the Boreal forest region of Canada (Wells and Blancher 2011), from northeastern B.C. and western Northwest Territories to Newfoundland, and south into northeastern US.

Yellow-rumped Warbler (*Setophaga coronata*)   Plate 47

Papiamento: *Chipe corona di oro, Chipe korona di oro*
Dutch: *Geelstuitzanger*

Description: One of the larger warblers. Has a thick, stubby, black bill; in all plumages, has a streaked back. In breeding plumage, bluish upperparts with white wingbars, black streaking underneath, yellow on sides, and bright yellow rump. Female and immature (plumages most likely seen in the islands) brown on upperparts but have white wingbars, bright yellow rump, rather coarse brown streaking below, and usually show yellow on sides. Thus far, all birds found on the ABCs appear to have been of the more easterly breeding "Myrtle" race, in which adults show a white throat, rather than "Audubon" race of western North America, in which adults show a yellow throat. **Similar species:** Magnolia Warbler, which also has a yellow rump, is entirely yellow below, with a striking tail pattern unlike that seen in Yellow-rumped Warbler. Cape May Warbler also shows a yellowish rump, and immature birds very brownish overall like Yellow-rumped Warbler. Cape May Warbler has thinner, more sharply pointed bill, is more finely streaked below, has an unstreaked back, and rump is a more muted yellow-green (not bright yellow). Palm Warbler also has a yellow rump but note yellow undertail coverts (white in Yellow-rumped Warbler); it regularly tips its tail up and down. **Voice:** Call is a very distinctive, low *chuck*. Song, unlikely to be heard on the islands, is a slow, sweet warble, typically rising or falling at the end. **Status:** Six records from Aruba: one male at Bubali, March 31, 2001 (Prins et al. 2009); one male and one female at Bubali, Apr 20, 2001 (Prins et al. 2009); at Bubali, two in female-immature plumage, Jan 12, 2002, and one in similar plumage on Jan 16, 2002 (our record); one in female-immature plumage, Jan 13, 2002 (our record); one in female-immature plumage at Bubali, Jan 14, 2003 (our record); one adult male, Apr 2, 2013, Bubali (N. Swick fide eBird). Five records from Bonaire: one at Dos Pos, Dec 1957; one near Gotomeer, Apr 4, 1977; one near the road to Sorobon, Apr 29, 1989; one at Dos Pos, Apr 16, 1992; one male at Dos Pos, Apr 24, 2001 (all from Prins et al. 2009). Curaçao: one at Hato in Jan 1954; one at Muizenberg on Dec 17, 1956 (Prins et al. 2009). **Range:** Winters through much of southern US, into Greater Antilles, Mexico, and Central America; in some years common as far south as Panama (Angehr and Dean 2010). Rare winter visitor to Colombia and Venezuela, where there are two records, one of each race (Hilty 2003). Breeding range of "Myrtle" race largely confined to Boreal forest region of Canada and Alaska, south to northeastern US. Breeding range of "Audubon" race from southern British Columbia south to western Mexico.

Prairie Warbler (*Setophaga discolor*)   Plate 48

Papiamento: *Chipe pradera*
Dutch: *Prairiezanger*

Description: A small songbird, similar in size to Bananaquit. Yellow underneath, with streaking on sides and flanks. Yellow extends to undertail coverts. Olive-greenish on upperparts, with yellow wingbars. Adults usually show chestnut stripes on back but these may be reduced in females or fall adults. Adults have distinctive facial pattern, with yellow eyebrow, dark line through eye, and dark loop below eye framing yellow. Immature birds can be quite drab but usually show faded, gray version of facial pattern. One of the best identification features is its habit of constantly bobbing its tail up and down. **Similar species:** Beware of molting resident Yellow "Golden" Warblers, with great variety of plumages, but note that these birds do not bob their tails and usually lack any facial pattern. Palm Warbler, which also bobs tail, does not show distinctive facial pattern, is browner above, and typically shows a contrast between the bright yellow undertail coverts and white, or paler yellow, belly; also note more contrasting, yellowish rump. Immature Magnolia could look very similar but does not bob tail, has white wingbars, white undertail coverts, contrasting yellowish rump, and has a distinctive tail pattern. **Voice:** Call described by Dunn and Garret (1997) as "a smacking *tsip* or *tchick*, quite similar to the call of Palm Warbler." Song, unlikely to be heard in the islands, is a series of rising, buzzy *zee* notes going up the chromatic scale. **Status:** Rare visitor to the islands, with 10 records from Aruba and one from Curaçao. Aruba: one at Lago Colony, Sero Colorado, on Nov 20, 1955 (Voous 1983); at Sero Colorado, presumably in Lago Colony area, one bird each winter (usually in Oct), in 1972, 1973, 1974, 1975, 1976, 1977, and 1978 (Voous 1983); one at Bubali, Jan 9, 1996 (our record); one at Spanish Lagoon, Apr 8, 2013 (S. Mlodinow fide eBird). Curaçao: one was seen at Rooi Sanchie, Knip, Apr 11, 1979 (Voous 1983). Note that some writers have considered the records of single birds sighted at Sero Colorado from 1972–1978 to be a single record. While this is possible, it is also entirely possible that they represent different individuals. Throughout the book we follow the protocol of considering sightings separated by long time periods, even at the same location, as independent records, unless there is some specific evidence that the sighting is of the same individual. **Range:** Breeds in eastern US and southernmost Ontario, Canada. Wintering range confined generally to southern Florida and West Indies, and to islands off southeastern Mexico, but rarely along Caribbean coast of Central America south to Panama. Several records from Colombia, but none listed for Venezuela in Hilty (2003).

Black-throated Green Warbler (*Setophaga virens*) Plate 47

Papiamento: *Chipe lomba bèrdè*
Dutch: *Gele Zwartkeelzanger*

Description: A small songbird, similar in size to Bananaquit. Breeding males have black throat, chin, and breast; also note black streaking along sides of chest. Face yellow; back and crown olive-green; underparts white. Has two white wingbars. Immatures and females in nonbreeding plumage (and some adult males in nonbreeding plumage) lack black on throat (or have reduced black) but still show streaking underneath, especially along sides. These birds always show a yellow face with pale, olive ear patch, greenish-olive back and crown, and white throat. **Similar species:** Some immature Blackburnian Warblers look similar but usually show orangish-yellow throat, much darker ear patch and back, and have pale lines on back. Although Townsend's Warbler has not occurred on the islands, it has occurred in Panama and Colombia. It always shows yellow on throat and/or upper breast, and a darker ear patch. Golden-cheeked Warbler is very unlikely in the islands but has occurred in Panama. That species lacks dark ear patch but would show a dark line through eye. **Voice:** Call is a quick *chip* or *tsip,* sometimes sounding similar to call of Yellow-rumped Warbler, but higher and quicker. Song, unlikely to be heard on the islands, includes a buzzy *zee-zee-zee-zoozee,* ascending on the last note. **Status:** Rare visitor with no published or archived photographic or video documentation of which we are aware. Two records from Aruba: Sero Colorado, Oct 1976; and Jamanota, Jan 26, 1979 (Prins et al. 2009). Five records from Bonaire: May 4, 1962 (location unspecified); Lima, Apr 27, 1976; Kralendijk, May 3, 1976; Pos Mangel, Apr 1999 (all from Prins et al. 2009); one at the Bonaire Sewage Treatment Plant, Nov 3, 2014 (J. Ligon fide eBird). One record from Curaçao: Klein Hofje, Nov 21, 1992 (Prins et al. 2009). **Range:** Winters throughout West Indies, in Central America, to northern Venezuela (where rare). Breeds across much of Boreal forest region of Canada, south into eastern US.

Canada Warbler (*Cardellina canadensis*) Plate 51

Papiamento: *Chipe canades*
Dutch: *Canadese Zanger*

Description: A small songbird, similar in size to Bananaquit. All sexes and ages are blue-gray above, with no wingbars and with yellow on underparts extending from throat to lower belly; and note white undertail coverts. All ages show an eye-ring and a yellow line that extends from above the eye to base of bill, though most striking in adult males. Adult males have a black "necklace" formed by streaks. Females and immatures have a gray "necklace" that can be quite faint in some immature birds. From underneath, tail itself is all dark. **Similar species:** Magnolia Warblers have a similar "necklace" but it usually extends onto flanks, and note the white wingbars and different facial pattern; also note the distinctive tail, which is half white and half black. **Voice:** Call described by Sibley (2000) as "a sharp, dry, slightly squeaky *tyup.*" **Status:** Only a single record. An immature male was found dead on Bonaire on Sept 23, 1991; it is preserved as a specimen at the Amsterdam Zoological Museum (Prins and Debrot 1996). **Range:** Breeding range extends across southern Boreal forest region of Canada, from southeastern Yukon and adjacent edge of southwestern Northwest Territories eastward to Quebec, and south into northeastern US, along Appalachian Mountains to northern edge of Georgia. Winters largely in foothills of northern Andes, from Venezuela south to Peru.

Family Coerebidae: Bananaquits

Bananaquit (*Coereba flaveola*)   Plate 45

Papiamento: *Chibichibi, barica geel, bachi pretu*
Dutch: *Suikerdiefje*

Description: A small, colorful bird. Solid black on top of head, back, wings, and tail; bright yellow breast; yellow-green rump. Black throat (white in Bonaire subspecies) and crown contrast with white eyebrow and reddish, fleshy base of otherwise black, down-curved bill. Legs black. White undertail coverts and undertail. Red fleshy base to bill more prominent in male during breeding season. No other seasonal changes, and sexes otherwise alike. In immature birds, black is replaced with dull greenish-black, yellow breast is duller, and eyebrow is yellow rather than white. Often referred to as the "chickadee of the Caribbean" for its feisty but friendly nature. **Similar species:** By sight, unlikely to be confused with other species. Yellow "Golden" Warbler is the only other year-round resident with bright yellow plumage but lacks black uppersides and throat, and has very small bill and reddish cap. Some migrant North American warblers, especially in spring, have bright yellow breasts but are only very superficially similar, and most occur very irregularly. **Voice:** Call note is a short *chip*. Song is a repetitive, sizzling *chiiIIII-bi-chibichibi* (hence the local Papiemento name of *chibichibi*). The longer note at the beginning of each phrase increases in volume (like breathing in through clenched teeth) and is followed by series of rapid chattery notes. Only species with somewhat similar song is Black-faced Grassquit, which lacks "sizzling" quality and starts with a sharp *chip*, followed by a quick, squeaky *for-each-of-you.* **Status:** One of the most familiar and beloved birds of the islands, occurring commonly in and around hotels, restaurants, suburban homes, and in shrubs and mangroves. Dozens (as many as 100 in some cases) are attracted to feeding stations offering sugar, fruit, or fruit juices. Often enjoy the "leftovers" (fruit juices, syrup, jam packets) at open-air restaurants within hotel complexes. Groups of 5–10 birds are also regularly found feeding on nectar from flowering trees and shrubs. **Range:** Occurs from southern Mexico and West Indies south through South America, to Argentina, with as many as 40 subspecies recognized by some authorities.

Family Thraupidae: Tanagers

Swallow Tanager (*Tersina viridis*)   **Plate 52**

Papiamento: *Tanagra swalchi, Tanagra souchi*
Dutch: *Zwaluwtangare*

Description: A very distinctive species that likes to perch upright in the open and in treetops. Adult male is unmistakable, with light blue body, black face, black barring on flanks, and white lower belly and undertail. Female has green upperparts and green on upper chest, barring on flanks, with yellowish lower belly and undertail. Immature male molting into adult plumage can show patches of blue and green but note barring on flanks and distinctive wide-mouthed, swallowlike bill shape. **Similar species:** Should be unmistakable. **Voice:** Call described by Ridgley and Tudor (2009) as "a sharp, unmusical *tzeep*, distinctive and often given in flight." **Status:** One record of an immature male photographed on Bonaire, at the Hilltop Apartments, 3 mi (4.8 km) north of Kralendijk, on Feb 25, 2008 (Prins et al. 2009). Known to be nomadic and at least partially migratory (Hilty 2003), so perhaps not entirely unexpected as a vagrant on the islands. **Range:** Wide ranging in South America, from southern Brazil and Paraguay north to Venezuela and Colombia (also Trinidad), and just across the border into Panama.

Red-legged Honeycreeper (*Cyanerpes cyaneus*)   **Plate 52**

Papiamento: *Barica blou, Barika blou*
Dutch: *Blauwe Suikervogel*

Description: Small, a bit larger than a Bananaquit, with long, thin, down-curved bill. Adult breeding male is a stunning violet-blue, with black back, wings, and tail; bright red legs, black around face, and light blue crown. Females greenish overall, darker on wings, back, and tail, with a light eyebrow line, lighter yellow-green throat, some light streaking on upper breast, and reddish legs. Nonbreeding male is similar to female but wings and tail are black and legs are brighter red. Underwing bright yellow in all ages and in both sexes but only visible in flight. **Similar species:** Breeding males of other similar violet-blue South American honeycreepers do not have black backs and/or do not have red legs and yellow underwing. Other similar female honeycreepers have more distinct streaking across underside, distinct whisker marks, no yellow in underwing, yellow or gray in face, and dark (not reddish) legs. **Voice:** Hilty (2003) describes "a high, wheezy, ascending *shree*" and "constant *tsip* notes." Angehr and Dean (2010) write that the "calls include a thin *dzt* and a piercing *chuweet!*" **Status:** Two records from Bonaire. One immature male, captured in Kralendijk, Jan 16, 1961 (specimen at Amsterdam Zoological Museum); one immature female at Lac Bay complex, Feb to March, 2004, and photographed on March 25, 2004 (Prins et al. 2009). Prins et al. (2009) note the existence of two specimens, reportedly from Curaçao, in the US National Museum, one without any data and the other said to have been collected in Sept 1917. Given that Voous (1983) did not mention these specimens, it seems likely he was skeptical that they were indeed from Curaçao. **Range:** Extensive range, with a number of disjunct populations from Brazil north to Mexico; also in Cuba.

Saffron Finch (*Sicalis flaveola*) Plate 52

Papiamento: *Parha hel, Saffraanvink, Kanari*
Dutch: *Gewone Saffraangors*

Description: A small, all-yellow finch (back, wings, and slightly darker tail) with stubby bill and orangish forehead. Females also all yellow but duller. Immatures gray-brown but with yellow on upper breast, nape, and rump. Immatures molting into adult plumage can be a patchwork of yellows, grays, or browns. **Similar species:** Yellow "Golden" Warbler is only other small, all-yellow bird on the islands, but it is much smaller, with a finer bill, and male has red cap and red streaks on breast. **Voice:** Hilty (2003) describes song as "a dry, chattery, rather monotonic set of notes, *weezip, weezip, tsit, tsit, weezip, ts-tsit, weezip, weezip, tsik, ta-sik, weezip*" and call as a "dry *chit.*" **Status:** A breeding resident on all three islands, established over the last 30–40 years, most likely from escaped cage birds. Apparently first established on Curaçao in the 1970s and now common throughout much of the island (Voous 1983; Prins et al. 2009). On Aruba, known to be breeding in the suburbs of Oranjestad by 1979 (Voous 1983). We counted at least 12, including several groups of recently fledged, begging young birds, at a feeder on Englandstraat in Oranjestad, Jan 12, 1999. Prins et al. (2009) suggest that the species no longer occurs on Aruba, but we have received relatively recent photos and reports from visiting birders, including three seen and photographed at La Cabana Resort, May, 2010, two somewhere between Oranjestad and California Lighthouse, Apr 2011 (Steve Abbott, personal communication), and one in the area near the high-rise hotels, Sept 2010 (John Galluzo, personal communication). The species is a more recent arrival on Bonaire, with early sightings near Kralendijk in 1994, and nesting confirmed in 2002; it is now found across much of the island (Prins et al. 2009). **Range:** In native South America, two disjunct populations. One is widespread across southern Brazil, south into Argentina. A second is in northern South America, across Colombia and Venezuela. Ridgely and Tudor (2009) suggest that the two populations may be separate species. There are also isolated populations in Peru and Ecuador. Populations have been established, likely from escaped caged birds, in Panama, Jamaica, and Puerto Rico. Birds established on Aruba, Curaçao, and Bonaire are probably from escaped caged birds, but the possibility of natural arrivals from the mainland cannot be discounted.

Blue-black Grassquit (*Volatinia jacarina*)

Plate 53

Papiamento: *Mòfi bachi blou*
Dutch: *Jacarinagors*

Description: Similar in size to the common Black-faced Grassquit. Adult males glossy-black, sometimes with a small spot of white visible on shoulder. White underwing visible only in flight. Females and immatures have brown upperparts and are streaked below. As they molt into full adult plumage, immature males can be mottled black and brown and/or show black feathers edged with brown. **Similar species:** Male Black-faced Grassquit similar to male Blue-black Grassquit but Black-faced has an olive-green back and tail. Female and immature Blue-back Grassquit is streaked below, unlike female and immature Black-faced Grassquit. Female and immature Indigo Bunting could easily be confused with female or immature Blue-back Grassquit, but Indigo Bunting shows less contrasting streaking below, is browner below, and has more contrasting, lighter brown wingbars. **Voice:** Call described by Howell and Webb (1995) as "A high, sharp to slightly liquid *tsik* or *sip*." **Status:** Exceedingly rare on the islands, unlikely to be seen. Two records. One male caught (said to be "in exhausted condition") photographed and released on Curaçao at CARMABI, Piscadera Bay, Sept 21–23, 1966 (Voous 1983). One male at Playa Grandi, Bonaire, March 18, 1976; and, with near certainty, the same bird was captured there by children on July 29, 1976, and kept in captivity until its death in June 1979 (Voous 1983). Although this species could have arrived on the islands naturally, since it is common on the Venezuelan mainland, there is also the distinct possibility that these birds are escaped cage birds. **Range:** From Mexico south through Central America, and South America to Argentina. Also in Trinidad, Tobago, and Grenada.

Lined Seedeater (*Sporophila lineola*)

Plate 53

Dutch: *Witsterdikbekje*

Description: A small, ground-loving bird, just a bit larger than a Black-faced Grassquit but with a stubbier, less pointed bill. Striking males black above and white below, with white crown stripe, white cheek, and black throat connected by a line to the nape. Small, white "handkerchief" on folded black wing, and black bill. Females are dull brown above with brownish-yellow throat, chest, and flanks contrasting with lighter whitish belly and undertail coverts; they show a pale, or mostly pale, bill. **Similar species:** Males unmistakable, although there are some other similar appearing seedeaters that occur in South America, so careful study of any unusual bird is always a good idea. Females very easily confused with female or immature Black-faced Grassquit or Indigo Bunting, but note very short, stubby bill on the seedeater. There are several other species of seedeater that occur in South America in which the females are difficult or impossible to differentiate by sight from Lined Seedeater, so any suspected female seedeater would require detailed study and documentation. **Voice:** Hilty (2003) notes that males regularly sing on Venezuelan (austral) wintering grounds and describes the typical song as "a few, short, chattery notes followed by a rattle, *jit, jit, jit, jit, d,d,d,d,d, d,d.*" **Status:** One male photographed near the sewage treatment plant on the outskirts of Kralendijk, on Feb 6, 2014, by Ben and Nelly de Kruijff (personal communication). They reported that the bird was wary and actively feeding on seeds. **Range:** Two disjunct breeding populations in South America. One occurs from northern Argentina across Paraguay to southern Brazil, and migrates north in the austral winter to western South America. Second population occurs in northeastern Brazil, and apparently migrates north in austral winter to Venezuela (Hilty 2003).

Black-faced Grassquit (*Tiaris bicolor*)   Plate 53

Papiamento: *Moffi,* Mòfi
Dutch: *Maskergrondvink*

Description: A small, often very confiding, bird that regularly feeds on the ground and allows close approach. Relatively short-tailed and proportionately thick-based bill. Males have olive-green back and black face, throat, and breast, fading to a lighter charcoal-gray on lower belly. Females and immatures brown or brownish-gray overall. **Similar species:** Males should be easily recognizable. Females and immatures nondescript, but are common and should allow close study. **Voice:** Song is a regularly repeated, high, squeaky "chip-for-each-of-you," almost always starting with a single short, sharp, raspy, sliding *chip* or *slick* note followed by a quick, jumbled series of sliding notes like "for-each-of-you" or some variation. Both males and females sing. Calls include a version of the same short, raspy *slick* note but also a high, thin *spit.* **Status:** One of the most common and well-known small birds on the islands, found on all three islands, just about everywhere, including in suburban and urban areas and around hotels and resorts. **Range:** Resident throughout most of the Caribbean except on Cuba; also resident along coast of Venezuela to eastern coastal Colombia and several isolated interior populations in Colombia and Venezuela.

Family Emberizidae: Sparrows

Grasshopper Sparrow (*Ammodramus savannarum*)   Plate 57

Papiamento: *Mòfi di sabana*
Dutch: *Sprinkhaangors*

Description: An extremely rare, elusive, and little-known species on Curaçao and Bonaire, found only in grassland or low savannah habitats. Very difficult to find unless males are singing or adults are carrying food to young. A small, short-tailed, and rather flat-headed sparrow that is buffy below, with no streaking, and has a light central crown stripe bordered by dark stripes. Juveniles show streaked upper breast. **Similar species:** Although a rather cryptic species, the combination of unstreaked buffy underparts, short-tail, flat-headed profile, lack of wingbars, light central crown stripe, and proportionately large bill should identify it. More often heard than seen and produces a very distincive song. **Voice:** Call heard most often is a rough *trrrrt*. Produces two distinct songs. One is a very high pitched *tick-bzzzzzzzz* or *tick-ta bzzzzzzzz*; the other is a long, high-pitched series of squeaky notes. Both songs are unlike the songs of any other bird found on the islands, and are likely to be dismissed by the unitiated as sounds from an insect. For North American birders, both songs will sound higher and thinner than those heard in the US. **Status:** Known only only Curaçao and Bonaire, with very few records, and absent from Aruba. Voous (1983) noted that the species was known from six locations on Curaçao, citing Hato, Ronde Klip, and Noordkant. In an earlier publication, he also cites Montagne, Santa Barbara, Suffisant, and Malpais as locations where it had been seen (Voous 1957). Apparently, the only documented record on Curaçao in recent years (perhaps since before 1984) was the sighting and audio recording by Jeff Wells and Gerard Phillips of two singing males at Koraal Tabak, Nov 24, 2003. On Bonaire, Voous (1984) reported that the species was known from eight locations, including Amboina and Kralendijk. He mentions that a specimen was brought to him by a school boy who caught it near Wanapa (Voous 1957). Prins et al. (2009) list six records since 1983, including four records of single birds in Kralendijk, and records of multiple birds at Seru Largu, including four pairs with recently fledged young, on Jan 20, 2004. We have received several recent reports of Grasshopper Sparrows from fields on the outskirts of Kralendijk, including three singing males in Jan 2009 (Jan van der Winden personal communication). **Range:** Has a fascinatingly extensive and fragmented breeding range, from southwestern Canada through much of US, and with populations in Greater Antilles, Mexico, and Central America, to Panama (many populations isolated and patchy and described as separate subspecies). Isolated, now very rare, populations in central Colombia and formerly Ecuador. Populations on Curaçao and Bonaire are far from any other populations and have been described as a separate subspecies.

Rufous-collared Sparrow (*Zonotrichia capensis*) Plate 57

Papiamento: *Chonchorogai*
Dutch: *Roodkraaggors*

Description: A small, ground-loving bird. Adults have rufous collar, gray head with black crown stripes, black line through eye and framing white throat, and conspicuous black mark on shoulder. Adults are unstreaked below and show two white wingbars. Immature birds are quite nondescript, showing fine streaking across breast and lacking rufous collar and gray head. **Similar species:** Adult should be quite unmistakable. Rare Dickcissel has rufous shoulder on wing but not on nape and lacks black mark extending from shoulder. Male House Sparrow has rusty-brown on back and back of neck but has black throat and lacks head stripes. Female House Sparrow is nondescript and might be confused with immature

Rufous-collared Sparrow, but lacks any streaking below and lacks dark crown stripes and dark whisker line. **Voice:** Song is series of sweet notes quite unlike songs of other species on the islands. Our audio recordings of males singing in Arikok in 1999 are of three- (occasionally four-) part whistled songs like *teer-cheer-cheer.* Voous (1983) described its song as *tsee-tssee-teerr,* ending in a distinctive terminal trill. Aruba birds we have recorded have not shown this terminal trill but we have heard recordings from Curaçao that, although also three-parted, have a trill on the final note. Call is a sharp *chink* or *chip.* **Status:** Common and widespread breeding resident on Curaçao, occurring both in wilder areas as well as suburban backyards. Restricted on Aruba and not occurring on Bonaire. To give a sense of abundance on Curaçao, here are some of our maximum counts: 20 at Christoffel NP, March 3, 2000; 20 at Malpais, Nov 22, 2003; 15 at Blue Bay Golf Course, Nov 22, 2003; and 46 at Christoffel NP, Nov 24, 2003. Three birds were known to have been introduced to Bonaire in the 1950s, but it is unclear whether a population became established (Voous 1983). A single bird was reported at Belnem, Bonaire, on May 19, 1978 (Voous 1983), but there are no subsequent records from Bonaire (Prins et al. 2009). On Aruba the species seems to have been much less abundant, with an ever more restricted range (Voous 1983). We have only encountered the species in Arikok NP (area near current visitor center), Cunucu Arikok, Jamanota, and once at Spanish Lagoon in Jan 2002, although we know of people who have observed the bird in the Frenchmen's Pass area. We have not found the species in the last few years at Arikok or elsewhere, nor has the species been reported to us anywhere on Aruba by any visiting birders in recent years, although it may still occur in some parts of the island that are not visited regularly by birders. It may be a species that has been impacted by introduced Boa Constrictors. Largest numbers we have counted on Aruba were: 8–10 in Arikok NP, Jan 13, 1996; 10+ including many birds singing and carrying nesting material, Arikok NP, Jan 9, 1999; and 7–8 in Arikok NP, Jan 14, 2002. **Range:** Very extensive South American range, occurring widely across southern South America and in highlands of northern South America. Also occurs from southern Mexico to Panama, and in highlands of Hispaniola.

White-throated Sparrow (*Zonotrichia albicollis*)   Plate 57

Papiamento: *Chonchorogai garganta blanco, Chonchorogai garganta blanku*
Dutch: *Witkeelgors*

Description: Adult birds have white throat; head with black and white stripes or tan and white stripes (two forms occur); yellow in front of eye. Adults have unstreaked underparts and rather reddish-brown wings, with two white wingbars. Immature birds similar to tan-striped adult form but show blurry streaks across breast. **Similar species:** Immature Rufous-collared Sparrow is quite similar to immature White-throated Sparrow but lacks rufous tones to upperparts, would not show any yellow above eye, and would have lighter underparts. **Voice:** Migrant birds typically give a long, high *seet* and a hard *chink* call. The species is famous for its melodious song on breeding grounds, rendered as

"Old-Sam-Peabody-Peabody-Peabody" or "Oh-Sweet-Canada-Canada-Canada." **Status:** Given that the species' wintering range extends only to northern Mexico, the single record from Aruba of a bird captured at an unspecified location, Jan 1964, is exceedingly unusual (Voous 1984; Prins et al. 2009). The bird is said to have died in captivity and is now a specimen in the Zoological Museum Amsterdam, but the age, sex, and color-morph of the specimen are not given in any publications we are aware of. The bird was said to have been "captured out of a group of six birds" (Prins et al. 2009), but it is unclear if there were six White-throated Sparrows (which stretches credibility) or if the captured bird was in a flock with other species. Voous, in a 1985 publication, writes that this bird probably arrived to Aruba via ship-assisted passage, though it is unclear whether or not he knew of more details or if this was conjecture based on the improbability of the record (Voous 1985). **Range:** Breeding range is largely confined to the Boreal forest region of Canada, from the Northwest Territories to Labrador and Newfoundland, and south to New York. Winters largely in southern US, south to extreme northern Mexico.

Family Cardinalidae: Grosbeaks, Buntings

Summer Tanager (*Piranga rubra*)   **Plate 54**

Papiamento: *Tanagra ala preto, Tanagra ala pretu*
Dutch: *Zomertangare*

Description: Roughly the same size as a Tropical Mockingbird but with noticeably large-looking bill. Adult male all red. Females and immatures can vary from yellow to golden to orangish or orange-red, with dark area in front of eye; wings and back only slightly darker than rest of body. In spring, molting males sometimes show patchy red and yellowish plumage. **Similar species:** Male breeding Scarlet Tanager has black wings and black tail. Female, immature, and nonbreeding male Scarlet Tanager have smaller bill; greener overall; more contrasting, dark wings and tail (quite black in nonbreeding adult male) and usually lack dark markings in front of eye. Female and immature Western Tanager have two, white wingbars and smaller bill. **Voice:** Call is a distinctive *pee-tuck* or *pee-tucky-tuck*. **Status:** Two records from Aruba, two records from Curaçao and two from Bonaire. <u>Aruba</u>: one photographed at Divi Links on Apr 23, 2013 (R. Wauben pers. comm.); one photographed at Balashi Gold Mill Ruins near Spanish Lagoon on Oct 5, 2014 (R. Wauben pers. comm). <u>Curaçao</u>: one bird "in golden-green plumage" photographed at Julianadorp, early fall of 1957; one "red male" at Seru Bientu, Sint Christoffel area, Apr 5, 1979 (Voous 1983). <u>Bonaire</u>: one immature male photographed at Peaceful Canyon, Apr 17, 2003; one female at Dos Pos, Apr 18, 2003 (Prins et al. 2009). Given these back-to-back dates, it seems a strong possibility that the same bird was involved in both sightings. The species is described as a "fairly common nonbreeding northern winter resident" in Venezuela (Hilty 2003), so it is curious that it has not been recorded more regularly on the islands. **Range:** Breeding range extends from southeastern and southwestern US, south into Mexico. Winters from Mexico south into South America, as far south as Bolivia and extreme eastern and northern Brazil.

Scarlet Tanager (*Piranga olivacea*)   Plate 54

Papiamento: *Tanagra còrá, Tanagra kòrá*
Dutch: *Zwartvleugeltangare*

Description: In breeding plumage, adult male quite unmistakable, with red body and black wings and tail. Females, immatures, and nonbreeding males are greenish-yellow, with contrasting darker wings and tail (black in nonbreeding adult male). **Similar species:** Male breeding Summer Tanager entirely red. Female, immature, and nonbreeding male Summer Tanager have larger bill, are more orangish or golden (versus green) overall, lack much contrast between the body and the wings and tail, and usually show dark markings in front of eye (lacking in Scarlet Tanager). Female and immature Western Tanager have two white wingbars and more pink or orangish bill. **Voice:** Distinctive call is a *chip-burr*, very different from call of Summer Tanager. **Status:** Despite the fact that Scarlet Tanager is much rarer in Venezuela than the Summer Tanager, on the islands there are many more records of Scarlet Tanager than of Summer Tanager. Voous (1983) describes the species as an "uncommon but regular migrant, more frequently seen in spring" and goes on to say that "occurrence substantiated by numerous records." In recent decades, we are aware of relatively few records. Prins et al. (2009) list only two records from Aruba: one at Oranjestad, Apr 3–10, 1952, and one at an unspecified location, Oct 11, 1977, and there is one recent record of two birds (at least one photographed) at Bubali, Sep 29–Oct 3, 2016 (M. Oversteegen fide eBird). There are five specimen records from Curaçao, three from Apr, one from Oct, and one unspecified (Prins et al. 2009). There are six records for Bonaire since 1983: one adult male in breeding plumage at Kralendijk, Feb 17, 1983; single birds at unspecified locations on Apr 1 and 2, 1998; a molting male at Washington-Slagbaai NP, Jan 13, 2001; and a male at Dos Pos, Apr 3 and 10, 2001, (Ligon 2006; Prins et al. 2009), one adult male in breeding plumage coming to a backyard birdbath near Kralendijk, May 10–11, 2015 (B. Pement, J. Ligon fide eBird). **Range:** Breeds across eastern North America, from southern Canada to southeastern US. Winters mostly in northwestern South America, from Colombia south to Bolivia; is a rare migrant in Venezuela (Hilty 2003; Ridgley and Tudor 2009).

Western Tanager (*Piranga ludoviciana*)   Plate 54

Papiamento: *Tanagra barica hel, Tanagra barika hel*
Dutch: *Louisianatangare*

Description: Adult male quite unmistakable, with bright yellow body, black back and tail, black wings with white wingbar, yellow shoulder bar, and reddish-orange on face and head (most extensive during breeding season). Females and immatures have grayish back, wings, and tail, two white wingbars, and an orangish or pinkish bill (top of bill is dark but the rest usually pale). **Similar species:** Adult male unmistakable. Females and immatures have two white wingbars, unlike unmarked wings of Summer and Scarlet tanagers. Note that occasionally immature or female Scarlet Tanagers can show very narrow and poorly defined wingbars that are completely unlike the bold wingbars of Western Tanager. Summer Tanager has a much larger bill and usually shows dark in front of eye (lacking in Western). Scarlet Tanager females have a darker bill. **Voice:** Call described by Sibley (2000) as "a quick, soft, rising rattle *prididit.*" **Status:** Single, rather extraordinary record from Bonaire of an adult male in breeding plumage that we photographed as it actively fed (seemingly largely on insects) at Playa Frans, on July 5, 2001 (Wells and Childs Wells 2002). **Range:** Breeds across western North America from southern Yukon and Northwest Territories south to northwestern Mexico. Winters from Mexico south regularly to Costa Rica and rarely to Panama.

Rose-breasted Grosbeak (*Pheucticus ludovicianus*)   Plate 55

Papiamento: *Pico grandi barica ros, Pik grandi barika ros*
Dutch: *Roodborstkardinaal*

Description: In breeding plumage, adult male is a striking black and white, with a blood-red bib and thick, ivory-colored bill; also note full, black hood, black back and wings, bold white wingbars, white rump, and black tail with white at corners. Females and immatures are brown above, streaked below, always with a bold, white eyebrow and boldly striped crown, pinkish bill, and bold, white wingbars. Wing linings are pinkish-red in males and yellow-orange in females. Some birds can be very buffy or pinkish below, especially as they progress through molt sequences, making for a variety of sometimes confusing plumage combinations. **Similar species:** Adult males generally unmistakable. Females and immatures could be confusing, but no other commonly occurring species from the islands has the large, light-colored bill, boldly striped head, distinctive eyebrow, and bold, white wingbars. Female and immature Black-headed Grosbeak can be exceedingly difficult to separate from Rose-breasted Grosbeak and should be carefully studied and photographed. Generally, female Rose-breasted Grosbeaks have thicker breast streaking and are less buffy underneath than female Black-headed Grosbeaks; immature male Rose-breasted Grosbeak can be buffy and show fine breast streaking, but should show pink, rather than yellow, underwing linings. **Voice:** Call is a sharp, squeaky *pik*, like the sound of sneakers stopping abruptly on a gym floor. **Status:** An uncommon migrant, with apparently a total of 34 records from the islands, though only three of these are from Aruba (Voous 1983; Prins et al. 2009; eBird 2016). Most records (27) come from fall (Oct–Dec), 6 from spring (March–Apr), plus one from June (Voous 1983; Prins et al. 2009). The three Aruba records are: two males and two females at Jamanota (now part of Arikok NP), Dec 19, 1976 (Voous 1983); one on the outskirts of Orenjestad, Oct 30, 2013 (F. Franken fide eBird); two female or immature plumaged birds photographed in a backyard near Calbas on Oct 25, 2015 (M. Oversteegen fide eBird).

Eight records from Bonaire listed by Prins et al. (2009) plus three additional from eBird and three from Observado: one male at Flamingo Beach Hotel, June 10, 1977, preserved as a specimen at the Amsterdam Zoological Museum; one male in Kralendijk, Oct 20, 1982; one male and one female at Dos Pos, Nov 9, 1987; one in the Hato region, Oct 1, 1995; a female in Washington-Slagbaai NP, Nov 12, 2002; a male said to be in winter plumage, near Seru Largu, Dec 11, 2004 (Prins et al. 2009); one seen at an unspecified location on Oct 22, 2013 (J. Ligon, J. and K. Keagle fide eBird); one immature male photographed as it came to backyard feeder near Kralendijk, Oct 9–16, 2014 (B. Pement fide eBird); female or immature seen in Rincon, Nov 7, 2014 (J. Ligon fide eBird); one immature male (photographed) at Bonaire Sewage Treatment Plant on Oct 25, 2015 (P. Schets fide Observado); one immature male (photographed) at Salina Slagbaai on Oct 23, 2015 (P. Schets fide Observado); three (two photographed) in Rincon, Feb 26, 2016 (D. van Straalen fide Observado). Three recent records for Curaçao listed by Prins et al. (2009): one female at Willemstad, Nov 16, 1992; one male at Willemstad, March 3, 2001; and one male at Julianadorp, Apr 23 and 26, 2004. We had a quick sighting of what appeared to be a female Rose-breasted Grosbeak at Klein Hofje, Curaçao, on March 29, 2000, but could not relocate the bird. **Range:** Breeding range extends from southern Northwest Territories south and east across southern Canada to New Brunswick and Nova Scotia, and then south in US to Oklahoma and down Appalachians to northern Georgia. Winters from Mexico south through Central America to South America, where it is found from western Venezuela through Colombia, to Ecuador.

Black-headed Grosbeak (*Pheucticus melanocepalus*) Plate 55

Papiamento: *Pico grandi kabez preto, Pik grandi kabes pretu*
Dutch: *Zwartkopkardinaal*

Description: In breeding plumate, adult male has orange body, black hood, black and orange stripes on back, and black wings with bold, white wingbars; also note black tail with white at corners and orange rump. Massive bill, dusky or lighter on lower mandible but not ivory as in male Rose-breasted Grosbeak. Females and immatures brown above, with bold head stripes and eyebrow line; bold, white wingbars, buffy-orangish below (strongest in first winter males), and sometimes with fine streaking on sides and across breast. **Similar species:** Adult males should be unmistakable. Females and immatures require careful study, as they can be almost indistinguishable from female and immature Rose-breasted Grosbeaks. See Rose-breasted Grosbeak account for more details and consult North American field guide references. **Voice:** Sibley (2000) describes the call as "a high, sharp *pik*, more wooden and less squeaky than Rose-breasted Grosbeak." **Status:** One record from Bullenbaai, Curaçao, of a male in "bright, unmistakable plumage," on Dec 17, 1978 (Voous 1983). **Range:** Breeds from southern British Columbia and Alberta south into Mexico. Winters in Mexico but with occasional records south to Belize and Costa Rica (Howell and Webb 1995; Jones 2003; Garrigues and Dean 2007).

Blue Grosbeak (*Passerina caerulea*)   Plate 55

Papiamento: *Pico grandi blou, Pik grandi blou*
Dutch: *Blauwe Bisschop*

Description: A bit larger than a House Sparrow. Blue Grosbeak has a large, broad-based bill, proportionately large head, and a relatively long and rounded tail that it often moves from side to side. Adult male is deep blue with rusty wingbars, lower mandible of bill is silvery. Black around base of bill and in front of eye. Females and immatures are a uniform gray-brown or rusty-brown, with two contrasting rusty or buffy wingbars and thick bill. **Similar species:** Most similar to Indigo Bunting, which is more likely on the islands. Indigo Bunting is smaller, with smaller, more pointed bill; it never shows rufous wingbars, although females and immatures can show weak, light gray-brown wingbars. Any sighting of a suspected Blue Grosbeak should be documented with photos and/or video. **Voice:** Sibley (2000) describes the call as "a very metallic, hard *tink* or *chink*." **Status:** One record of a female from Bonaire, photographed at an unspecified location, Nov 14–20, 1983 (Prins et al. 2009). **Range:** Breeds across southern US and south through Mexico, to Costa Rica. Wintering migrants known south to Costa Rica and Panama; exceptional vagrant records in Colombia and Ecuador (Garrigues and Dean 2007; Ridgely and Tudor 2009; Angehr and Dean 2010).

Indigo Bunting (*Passerina cyanea*)   Plate 55

Papiamento: *Parha indigo, Para indigo*
Dutch: *Indigogors*

Description: A small bird, slightly smaller than a House Sparrow. Males in breeding plumage are a deep, stunning blue but can appear entirely dark in bad light. Females and immatures are fairly uniform buffy-brown, with faint streaking on breast and sometimes with dull wingbars. Molt sequences are rather complicated in this and related species (Howell 2010), so it can be hard to pinpoint the exact age of some birds; males in nonbreeding season often a patchwork of blues and browns. **Similar species:** Indigo Buntings of all ages differ from exceedingly rare Blue Grosbeak; they are smaller, have a smaller bill, and lack rusty wingbars. Females and immatures could be mistaken for a female Black-faced Grassquit, but Indigo Bunting is larger, longer-tailed, has a different bill and head shape, and usually shows weak wingbars and at least faint breast streaking. Also note very different call notes from Black-faced Grassquit. **Voice:** If males sing, as we heard on Aruba in Apr 2000, they sound distinctly different from any resident bird species on the islands. Song is a series of sweet, finchlike, warbled notes, each repeated two-times, like *sweet-sweet, chew-chew*, etc. Flight call is a relatively long, resonant, springy or buzzy *bzeet*. Alarm call is a sharp, dry *spik* or *tsik*. **Status:** There are at least 15 documented occurrences of Indigo Bunting across the three islands: five from Aruba, seven from Bonaire, and three from Curaçao. On its status, Voous (1983) wrote: "apparently less rare than supposed…" and "single birds and small flocks have been observed, particularly in March and April, on harvested fields of millet…"

Unfortunately, Voous (1983) did not provide any detailed summary of records for the islands but noted only that the species had been seen between Nov 7 and Apr 23, and that it had been documented from Curaçao (at Malpais and Santa Cruz) and Bonaire, but not from Aruba. Prins et al. (2009) list a single record from Aruba of a female at Bubali, on March 12, 2005 (Mlodinow 2006). We have documented the species four times on Aruba: one or two female or immature birds at Spanish Lagoon, on Jan 7 and 11, 1997; one molting male at Arikok NP (near current head-quarters building), Apr 13, 2000; five or six birds, including at least two singing males, at Spanish Lagoon, Apr 14, 2000; one molting nonbreeding male (plumage a patchwork of blue and brown) and one or two female or immatures bird at Spanish Lagoon, Jan 13–16, 2002. Additionally there are two eBird records for Aruba: 30 birds (20 males/10 females, some photographed) at Balashi Gold Mill Ruins near Spanish Lagoon, Mar 15, 2015 (R. Wauben fide eBird); 21 birds (13 males/8 females) at Spanish Lagoon, Apr 12, 2015 (G. Peterson fide eBird). Prins et al. (2009) list seven records from Bonaire: female that died after hitting a window at Kralendijk, Jan 5, 1977, and preserved as specimen at Amsterdam Zoological Museum; two at an unspecified location, from Oct 1980 until Jan 26, 1981, with one of the birds until Feb 3, 1981; one at Fontein, Apr 4, 1981; one molting male at Kralendijk, Feb 4, 1983; one at Tolo Trail, Oct 1, 1997; one male in full breeding plumage at Sabadeco, Apr 15, 2002; one molting male at Fontein, Oct 28, 2003. **Range:** Breeds from southern edge of Manitoba east to very southern edge of Quebec, south through eastern US, and in southwestern US, into Arizona. Winters regularly from Mexico south to Panama and rarely to Colombia and Venezuela (one record, Hilty 2003).

Dickcissel (*Spiza americana*)   Plate 54

Papiamento: *Arozero, Para di aña dashi gris*
Dutch: *Dickcissel*

Description: A sparrowlike bird, similar in size to House Sparrow. During migration, often occurs in flocks. Dickcissels have a wide-based, triangular bill and chestnut shoulder patch in all plumages. Adult breeding males are bright yellow on breast and sides of throat, with a black V on the upper breast connected to base of bill by two thin whisker lines. Head is gray with a light eyebrow tinged with yellow. Females and immatures are duller but usually with at least a tinge of yellow on the breast, streaking on the upper breast (immatures), a whisker mark, light eyebrow line, and streaked back. **Similar species:** Male House Sparrows have chestnut on nape and back and have black on throat, but lack any yellow, have a different facial pattern, and have noticeable white wingbars. Female House Sparrow has noticeable wingbars, has smaller, less triangular bill; lacks chestnut shoulder, lacks whisker mark, and lacks dark cheek. **Voice:** Distinctive flight call is a low *brrrt*, sometimes likened to the sound of someone "passing gas," and somewhat similar to calls of Northern and Southern rough-winged swallows. **Status:** Although Voous (1983) wrote of the species that it was an irregular fall migrant, there seem to be very few documented records from the islands. He noted that they were numerous birds on Aruba, in Sept and Oct of 1954, and on Curaçao, in late Oct and early Nov of 1955, when they occurred "in flocks of 10-50 birds on roadsides and in acacia scrub." He also notes one spring record of a bird on Aruba, on Apr 24, 1908 (Voous 1983). Prins et al. (2009) do not list any other records but note that two specimens were taken on Curaçao in 1954 and two in 1955. Perhaps the species was more regular historically than this published record indicates, but we would still categorize the species as a rare visitor. There were reports of single birds submitted to eBird for two Aruba locations (Tierra del Sol and Westpunt) in Oct 2015 and of up to three birds at Divi Links (some photographed) in Oct 2015 (G. Peterson, R. Wauben et al. fide eBird). A flock of 30 was seen (some photographed) at Bubali on Oct 4, 2016, flying intermixed with a flock of migrating Bobolinks (M. Oversteegen fide eBird). Two were photographed on Bonaire on Oct 2, 2016, and three seen on Oct 4, 2016, at the Bonaire Sewage Treatment Plant (P. Schets fide Observado). We encourage birders to report any sightings from the islands and to document sightings with photographs and/or video. **Range:** Breeds in grasslands, from very southern edge of Saskatchewan and Manitoba south to Texas and Gulf Coast states. Winters from Mexico south to Colombia and Venezuela; formerly, a large proportion of the world population was thought to winter in Venezuela (Hilty 2003).

Family Icteridae: Blackbirds, Orioles

Bobolink (*Dolichonyx oryzivorus*)   **Plate 58**

Papiamento: *Parha di aña, Para do aña*
Dutch: *Bobolink*

Description: A small, migrant blackbird in which male undergoes a dramatic plumage change between nonbreeding season and breeding season. Perhaps more likely to be seen on the islands in fall, when males, females, and immatues are clad in similar, nondescript, streaky brown plumage. Bill pink or pink-based, head with light central crown stripe bordered by dark stripes on either side. Back strongly striped, sometimes with two brighter, whitish stripes. Sides of face and nape pale, with variable distinct dark line extending back from eye. Underside often yellowish or buffy, with streaks confined to sides of upper breast and darker streaks along flanks. Tail has sharply pointed feathers. Male in breeding plumage has black underside, face, and back, with bold, cream-colored patch on back of head; bold, white wing patches and white rump. Males moulting from nonbreeding plumage into breeding plumage can show a variety of confusing intermediate stages. **Similar species:** Male Bobolink in full breeding plumage should be quite unmistakable. The Bobolink tends to travel in flocks, while most similar species do not. Listen carefully for any vocalizations, as the *bink* call of the Bobolink is very distinctive. Female or immature Red-breasted Blackbird could be confused with nonbreeding or immature Bobolink. Red-breasted Blackbird usually shows obvious pinkish wash on breast; largely dark (not pink) bill; has a barred and squared-off tail (not pointy); shows more obvious streaking on the nape and across the upper breast, and possibly more obvious barred flight feathers (versus unbarred in Bobolink). Female or immature Yellow-hooded Blackbird is more dark overall, with contrasting yellowish throat and eyebrow, and dark bill. Female House Sparrow and Indigo Bunting, both drab brownish, might be confused, but neither shows any streaking underneath. Immature Dickcissel has dark, thick-based bill, black stripes framing throat, and chestnut patch in wing. **Voice:** Call most likely to be heard is the very distinctive *bink* flight call, given regularly by migrant birds. Song that presumably might be made by migrating males in spring is a long, exuberant jumble of reedy notes (sounding remarkably like the voice of R2D2 in the old Star Wars movies). Song sometimes rendered as *bob-o-link, bob-o-link, spink-spank-spink*. **Status:** Apparently once relatively regular during fall and spring migration, although we have never found any detailed information about its status in any publication, not even in the normally very detailed accounts of Voous (1957, 1983). Voous (1983) wrote that Bobolinks "appear singly, or more often in flocks of up to 50 or more, during short periods of autumn and spring migration" and notes that fall records span from Sept 13 to Nov 21, and spring dates range from Feb 25 to May 14. Prins et al. (2009) wrote that there was a "notable decline in numbers in the last decade." Ligon (2006) reported that the species was seen on Bonaire in Sept or Oct almost every year between 1994 and 2000, and that a flock of 50 were seen at Fontein, Oct 21, 2000. We had never observed the species on the islands until, during Oct 6–10, 2016, we saw and heard small flocks of 5–10 at multiple locations daily on Aruba. In Sept 2014, up to 29 birds were seen (some photographed) at Tierra del Solf golf course on Aruba by Ross Wauben and Luis Matheus (fide eBird). In Oct 2015, up to 100 birds were seen at the Divi Links, Aruba (many photographed), by Ross Wauben and Greg Peterson, and 300 at Tierra del Sol on Oct 11, 2015, by Greg Peterson (fide eBird). On Bonaire, flocks of Bobolinks were seen (some photographed) throughout Oct 2016 at the Bonaire Sewage Treatment Plant, with a maximum reported of 75 on Oct 13; 50 were seen there (some photographed) on Oct 25, 2015 (P. Schets fide Observado). **Range:** Breeds across southern Canada and northern half of US. Undertakes an amazingly long migration to its wintering grounds in the grasslands of southern South America, from southwestern Brazil and eastern Bolivia south to northern Argentina.

Eastern Meadowlark (*Sturnella magn*)   Plate 58

Papiamento: *Chuchubi dashi preto, Chuchubi dashi pretu*
Dutch: *Witkaakweidespreeuw*

Description: A chunky, short-tailed, large-billed, and very distinctive ground-loving blackbird. Bright yellow on throat and belly, with a striking black V across the breast. Upperparts mottled brown and black; black crown stripes and eye stripe; barred wing feathers. Tail is white edged with barred central tail feathers. **Similar species:** Should be unmistakable. **Voice:** Alarm call is a short buzzy *zreet* sometimes followed by a high rattle or chatter. Flight call is a high rising *weet*. A vagrant individual would seem unlikely to sing; however, Voous (1983) reported that the observers of the individual on Bonaire were alerted to its presence by a song similar to that made by Eastern Meadowlarks in their home state of Wisconsin. The songs of Venezuelan birds are described by Hilty (2003) as resembling "those of North American birds but are flatter, more run together, and often with more than 4 slurred whistles." Hilty (2003) renders one version of the song as "a clear, slurred *cheewa-seea, chewa-chorra.*" **Status:** One sight record from Bonaire of an adult seen at Playa Pabau, Kralendijk, on Nov 2, 1977 (Voous 1983). **Range:** Extends from southern fringes of Ontario and Quebec in Canada, south through eastern and southwestern US, through Mexico and Central America, to northern South America and as far south as northeastern Brazil. Also in Cuba. Birds from southern Canada and northern US migrate south in winter, apparently largely to southern US. Birds in remainder of range are thought to all be non-migratory residents.

Red-breasted Meadowlark (*Sturnella militaris*)   Plate 58

Papiamento: *Chuchubi pecho còrá, Chuchubi pechu kòrá*
Dutch: *Zwartkopsoldatenspreeuw*

Description: A small blackbird. Male is striking, with red throat and breast contrasting with black on the rest of its body. Female is brown with streaked back, light eyebrow stripe, pink tone to upper breast; streaks on nape, upper breast, along flanks, and vent, but center of belly unstreaked, and tail barred. Immature similar to female but sometimes fully streaked below. **Similar species:** Bobolink in non-breeding plumage is probably the most similar species. It has a pinkish bill (dark in Red-breasted Blackbird), usually some whitish streaking on back, no streaks on nape, no pink wash on breast, less streaking across breast, less distinct eyebrow line, and more pointy and unbarred tail. Also see female Yellow-hooded Blackbird (p. 284) and Dickcissel (p. 278). **Voice:** Hilty (2003) writes that the "calls include a hard *pleek* and a dry rattle." These should be readily distinguishable from the distinctive *bink* call of the Bobolink. **Status:** One female or immature photographed at Tierra del Sol golf course, Aruba, on March 28, 2003 (S. Mlodinow in Prins et al. 2009). **Range:** Found in open, non-forest habitats, from northern Brazil north to Costa Rica. Range has been expanding as more forests are cleared for pasture and agriculture.

Great-tailed Grackle (*Quiscalus mexicanus*)   Plate 58

Papiamento: *Zenata rabo largo, Zenata rabu largu*
Dutch: *Langstaarttroepiaal*

Description: A very large blackbird with long, keel-shaped tail that widens at the tip, like a spoon. Males are glossy purple-black, with pale eye and large bill. Females similar in shape, and with pale eye, but have dark brown back and tail, and lighter, buffy-brown underparts, throat and face, though this can be quite variable. **Similar species:** Much larger and longer tailed than the more likely Carib Grackle. Female and immature Great-tailed Grackles usually a warmer, buffier brown on breast and throat. **Voice:** Like Carib Grackle, a very noisy species with many vocalizations. Hilty (2003) writes that "both species have a variety of calls including a rough *chuk* and sharp series of *krit* or *quit* notes given rapidly" and that males have a "shrill, quavering *kuuueeeeeee*, drawn out." Nevertheless, most vocalizations are quite different from those of Carib Grackle, with more liquid notes and rising and drawn-out notes. **Status:** Presumably a vagrant from nearby coastal Venezuela, around the mouth of Lake Maracaibo, where it reaches its easternmost limits in South America. Possibility of ship-assisted passage cannot be ruled out, especially since the records on Curaçao have been from the areas near active shipping docks. Three early records from Aruba, where now a small number seem established. Eight early records from Curaçao, where the species is thought to have bred and is likely established (Prins et al. 2009). No records from Bonaire. Aruba: One bird at Bubali, May 27, 1990; one male at unspecified location, Nov 1991 (Prins et al. 2009); two males, Dec 7, 2013, joined by a female on Dec 15, 2013, at the Marriot Hotel, and all photographed (Erik Kramshøj, as reported to us through our Aruba Birds website www.arubabirds.com). Since 2013, reports of 1–4 birds have been regularly submitted to eBird, some with photographs, from the area extending from Bubali to Malmok. This would seem to indicate that a small resident population may be established on Aruba. Early Curaçao records, in chronological order: two adults and five juveniles at Zeelandia and Groot Davelaar, July 23, 1991; five birds near the docks, Apr 18, 1992; one male near the docks, May 3 and 5, 1992; four birds near Dok, May 4, 1997; one bird near the docks, Oct 31, 1998; 2–5 birds in female or immature plumage, one photographed at Schottegat, near Zeelandia, Nov 23, 2003, by Jeff Wells and Gerard Phillips; several females with nesting material in the Schottegat area, in 2003 (Prins et al. 2009); in March 2014, 11 birds were seen near Jan Thiel, Curaçao (P. van Rooij fide Observado) and up to four birds were seen regularly in Aug 2016 at Klein Hofje Curaçao (E. Lenting fide Observado). **Range:** Extends from center of US, south through Central America, to coastal northern South America, from extreme northwestern Peru along coast through Ecuador and Colombia to extreme northwestern Venezulea, around the mouth of Lake Maracaibo.

Greater Antillean Grackle (*Quiscalus niger*)   **not illustrated**

Papiamento: *Zenata antiano*
Dutch: *Antilliaanse Troepiaal*

Description: Very similar to Carib Grackle, but slightly longer-tailed and longer-billed. Unlike in Carib Grackle, females in this species are also black, not brown. **Similar species:** Any sighting of a grackle obviously larger and longer-tailed than Carib Grackle should be checked very carefully to eliminate the more likely possibility of Great-tailed Grackle. This species occurs almost exclusively on the Greater Antilles and is a highly unlikely candidate as a vagrant on Aruba, Bonaire, and Curaçao. **Status:** On Aruba, there is a claimed sight report of a bird that appeared after the passing of a hurricane, at Savaneta, Sept 18–19, 1999; subsequently observed with flocks of Carib Grackles and Shiny Cowbirds at

Savaneta and Bubali until Dec 2002 (Prins et al. 2009). There appears to be no photo or video documentation of the bird, nor any explanation of how it was distinguished from Carib Grackle or Great-tailed Grackle. Without formal published documentation, we think it is premature to include the species on the official list of birds for Aruba, Bonaire, and Curaçao. **Range:** Resident of Greater Antilles.

Carib Grackle (*Quiscalus lugubris*)   Plate 58

Papiamento: *Zenata caribeña, Zenata karibeño*
Dutch: *Caribische Troepiaal*

Description: Male all black, with glossy, purple sheen, pale eye, fairly long, keel-shaped tail with graduated tail feathers; bill stout but fairly long, pointed. Female and immature brown. Juveniles have a dark eye that gradually lightens; also sometimes has shorter tail feathers until fully grown. **Similar species:** Pale eye differentiates Carib Grackle from Shiny Cowbird. Shiny Cowbird has a shorter tail; a shorter, more triangular bill; and flatter head profile, but this can be easily overlooked. Juvenile Carib Grackle with dark eye can be very confusing but look for differences in bill shape and size, tail length, and head profile. Rare Great-tailed Grackle may be expanding its range and could begin occurring more regularly; it is larger, larger billed, and has a tail that is about twice as long as that of Carib Grackle. **Voice:** Males have an ear-splitting, screechy song given frequently, sometimes even while foraging on lawns and around open-air restaurants: makes a quickly repeated *REET-REET-REET-REET-REET*, along with a chatter and a repeated single high *teer*. Males often display while singing, with tail up and feathers fluffed out. Both males and females have a typical blackbird *chuck* call. **Status:** Common on all three islands. Introduced to Aruba. where it is now abundant; it is common to see birds eating scraps in and around open-air restaurants (sometimes even jumping onto unattended plates of food) and on lawns of hotels and resorts. Twelve birds from Maracay, Venezuela, were apparently released on Aruba, close to Lago Colony near San Nicolas, 1981–1988, and four in downtown Oranjestad (Reuter 1999). We know of no published records between 1981 and 1993, necessary in order to trace popuation growth on Aruba. We observed up to three birds on the grounds of the Divi Tamarijn Resort, near Oranjestad, in Jan 1993; up to six birds at Bubali, Jan 1994; at least 15 in Oranjestad, near the former Sonesta Suites hotel (across from the Parliament building), Jan 1995; and at least 10 at Savaneta, Jan 1996. Maximums we have recorded were at least 60 at Bubali, Aruba, Jan 6, 1999 (during a very wet year) and at least 60 coming to roost at the Divi Tamarijn Resort, near Oranjestad, Jan 8, 2003. It is now widespread and abundant throughout many settled area of Aruba, and likely numbers in the thousands. It is unclear whether the species arrived on Bonaire and Curaçao naturally, but it is now established on both islands. First records on Bonaire: one female near the Salt Company, March 7, 1980; one male in Kralendijk, June 23–24, 1984; one male at Simon Bolivar, March 18, 1985, and at same location, May 13, 1985 (Prins et al. 2009). We heard what we thought to be a Carib Grackle in Kralendijk, Bonaire, on Jan 14, 1998. The species was found nesting in Kralendijk, in March, 2001 (Ligon 2006). We saw at least four in Kralendijk, July 4, 2001. The species is now found in suburbs around Kralendijk. First records on Curaçao: five birds at Klein Hofje, Sept 7, 1991; two males and a female at Dokweg, May 3, 1997; one bird at the Holiday Beach Hotel, May 4, 1997; four birds at Dok, May 11, 1997; one bird in the mangroves of Shottegat-east, Dec 1, 2001 (all Prins et al. 2009). Breeding colonies were found at Rif, Otrabanda, 2003–2005, and at Jan Sofat in 2006 (Prins et al. 2009), and the species is now common across many urban and suburban parts of the island. **Range:** Resident of northeastern South America, from the mouth of the Amazon in Brazil northwards along coast to Venezuela and central eastern Colombia. Populations occur north, through Trinidad and Lesser Antilles to Anguilla.

Oriole Blackbird (*Gymnomystax mexicanus*)   Plate 59

Papiamento: *Zenata mexicano*
Dutch: *Wielewaaltroepiaal*

Description: Large, Troupial-sized blackbird with large, pointed bill, bright yellow underside and head; black back, wings, and tail. Yellow shoulder on wing, black between eye and base of bill, and black line extending down from bill base to form whisker-mark. Immature has black cap. **Similar species:** Yellow Oriole is smaller, has yellow back, white in wings, and black throat. Troupial has black hood; bold, white patch in wing, blue skin around eye. Any possible sightings of this species should be carefully documented with photos and/or video. **Voice:** Jaramillo and Burke (1999) describe the common call as "a long drawn-out screech, resembling the sound of a rusty gate." **Status:** One apparently undocumented sight record from Bubali, Aruba, on May 27 and 29, 1990 (Prins et al. 2009). **Range:** Occurs as an apparently non-migratory resident in two disjunct regions of northern South America. Northern range from Colombia through Venezuela to Guyana. A separate population occurs in a linear band along the southern edge of the Amazon River, from Ecuador eastward to the mouth of the Amazon.

Yellow-hooded Blackbird (*Chrysomus icterocephalus*)   Plate 59

Papiamento: *Trupial preto kabez hel, Trupial pretu kabes hel*
Dutch: *Geelkaptroepiaal*

Description: A very rare, small blackbird. Males have a bright yellow hood and black body, with unmarked wings and a relatively short tail; also note black between base of bill and eye. Females and immatures have dark brown upperparts with streaking on back, lighter gray-brown underparts, and yellow throat and eyebrow. **Similar species:** Male distinctive. Oriole-Blackbird has yellow underparts, not all black. Although never recorded in the islands, Yellow-headed Blackbird has occurred in Panama; it is larger, and males have conspicuous white in wings. Female Yellow-headed Blackbird has darker, unstreaked back and darker underparts. Female Yellow-hooded Blackbird might conceivably be confused with female or immature Red-breasted Blackbird but note flatter head profile, lack of obvious yellow in throat and eyebrow, and more distinct streaking on underparts, including along flanks and vent area. Female Yellow-hooded Blackbird could also be confused with nonbreeding adult or immature Bobolink. Note Bobolink's pink bill (versus dark), usually some white stripes on back, lack of obvious contrasting yellow throat and eyebrow, black stripe behind eye and obvious white crown strip, and streaking on flanks. **Voice:** Call described by Hilty (2003) as "a low, harsh *check.*" Song, unlikely to be heard on the islands, is reminiscent of song of North American Red-winged Blackbird, rendered by Hilty (2003) as *jur-gul-ZLEEE.* **Status:** Three reports from the 1970s and then none until 2015 and 2016, when birds were found on all three islands. All records of which we are aware to date are listed here. <u>Aruba</u>: one adult male photographed on Jul 8, 2016 at Bubali (R. Halff fide Observado) but originally mistakenly input into Observado as Yellow-headed Blackbird. On Aug 21, 2016, Ferdinand Kelkboom independently discovered and photographed a single male Yellow-hooded Blackbird at Bubali (fide eBird). Up to three birds were seen intermittently at Bubali following this, with three birds seen and photographed by Michiel Oversteegen on Oct 8, 2016 (fide eBird). <u>Curaçao</u>: One male was seen at Malpais, Aug 14, 1971; one male was seen at Klein Hofje, Oct 11, 2015 (J. Stumpel fide Observado). <u>Bonaire</u>: one male was seen at Tera Corá, Bonaire, Jan 18, 1977; one male was seen at Kralendijk, Bonaire, March 9, 1979 (Voous 1983; Prins et al. 2009); one male photographed at Bonaire Sewage Treatment Plant, Oct 24, 2015; three photographed at Bonaire Sewage Treatment Plant, Mar 3, 2016, and two reported there until Mar 23, 2016 (P. Schets, D. van Straalen, H. Sieben fide Observado). **Range:** Occurs along Amazon, from Peru eastward to mouth of Amazon in Brazil; also north along coast and west through Venezuela and northern Colombia, to Panama.

Shiny Cowbird (*Molothrus bonariensis*)   Plate 58

Papiamento: *Parha vakero lustroso, Para vakero lustroso*
Dutch: *Glanskoevogel*

Description: Male is all black, with glossy purple sheen; female immature brown, with some streaking below. Dark eyes, short tail, and relatively short, wide-based bill. **Similar species:** Most similar to Carib Grackle, but it has longer tail and yellow eye. Juvenile Carib Grackles have a dark eye and may show a short tail if recently fledged, but note the longer, slimmer bill. **Voice:** Call is a low *chuck*, perhaps a bit lower and more resonant than in Carib Grackle. Male song is series of liquid whistles and purrs. Also give whistled flight calls and rattles. **Status:** Recently established breeding resident in all three islands. Fairly common and widespread in Aruba and Curaçao, where it is usually found in settled areas and near hotels and resorts; no recent records from Bonaire (Prins et al. 2009). First documented on Aruba, when three birds were seen at Bubali, on Aug 21, 1997, but now widespread, with maximum a roost of 300 seen near the Bucuti Beach Resort, Oct 2007 (Prins et al. 2009). Our first sighting of the species on Aruba was of 25 at a feeder on Englandstraat, Oranjestad, on Jan 12, 1999. The person who kept the bird-feeding stations told us that she had first seen them about 12 months previously. They had reached Savaneta by 1998 (Prins et al. 2009). We observed a single female near the Lago Colony complex near Baby Beach on Apr 13, 2000, and saw a flock of 13 flying to roost past the Divi Resort near Oranjestad on Jan 16, 2002. We have not documented the species in Arikok NP, but it may now occur there. On Curaçao, the species was documented by 1985, and had become widespread on the island by 2005, including in Christoffel NP (Prins et al. 2009), although it is most common in settled areas. We saw a female at Klein Hofje, March 29, 2000, a male at Piscadera, March 30, 2000, and a male at Malpais, Apr 1, 2000; Debrot (personal communication) reported that the species was common in many urban and suburban areas of the island by 2000. First record on Bonaire was of a male and female at Kralendijk, Aug 10, 1983, and a juvenile being fed by a Yellow Oriole in Kralendijk, late in 1983, but no further reports were known to Prins et al. (2009). Six reports of 1–4 birds have been submitted to Observado from Kralendijk and the Bonaire Sewage Treatment Plant since 2009, which may indicate a small population persists in the suburbs of Kralendijk. **Range:** Resident throughout South America, north to Panama; also occurs throughout Caribbean, to southern Florida.

Orchard Oriole (*Icterus spurius*)   **Plate 59**

Papiamento: *Trupial shoura*
Dutch: *Tuintroepiaal*

Description: A small oriole, much smaller than the Troupial. Adult males are very unique, with brick-red body; black hood, back, and tail; white wingbars and edges of flight feathers; and brick-red shoulder. Females are fairly nondescript, with greenish-yellow underparts, darker olive upperparts, and two white wingbars. First-year males similar to females but have a black throat patch and face. **Similar species:** Given that there is only a single, undocumented report of the species on the islands, not likely to be seen. Adult male should be unmistakable. One could mistake an immature Yellow Oriole for a female or immature Orchard Oriole but Orchard Oriole would be much smaller, greener rather than yellow below, and darker olive on back (versus more yellowish). **Voice:** Call a sharp *chit or chit-chit,* sometimes given in a series, quite unlike the more common, screechy calls of Yellow Oriole. **Status:** One sight report, apparently accompanied by drawings, of a female seen at Boka Santa Marta, Curaçao, Jan 16, 1995 (Prins et al. 2009). **Range:** Breeds across eastern US north to extreme southern Saskatchewan, Manitoba, and Ontario, and south into Mexico. Wintering range extends from southern Mexico to northern Colombia and, rarely, to northwestern Venezuela.

Troupial (*Icterus icterus*)   **Plate 59**

Papiamento: *Trupial, trupial sabi*
Dutch: *Oranje troepiaal*

Description: North American birders familiar with the bright-orange male Baltimore and Bullock's orioles will immediately recognize this species as a member of the genus *Icterus.* The bird's black hood contrasts strongly with its bright orange underparts, rump, and upper back. Remainder of back and tail black. Wings black with large white patches and orange shoulder patch. Close views reveal striking, bare blue skin around and behind yellow eye. Rather long, pointed bill is black, with base of lower mandible bluish. Male and female similar in appearance. **Similar species:** No other species shares combination of bright orange body, black hood, and white patches in otherwise black wings. Only the migrant Baltimore Oriole is similar, and it is smaller, has a black back, less white in wing, orange corners to black tail, dark eye, and lacks blue skin around eye. **Voice:** Both sexes sing. Song is a series of loud, piping whistles, with variations that include a rich, three-part *truuu-PE-al, truuu-PE-al*, a two-part *truuu-PEA, truuu-PEA*, and sometimes a single note. No other species on the islands has a song with such a loud, piping, whistled quality. Yellow Orioles have a weaker, whistled song and Tropical Mockingbirds sing repeated phrases, but again without the Troupial's loud, piping quality. **Status:** Historically, resident on Aruba and Curaçao. Introduced to Bonaire in 1973, where now established. Common bird of the arid, cactus-shrub habitats but also occurs infrequently on grounds of landscaped hotels and suburban yards near shrub-scrub habitat. Aruba: Most common in backcountry cactus-scrub habitat, such as in Arikok NP, Sero Colorado, but small numbers also noted around Bubali and Spanish Lagoon; most birds at a single sighting was 15, at Arikok NP, Jan 10, 1998. Bonaire: Found throughout most of the island, including in Washington-Slagbaai NP, but not in southern part of island, around Pekelmeer, where there is little vegetation. Curaçao: Found throughout the island, especially common in Christoffel NP. **Range:** Northern South America, from extreme northeastern Columbia through northern Venezuela. Introduced to Puerto Rico and Virgin Islands.

Yellow Oriole (*Icterus nigrogularis*)   Plate 59

Papiamento: *Gonzalito, Trupial kachó*
Dutch: *Gele Troepiaal*

Description: A resident oriole, smaller than Troupial. Adult males and females both bright yellow or yellow-orange, with black throat (female slightly duller on back) and black between bill and eye; black tail and wings, with a white wingbar and white edgings to flight feathers. Immatures are duller yellow, lack yellow throat, have gray-olive wings (rather than black), and tail is dull grayish, often with yellow sides. Immature males develop a black throat but still show the duller immature-type plumage. **Similar species:** Troupial is much larger, with black hood, blue skin near eye, and bold white wingstripe. See Baltimore Oriole and Orchard Oriole. While there are some very similar oriole species that occur in northern South America, odd-looking birds are most likely worn Yellow Oriole birds or reflect variations in pattern. **Voice:** Regularly heard calls are a series of screechy alarm notes: *reet-reet-reet-reet*. Song, not often heard, is a sweet, full melody with short, repeated phrases. One song we recorded at Arikok NP, Aruba, Jan 1999, begins each phrase with a single note, like a small hammer striking solid metal, sometimes doubled, sometimes followed by one or two lower, richer, ringing notes: *TINK TINK-chur TINK-chur chur TINK-chur TINK*

TINK-chur. Voous (1983) renders the song as *tjee-tju tjee-tju-tju tjee-tjee*. Hilty (2003) describes the song of Venezuelan birds as *tur-a-leet, tur-sweet, tuur... tweet, tweet*. **Status:** Although not as common as the Troupial, the Yellow Oriole is relatively widespread (but never seemingly abundant) on Bonaire and Curaçao (perhaps more common on Curaçao); it will come to fruit left out on porches and backyards with sufficient brushy habitat nearby. The species has seemingly remained more restricted on Aruba. Voous (1983) commented that it was "decidedly scarce in Aruba, where almost restricted to hills and valleys of Arikok and Yamanota and the mangroves of Spanish Lagoon." Although it must occur at other locations on Aruba, we have only found it and its long, hanging nests at Spanish Lagoon and in Arikok NP. We have most seen groups of 2–4, though we found about 10 birds in Arikok NP on Jan 9, 1999, in a very wet year. **Range:** Confined to northern South America, from northeastern Colombia eastward to the coastal section of Brazil, north of the mouth of the Amazon. Occurs on Aruba, Bonaire, Curaçao, Isla Margarita, and Trinidad, but apparently not on Tobago (Jamarillo and Burke 1999).

Baltimore Oriole (*Icterus galbula*)   Plate 59

Papiamento: *Trupial di Baltimore*
Dutch: *Baltimoretroepiaal*

Description: Slightly smaller than the resident Yellow Oriole and much smaller than Troupial. Adult males have a bright orange body, black back, white wingbar, orange-edged tail, and no blue around eye. Adult females can be bright orangish but lack black hood and back. Immatures have an orange-yellow throat, paler whitish belly, and two white wingbars. **Similar species:** Adult male Baltimore Oriole distinguished from Troupial by smaller size, lack of blue face, black back (orange in Troupial), orange edges to tail (all black in Troupial), and white wingbar and white in flight feathers (bold white wingstripe in Troupial). Adult Yellow Oriole similar in size to Baltimore Oriole but lacks black hood (black throat only in Yellow Oriole) and is bright yellow, not orange. Immatures could be confused with female Orchard Oriole but note that Orchard Oriole is smaller and more greenish or greenish-yellow, as compared to orange tones of Baltimore Oriole. Immature Yellow Oriole is a bit larger; much yellower, and with yellow evenly distributed (not patchy); and much lighter on back. **Voice:** Often gives a loud chatter and sometimes a sharp, rising *eeah*. **Status:** At least two records for Aruba and six for Bonaire, but none for Curaçao (Prins et al. 2009). On Aruba, this species was said to have been found every Sept, over of the course of several years, at Sero Colorado (Voous 1983), and there is one additional record from an unspecified location on Aruba, from Nov 25, 1971 (Voous 1983). Bonaire records as follows: one photographed at unspecified location, May 4, 1962; one male at Kralendijk, Oct 21, 1978; one male at Kralendijk, March 31–Apr 21, 1985; one first-year female at Hato, Oct 16, 2001; and one near Divi Flamingo Dive Center, Oct 28, 2003 (all from Prins et al. 2009); one coming to backyard feeder in Kralendijk through at least Nov 2, 2014 (J. Ligon fide eBird). **Range:** Breeding range extends from central Alberta to southernmost Quebec; also in eastern US to Texas. Wintering range extends from southern Mexico to northwestern Colombia and northeastern Venezuela, where uncommon to rare (Hilty 2003).

Family Passeridae: Old World Sparrows

House Sparrow (*Passer domesticus*) Plate 56

Papiamento: *Parha di Joonchi, Para di Jonchi*
Dutch: *Huismus*

Description: Well-known bird from urban areas around the world, where it has been introduced over the centuries from its original European and Asian breeding range. Males have chestnut-brown upperparts with a white wingbar, gray crown, black bib extending to eye, and pale cheek. Nondescript females could be confused with a variety of other species, but they tend to stay in flocks in urban areas, so best key to identification would be males in the same flock. Females show brownish-tan and black streaks on back and wings; also note brownish crown, light eye stripe, pinkish bill, and paler gray underside. **Similar species:** Female or immature Dickcissel resembles female House Sparrow, but on Dickcissel note chestnut wingbar and black stripe on sides of throat. Female and immature Indigo Bunting and Blue Grosbeak are more uniformly warm brown and lack lighter underside. Blue Grosbeak has rusty wingbars. **Voice:** Worldwide, the loud "*chirup, chirup*" or "*cheep, cheep*" of House Sparrows has become the background sound in many urban and suburban neighborhoods. The male song is a repeating series of loud *chirp* notes. **Status:** Introduced and now common on all three islands. Introduced on Curaçao in 1953, at Mundo Nobo, Willemstad (Voous 1983); founder population said to be of eight birds (Buurt and Debrot 2012), remaining largely confined to this neighborhood for about 10 years (Voous 1983) before spreading. This species is now common in many parts of Curaçao (Prins et al. 2009). On Aruba, the origin of the first birds on the island is unknown, but the first sightings were at Oranjestad, in Feb 1978. A colony of 5–10 birds, with nests, was found in Oranjestad, in Oct 1979 (Voous 1983). It appears that they may have remained largely confined to parts of Oranjestad for many years, but there is scant documentation in the period between 1983 and early 1990s. Three birds were found at Tierra del Sol, March 1997 (Prins et al. 2009). After six consecutive years of visits to Aruba, we found our first House Sparrows in Jan 1999, in the western suburbs of Oranjestad (Englandstraat), where we counted more than 30 coming to a bird-feeding station on Jan 12, 1999. We found a flock of 5–10 near Baby Beach, Aruba, on Apr 13, 2000, and heard one as we drove through San Nicolas on Apr 14, 2000. In Jan 2002, we noted the presence of House Sparrows at Aruba Phoenix Resort and Wyndham Resort near Bubali, but we did not get a thorough estimate of numbers. We were somewhat surprised to find at least four House Sparrows at Fontein, Arikok NP, on Jan 14, 2002. These birds were likely finding food at the nearby restaurant at Boka Prins. Our first sightings of House Sparrows at the Divi Divi resort near Oranjestad were of five on Jan 9, 2003, despite the fact that the species had been common for at least four years only about 500 meters away. We found at least three House Sparrows at Savaneta, Aruba, on Jan 17, 2003; Prins et al. (2003) reported 10 birds there from Aug 3–16, 2003. In recent years, the species has become much more widespread and abundant over most of the inhabited parts of Aruba, and even occurs at the headquarters of Arikok NP. It is unclear when House Sparrows first arrived on Bonaire; some speculate that birds from Curaçao hitched a ride on a ship transporting rice between the islands. Voous (1983) does not include the species as having been documented as of 1983. The earliest records for Bonaire listed by Prins et al. (2009) are: six birds seen near the airport on Nov 7, 1989; "several birds" near airport in Nov 1990; one bird at 1000 Steps, Dec 8, 2000; one bird at Boca Slagbaai, Jan 13, 2001. Although we did not detect any House Sparrows on a short visit to Bonaire in 1998, our observations show the birds were quite widespread by 2001. We found about 20 birds near the airport on July 4, 2001; four birds at Playa Frans, July 5, 2001; two birds near Sorobon, March 24, 2001; two birds at the headquarters building of Washington-Slagbaai NP, March 25, 2001; at least two at Rincon, March 25, 2001; one at pond near entrance to Washington-Slagbaai NP, March 26, 2001; four at Marcultura Project, near Pekelmeer, March 28, 2001; at least five near a plant nursery in Guatemala, March 28, 2001; at least four at Hato, March 28, 2001; at least two birds at Flamingo Airport, March 29, 2001. The species is now widespread and common across most inhabited parts of Bonaire. **Range:** Originally native to Europe and Asia. In Western Hemisphere, now established across much of North America, Central America, the Caribbean, and southern South America, and continuing to spread in South America. As of 2003, known only from a single location in Venezuela, La Guaira (Hilty 2003).

Family Ploceidae: Weavers

Village Weaver (*Ploceus cucullatus*)   **Plate 56**

Papiamento: *Flègtudó hel*
Dutch: *Grote Textorwever*

Description: An introduced, non-native species resident only in a suburb of Willemstad, Curaçao. Males are yellow, with a black hood, chestnut nape, black spotting on back and wings; also note large dark bill with thick base, dark red eye, and short tail. Females are remarkably drab; yellowish underneath; brownish back and wings with yellowish wingbars; like males, they also have a red eye and a bill with a thick base. **Similar species:** Given that the species is currently limited to a small region of Curaçao, identification should not be a problem, at least there. However, the species has occurred as an escaped cage bird on Aruba, and more could show up in other areas. Males should be easy to identify. Females could easily be mistaken for female House Sparrow but note Village Weaver's obvious yellowish cast overall, red eye, larger size, and larger bill. Female could also be confused with Bobolink, but note Bobolink's dark head stripes, browner upperparts, streaking on sides, and pinkish bill. Also compare with Baya Weaver. **Voice:**

Described by Latta et al. (2006) as a "steady high-pitched chatter with musical whistling calls." **Status:** Small breeding population established by 1982 or 1983 in Damacor (near the Botanical Gardens), Curaçao, from escaped cage birds (Prins et al. 2009; Buurt and Debrot 2012). An escaped cage bird was seen on Aruba, on the grounds of what was then known as the Americana Hotel, on Dec 7, 1983 (Prins et al. 2009). **Range:** Native to Sub-Saharan Africa. Introduced populations resident on Hispaniola and Martinique (Raffaele et al. 2003). One small introduced population at Mariara, Venezuela (Hilty 2003).

Baya Weaver (*Ploceus philippinus*)   **Plate 56**

Papiamento: *Flègtudó skur*
Dutch: *Bayawever*

Description: Introduced, non-native species resident only in a suburb of Willemstad, Curaçao. Males are yellow, with black face, cheek, and throat, and yellow crown. White on lower belly. Large, thick-based, dark bill; brown eye, very short tail. Females, immatures, and nonbreeding males remarkably drab, but note fine streaking on upper breast and sides, contrasting crown and cheek, and brown eye. **Similar species:** Given that this species is currently only found in a limited area on Curaçao, identification should not be a problem. Males should be easy to identify. Females could be easily mistaken for female House Sparrow but note Baya Weaver's larger size, larger bill, and fine streaking on upper breast and sides. Female could also be confused with Bobolink but note Bobolink's darker head stripes, browner upperparts, and smaller, pinkish bill. Female very similar to female Village Weaver, but on female Baya Weaver note streaking on breast, brown eye (red on Weaver), and greater contrast between crown and cheek. **Voice:** Makes chattering noises and scratchy whistles, not unlike Village Weaver. **Status:** Small breeding population established by early 1980s in Cas Cora area (near the Botanical Gardens), Curaçao, from escaped cage birds (Prins et al. 2009). **Range:** Native to southern Asia.

Color Plates

On a given page, perched birds are nearly always represented at the same scale. Sometimes, however, there is a change of scale on a page; this is indicated with a horizontal line that spans the page. For design reasons, the scale can change from page to page, even within a family.

PLATE 1
Ducks

Anatidae: Ducks

1. White-faced Whistling-Duck (*Dendrocygna viduata*), p. 99
L 15–18 in (40–45 cm). Rare visitor. Stands erect. Very distinctive, with white face and black nape and neck, dark bill, reddish upper breast and lower neck, and barred sides. Long neck.

2. Black-bellied Whistling-Duck (*Dendrocygna autumnalis*), p. 99
L 20–22 in (51–56 cm). Resident breeding bird. Stands erect. Long neck; orange-pink bill contrasts with gray head. Chestnut back, upper breast, and lower neck; black belly; pink legs.

3. Fulvous Whistling-Duck (*Dendrocygna bicolor*), p. 100
L 18–21 in (46–53 cm). Rare visitor. Stands erect. Long neck; buffy-rust head and underparts white along flanks; dark back and upperwing; rust-edged scalloping to back feathers. Dark bill and legs.

4. Greater White-fronted Goose (*Anser albifrons*), p. 100
L 28 in (71 cm). Exceptionally rare, with only one previous record. Large goose. Note pink bill, white face, bright orange legs, and dark markings on belly.

5. Comb Duck (*Sarkidiornis melanotos*), p. 101
Male L 30 in (56 cm); female 22 in (55 cm). Rare visitor. Large, heavy-bodied, short-legged duck. White breast, gray flanks, iridescent blue-green back and wings. Neck and head white with dark speckles. Males have prominent black knob on bill.

6. American Wigeon (*Anas americana*), p. 101
L 20 in (51 cm). Rare to uncommon winter visitor. Medium-sized duck with distinctive, rounded head. Green face patch and large white wing patch. Black rear, with white on sides. Males show whitish crown in breeding plumage.

7. Mallard (*Anas platyrhynchos*), p. 102
L 23 in (58 cm). Very rare winter visitor. Male in breeding plumage unmistakable, with green head, white neck ring, and yellow bill. Female-plumaged birds similar in appearance to female-plumaged Blue-winged Teal and Northern Pintail but note orangish rather than dark bill and larger size. Bill on female-plumaged Northern Shoveler has similar color but is much larger and spoon-shaped; Mallard is larger, has dark blue rather than green speculum, and lacks powder blue patch on leading edge of upperwing.

immature

1

2

immature

3

immature

4

♀

♂

5

♂

6

♀

♂

7

♀

♀

♂

PLATE 2
Ducks

Anatidae: Ducks *continued*

1. **Blue-winged Teal** (*Anas discors*), p. 102
L 14-16 in (36–41 cm). Commonest of wintering ducks on the islands. Small duck; mottled brown with dark bill. The large light-blue patch on the front of the wing and smaller dark green patch on trailing edge of wing are best seen in flight, but sometimes a small bit of blue is apparent in the folded wing. Males in breeding plumage show a striking white crescent in front of eye and a black rear contrasting with small, white flank patch.

2. **Cinnamon Teal** (*Anas cyanoptera*), p. 103
L 14-16 in (36–41 cm). Very rare with only a single record from Aruba. A full breeding plumage male should be unmistakable with its reddish head and body and red eye.

3. **Northern Shoveler** (*Anas clypeata*), p. 103
L 17-20 in (43–51 cm). Rare but regular winter visitor. Medium-sized duck with distinctive shovel-like bill. Males in breeding plumage have iridescent green head, and bright yellow eye, chestnut flanks and belly; and white breast. Females are mottled buffy-brown overall; note large, orange-based, spatulate bill.

4. **White-cheeked Pintail** (*Anas bahamensis*), p. 104
L 15–19 in (38–48 cm). Resident breeding bird. Distinctive, buffy-brown duck with sharply contrasting white cheeks and reddish-based bill.

5. **Northern Pintail** (*Anas acuta*), p. 104
Male 25–29 in (64–74 cm); female 21–23 in (53–58 cm). Rare to uncommon winter visitor. Breeding male has light silvery-gray on body, with brown head; white on breast extends up back of neck to form distinctive white neck stripe. Male has distinctive long tail in breeding plumage. Female, immature, and nonbreeding male buffy-brown overall but show long-neck and dark gray bill and legs. All plumages show a white-edged green or brown bar on trailing edge of wing.

6. **Green-winged Teal** (*Anas crecca*), p. 105
L 13-14 in (33-36 cm). Rare winter visitor. Males in breeding plumage are unmistakable, with reddish heads and large patch of green behind the eye, gray sides, and vertical white stripe on the side. In flight, female Green-winged Teal lack the large blue patch on the upperwing shown by Blue-winged Teal. Female plumaged Green-winged Teal are smaller than Blue-winged Teal with a smaller, thinner bill, darker body and head (often lacking light at base of bill shown by Blue-winged Teal), dark gray rather than yellow legs, and a whitish streak along the sides of the tail.

1

2

breeding

breeding

3

4

5

6

PLATE 3
Ducks, Grebes

Anatidae: Ducks *continued*

1. Ring-necked Duck (*Aythya collaris*), p. 105
L 14–18 in (36–46 cm). Rare winter visitor. Diving duck with distinctive, peaked head. Male has dark head, breast, back, and rear; gray sides. Characteristic band of white separates gray sides from dark breast. Both sexes have gray bill; white ring separates gray bill and black tip (more obvious in male). Female and immature mottled brown; white eye-ring and (sometimes) a thin white line extending back from eye; white at base of bill; may show indistinct finger of white on side of breast, as in male.

2. Lesser Scaup (*Aythya affinis*), p. 106
L 15–18 in (38–46 cm). Rare to uncommon winter visitor. Diving duck with slightly peaked head (less peaked than Ring-necked Duck). Males show dark head, breast, and rear; white sides; and gray barring on back. Females and immatures mottled brown; white at base of bill but no white around eye.

3. Bufflehead (*Bucephala albeola*), p. 107
L 12.5 in (32 cm). Exceptionally rare winter visitor, with only a single record. A tiny duck; even smaller than Blue-winged Teal. Male in breeding plumage unmistakable, with bold white patch on dark head, black back, and white sides. Female-plumaged birds show white stripe on cheek behind eye.

4. Masked Duck (*Nomonyx dominica*), p. 107
L 13 in (33 cm). Rare visitor. Medium-sized, rust-colored duck; blue bill with black tip; black mask and thin white eye-ring. Females and immature brown, with black barring and black lines on face. Stiff tail.

Podicipedidae: Grebes

5. Least Grebe (*Tachybaptus dominicus*), p. 109
L 9.5 in (24.1 cm). Small numbers resident on permanent or semi-permanent water bodies. Small waterbird; yellow eyes; thin, black, pointed bill. Thin gray neck; black on crown and throat.

6. Pied-billed Grebe (*Podilymbus podiceps*), p. 108
L 13 in (33 cm). Small numbers resident on permanent or semi-permanent water bodies. Small brownish waterbird; short, whitish, rounded bill with black ring in breeding season. Cottony white rear end.

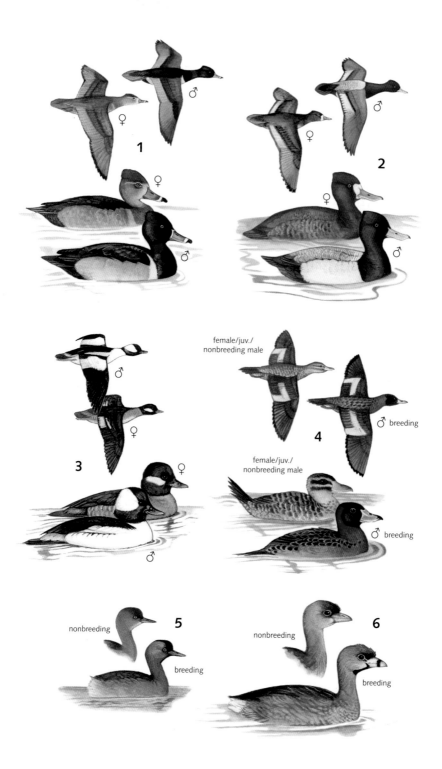

1

2

♀
♂
♀
♂

♀
♂
♀
♂

3

female/juv./
nonbreeding male

4

♂
♀

♂ breeding

female/juv./
nonbreeding male

♀

♂

♂ breeding

5

nonbreeding

breeding

6

nonbreeding

breeding

PLATE 4
Petrels, Shearwaters

Procellaridea: Petrels, Shearwaters

1. Black-capped Petrel (*Pterodroma hasitata*), p. 110
L 16 in (40.6 cm). Very rare and likely seen only far offshore. Long, pointed wings; dark gray on upperwing and back, with black M on wings; black cap and short, black, thick bill; white rump.

2. Bulwer's Petrel (*Bulweria bulwerii*), p. 111
L 10.2 in (26 cm). Very rare and likely seen only far offshore. Dark seabird with long wings and long pointed tail; on upperwing, note pale buffy bar.

3. Great Shearwater (*Ardenna gravis*), p. 112
L 18 in (45.7 cm). Very rare and likely seen only far offshore. Dark cap set off from back by white nape. Black tail separated from brown rump by narrow white border. Back and upperwing brownish, with scaly appearance, and lacking dark bars as in Black-capped Petrel. Black bill (yellowish in similar Cory's Shearwater). Dark belly patch and dark marks on underwing.

4. Cory's Shearwater (*Calonectris diomedea*), p. 112
L 18 in (45.7 cm). Very rare and likely seen only far offshore. Light brown upperparts extending on to head but lacking a contrasting dark cap as seen in Great Shearwater or Black-capped Petrel. Darker tail separated from light brown rump by narrow white border. Thick, yellow bill. White underneath lacking the dark belly patch of Great Shearwater or the dark underwing markings of Great Shearwater or Black-capped Petrel.

5. Audubon's Shearwater (*Puffinus lherminieri*), p. 113
L 12 in (30.5 cm). Rarely seen from shore but regular in small numbers offshore and may breed. Smallest shearwater. Short wings, long tail; undertail coverts partially dark. Black upperparts; white underwing bordered with black; white mark near eye.

PLATE 5
Storm-Petrels, Tropicbirds, Frigatebirds

Hydrobatidae: Storm-Petrels

1. Wilson's Storm-Petrel (*Oceanites oceanicus*), p. 114
L 7.25 in (18.3 cm). Rare and likely seen only far offshore. Square tail, white rump. Feet extend beyond edge of tail. Stiff, shallow wingbeats and fluttery flight pattern.

2. Leach's Storm-Petrel (*Oceanodroma leucorhoa*), p. 115
L 8 in (20.3 cm). Very rare and likely seen only far offshore. Notched tail, white rump. Long, pointed, angular wings. Deep, nighthawklike wingbeats and erratic flight pattern.

Phaethontidae: Tropicbirds

3. Red-billed Tropicbird (*Phaethon aethereus*), p. 115
L 36 in (91.4 cm). Uncommon and likely only seen far offshore. Has extensive black on outer wing but lacks black on inner wing and shows barring on back. Reddish bill.

4. White-tailed Tropicbird (*Phaethon lepturus*), p. 116
L 29 in (73.7 cm). Uncommon and likely only seen far offshore. Black on both outer wing and inner wing, and lacks barring on back (except in juvenile plumage). Yellowish bill.

Fregatidae: Frigatebirds

5. Magnificent Frigatebird (*Fregata magnificens*), p. 117
L 40 in (102 cm). Common resident seen soaring along shoreline. Very large, dark seabird; long, pointed wings with bend in middle; long, forked tail. Short neck; long, hooked bill. Adult male all black with red throat pouch. Adult female with white belly. Immature has white head and belly.

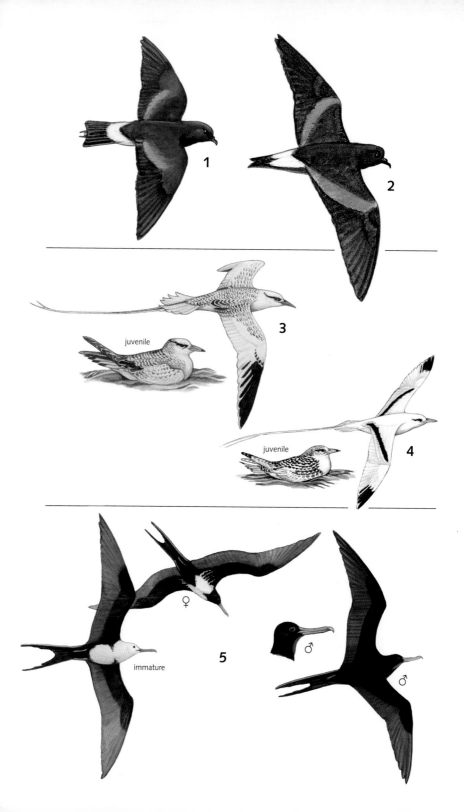

1

2

3

juvenile

4

juvenile

5

♀

immature

♂

♂

PLATE 6
Boobies

Sulidae: Boobies

1. Masked Booby (*Sula dactylatra*), p. 118

L 32 in (81.3 cm). Quite rare and likely seen far offshore. Large, white waterbird with long, pointed, black tail. Black wingtips and trailing edge of wing. Large, pointed, yellow bill. Black face mask. Immature with brown back and upper-wing, brown tail and head, white rump.

2. Brown Booby (*Sula leucogaster*), p. 119

L 30 in (76.2 cm). Only booby regularly seen from shore. Adult choco-late-brown above and on upper breast, contrasting with clean white under-parts. Large, yellow bill. Immature all dark but with pale underwing.

3. Red-footed Booby (*Sula sula*), p. 120

L 28 in (71.1 cm). Quite rare and likely seen far offshore. White form lacks black tail of Masked Booby and has more extensive black on underwing. White-tailed brown form is unmistakable. Brown morph is like immature Brown Booby but smaller, with smaller gray bill and darker underwing. Immatures like immature Brown Booby but with dark underwing contrasting with pale body.

1

juv.

2

juv.

brown
morph

white-tailed
brown morph

3

white
morph

brown
morph

PLATE 7
Cormorants, Pelicans

Phalacrocoracidae: Cormorants

1. Double-crested Cormorant (*Phalacrocorax auritus*), p. 122
L 33 in (83.8 cm). Exceptionally rare. Much larger and bulkier than common Neotropical Cormorant, with orange (not dark) lores. Adult lacks white border to throat pouch.

2. Neotropic Cormorant (*Phalacrocorax brasilianus*), p. 121
L 25 in (63.5 cm). Common at permanent and semipermanent water bodies on Aruba and Curaçao and becoming increasingly regular on Bonaire. Black body; yellowish throat and bill pouch. Adult has thin white border to throat pouch. Immature dirty brownish, but paler on undersides. All ages show dark lores.

Pelecanidae: Pelicans

3. Brown Pelican (*Pelecanus occidentalis*), p. 122
L 51 in (129.5 cm). Common resident. An unmistakable large, dark waterbird; has large, pouched bill.

1

nonbreeding adult

breeding adult

immature

2

immature

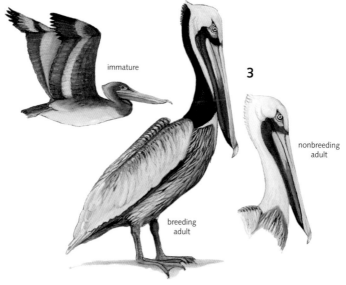

immature

3

nonbreeding adult

breeding adult

PLATE 8
Herons, Egrets, Bitterns

Ardeidae: Herons, Egrets, Bitterns

1. **Pinnated Bittern** (*Botaurus pinnatus*), p. 123
   25–30 in (63–76 cm). Exceptionally rare with only a single record from Aruba. Back buffy with black speckles. White throat. Foreneck streaked brown; back of neck buffy with black barring. Breast and belly white with brown streaks. Broad bill, yellowish overall with darker upper mandible.

2. **Least Bittern** (*Ixobrychus exilis*), p. 124
   L 10–11 in (25–28 cm). Rare visitor to Bubali, Aruba, where numbers have increased in recent years. Tiny heron with dark cap and back contrasting with buffy sides of face, neck, and nape; large buffy wing patches. Long yellow bill and yellow legs.

3. **Green Heron** (*Butorides virescens*), p. 129
   L 15-17 in (38–43 cm). Common breeding resident. Small, dark, and rather short heron; bright yellow legs, dark rufous neck and face, darker on cap; greenish-gray back and upperwing. Immature with brownish stripes on neck and cheek.

4. **Striated Heron** (*Butorides striatus*), p. 130
   L 15–17 in (38–43 cm). Vary rare visitor from South America. Similar to Green Heron but rufous replaced with gray on neck and face. Immature may not be distinguishable from Green Heron.

5. **Great Blue Heron** (*Ardea herodias*), p. 125
   L 40–50 in (102–127 cm). Commonly seen in small numbers year-round but not known to breed. Large heron with blue-gray body; grayish-purple neck; wide, black stripe over eye; rufous thighs. Immature dingier, with black cap.

PLATE 9
Herons, Egrets, Bitterns

Ardeidae: Herons, Egrets, Bitterns *continued*

1. Black-crowned Night-Heron (*Nycticorax nycticorax*), p. 131

L 23–28 in (58–71 cm). Common to uncommon resident. Medium-sized, stocky heron with short neck and thick, black bill. Black cap and back. Gray wings, white underparts. Immature brown, with yellowish bill and large white spots on wing coverts; has shorter legs than immature Yellow-crowned Night-Heron.

2. Yellow-crowned Night-Heron (*Nyctanassa violacea*), p. 131

L 22–27 in (56–69 cm). Common to uncommon resident. Medium-sized, stocky heron. Black face and bill, white cheek. Gray body, yellowish crown stripe. Immature with all-black bill and only small white spots on wing coverts, with longer legs than immature Black-crowned Night-Heron.

3. Boat-billed Heron (*Cochlearius cochlearius*), p. 132

L 18–20 in (46–51 cm). Very rare visitor from South America to Bonaire. Medium-sized heron with extremely large, broad dark bill, black cap and nape. Pale gray upperparts; white face, forehead, throat, and breast. Lower underparts pale rufous, with black flanks. Immature has all-brown upperparts.

4. Whistling Heron (*Syrigma sibilatrix*), p. 130

L 21-23 in (53-58 cm). Very rare visitor from South America to Aruba and Bonaire. Very distinctive heron with tan neck, black cap extending down and around colorful blue skin surrounding eye. Bill pink with black tip.

5. Tricolored Heron (*Egretta tricolor*), p. 127

L 22-26 in (56–66 cm). Common resident. Only common heron in the islands with dark upperparts and neck contrasting with white lower belly. Immature has rusty neck and upper back.

immature

immature

immature

immature

1

2

3

4

5

immature

PLATE 10
Herons, Egrets, Bitterns

Ardeidae: Herons, Egrets, Bitterns *continued*

1. Little Blue Heron (*Egretta caerulea*), p. 127
L 22–29 in (56–74 cm). Common resident. Adults are dark blue-gray, with purplish neck. Black-tipped bill with lighter gray base. Legs greenish-yellow. Immature birds start off all white but become mottled.

2. Reddish Egret (*Egretta rufescens*), p. 128
L 30 in (76 cm). Common resident on Curaçao and Bonaire, occasional on Aruba. Dark gray overall, with reddish neck and dark gray legs. Bill pink at base, with black tip. White form all white with two-toned bill and dark gray legs.

3. Great Egret (*Ardea alba*), p. 125
L 35–41 in (89–104 cm). Common resident. Large all-white heron with yellow-orange bill and black legs.

4. Little Egret (*Egretta garzetta*), p. 126
L 20–27 in (51–69 cm). Very rare with only a single confirmed record as of 2016 but perhaps to be expected more regularly. Very similar to Snowy Egret but note two long head plumes in breeding plumage as compared to shorter tuft of many feathers in plume of Snowy Egret. Little Egret tends to show darker lores than Snowy but this is variable depending on breeding condition. The bill of Little Egret usually appears longer than in Snowy Egret.

5. Snowy Egret (*Egretta thula*), p. 126
L 20–27 in (51–69 cm). Common resident. Slender all-white heron with dark bill. Lores usually yellowish. Adult has black legs and yellow feet. Immatures and nonbreeding adults have yellowish legs, but usually with black running down the front side of the legs; bill with varying amounts of pale yellow except for black tip. Breeding adults lack long head plumes (very rare Little Egret has two long head plumes in breeding plumage).

6. Cattle Egret (*Bubulcus ibis*), p. 128
L 20 in (51 cm). Common to uncommon resident. Medium-sized heron with short, thick neck. All-white plumage contrasts with yellow bill and (in nonbreeding plumage) dark legs and feet. In breeding season, bill and legs turn yellow-orange; buffy on crown, breast, and back.

immature

first
summer

1

immature

2

adult
dark morph

adult
pale morph

3

4

5

nonbreeding
adult

6

breeding
adult

nonbreeding
adult

PLATE 11
Limpkin, Storks, Flamingos

Aramidae: Limpkin

1. Limpkin (*Aramus guarauna*), p. 143
L 25–28 in (64–71 cm). Very rare visitor with only a single confirmed record. Tall, long-legged heronlike bird (not actually a heron). Brown overall, but head, neck, and back streaked and spotted with white. Long, stout, dull yellow bill; dark legs.

Ciconiidae: Storks

2. Wood Stork (*Mycteria americana*), p. 117
L 40–44 in (101–112 cm). Extremely rare, with only a single record. Very large, long-legged white bird with bald, black head. Long, thick, down-curved bill. White wings, black flight feathers, and black tail. On immature, neck and most of head whitish; pale bill.

Phoenicopteridae: Flamingos

3. American Flamingo (*Phoenicopterus ruber*), p. 109
L 48 in (122 cm). Abundant resident breeder on Bonaire. Common on Curaçao. Small numbers have become regular on Aruba in recent years. Unmistakable large, pink wading bird; thick, down-turned bill with black tip. In flight, note black on trailing edge of wings. Immature paler overall.

1

2

immature

3

immature

PLATE 12
Ibises, Spoonbills

Threskiornithidae: Ibises, Spoonbills

1. White Ibis (*Eudocimus albus*), p. 132
L 22–24 in (56-61 cm). Rare visitor. Medium-sized white wading bird with long, red legs and long, down-curved, bright red bill. Black on tips of wings. Immature with brown upperwing, back, and tail tip contrasting with white rump and white underparts; in certain light, indistinguishable from immature Scarlet Ibis. Immature Scarlet Ibis eventually shows pink tones to white rump and underparts.

2. Scarlet Ibis (*Eudocimus ruber*), p. 133
L 22–24 in (56-61 cm). Rare visitor. Adult quite unmistakable, with red plumage and legs; and long, down-curved bill. Immature sometimes indistinguishable from White Ibis, but eventually shows pink tones to white rump and underparts.

3. Glossy Ibis (*Plegadis falcinellus*), p. 133
L 22–24 in (56-61 cm). Rare visitor. Medium-sized, dark-bodied wading bird with long, down-curved bill and long, darkish legs. Metallic green sheen on back and wings. Difficult to distinguish from White-faced Ibis without clear view of face. In breeding plumage, Glossy Ibis shows dark area between eye and bill with thin white lines that extend from bill to eye, but not around eye; iris is dark (reddish in White-faced Ibis). Immature similar to adult, but body duller; almost indistinguishable from White-faced Ibis.

4. White-faced Ibis (*Plegadis chihi*), p. 134
L 22–24 in (56-61 cm). Very rare visitor, no recent records. Medium-sized, dark-bodied wading bird with long, down-curved bill and long, dark legs. In breeding plumage, note reddish skin between eye and base of bill and bold white lines extending from bill base to and around eye. Iris reddish (dark in Glossy Ibis). Immature similar to adult, but body duller; almost indistinguishable from Glossy Ibis.

5. Roseate Spoonbill (*Platalea ajaja*), p. 135
L 30–32 in (76–81 cm). Rare visitor but perhaps becoming somewhat more regular on Aruba. Unmistakable, pink wading bird with a bald head, reddish legs, and distinctive long, spoon-shaped bill. Immature paler than adult, with white, feathered head.

1

immature

immature

2

immature

3

immature

4

immature

breeding
adult

breeding
adult

5

immature

PLATE 13
Osprey, Hawks, Kites

Pandionidae: Osprey

1. Osprey (*Pandion haliaetus*), p. 136
L 21–24 in (53–61 cm). Common winter visitor and occasional throughout the year. Large, long-winged hawk. Dark upper back, wings, and tail contrast with white head and breast. White head shows dark mark through eye. In flight, note distinctive "kink" in wings; underside of wings light, with black patches at "wrist," wingtips, and trailing edge of inner wing. Tail narrowly barred.

Accipitridae: Hawks, Kites

2. White-tailed Kite (*Elanus leucurus*), p. 137
L 15–16 in (38–41 cm). Very rare visitor (no recent records). Small, whitish, graceful hawk with long tail, pointed wings. All-white head, tail, and underside; pale gray back and upper wings sharply contrast with black bar on shoulders. In flight, undersurface of wing shows black outer wings with black spot at wrist.

3. Swallow-tailed Kite (*Elanoides forficatus*), p. 137
L 22-26 in (56-66 cm). Very rare stop-over migrant. Unmistakable gentle-looking hawk with white head, neck, and underside; black upper wings and tail with a deeply forked tail unlike any other hawk.

4. White-tailed Hawk (*Geranoaetus albicaudatus*), p. 139
L 21–23 in (53–58 cm). Resident breeder on Curaçao and perhaps still on Bonaire but no longer present on Aruba. Stocky hawk with broad, very short tail and broad, bulging wing-shape in flight. White below, dark above, with rufous shoulder patch. White tail with distinctive black tail band. Immature dark chocolate-brown, with yellowish or white on upper breast and on side of head.

5. Long-winged Harrier (*Circus buffoni*), p. 138
L 18–24 in (46-61 cm). Very rare visitor from South America. Long winged and long tailed hawk with a white rump. Often glides low over wetland areas looking for prey. Much variation but typically with dark upper breast band, strong contrasting facial pattern, and contrasting lighter patches in outer part of upperwing. More barring in underwing than in Northern Harrier.

6. Northern Harrier (*Circus cyaneus*), p. 138
L 18–21 in (46-53 cm). Apparently very rare vagrant from North American migratory population based on a single record. Lacks strong contrasting pattern in upperwing of Long-winged Harrier. Immature Northern Harrier has strong cinnamon tones in breast and underwing linings lacking in Long-winged Harrier.

1

2

immature

3

immature

4

5

♂
pale morph

♀
pale morph

immature

♂

immature

6

♀

PLATE 14
Falcons, Caracaras

Falconidae: Falcons, Caracaras

1. Yellow-headed Caracara (*Milvago chimachima*), p. 208
L 16–18 in (41–46 cm). Very rare with only a single historical record. Buffy-white below, dark above. Head and neck pale, with dark stripe behind eye. White patch at base of outer primaries conspicuous in flight.

2. Crested Caracara (*Caracara cheriway*), p. 207
L 20–22 in (51–56 cm). Resident breeding bird on all three islands but likely declining on Aruba. Large, long-legged raptor; dark cap, body, and wings contrast with white face, neck, and throat; large, brightly colored bill. In flight shows flashing patches of white at wing tips and white tail with black terminal band.

3. American Kestrel (*Falco sparverius*), p. 208
L 9–12 in (23–30 cm). Common on Aruba and Curaçao, rare on Bonaire. Small hawk with long tail and pointed wings; often seen perched on wires or hovering. Male has rust-red back and tail, with black terminal band on tail, bluish wings, black teardrop mark coming down from eye, largely unstreaked underparts. Female similar but bluish wings replaced with rust-brown; also note streaking on underparts.

4. Merlin (*Falco columbarius*), p. 209
L 10–14 in (25–36 cm). Regular winter visitor in small numbers. Small hawk with long tail and pointed wings. Male with dark bluish-gray upperparts; female and immature with dark brown upperparts. Note heavily streaked underparts; tail with dark, wide bands; also note dark on crown and weak teardrop mark below eye.

5. Peregrine Falcon (*Falco peregrinus*), p. 209
L 15–21 in (38–53 cm). Regular winter visitor in small numbers. Large hawk with long tail and pointed, but broad, wings. Adults have dark bluish-gray upperparts and pale, barred underparts, with dark "helmet" and strong mark that extends down from eye and contrasts with white cheek. In immature, dark blue-gray is replaced with brown; also note paler crown.

1

immature

2

immature

3

♂

♀

♂

5

4

♂

♂

♀
immature

immature

PLATE 15
Vultures, New World Quail, Sungrebe

Cathartidae: Vultures

1. Turkey Vulture (*Cathartes aura*), p. 136
L 26-30 in (66–76 cm). Very rare vagrant to Aruba. Massive black bird with a bare, red head. Note silvery underwing flight feathers contrasting with darker wing linings. In flight holds its wings in a shallow V.

2. Black Vulture (*Coragyps atratus*), p. 135
L 22-26 in (56–66 cm). Sightings on Aruba apparently pertain to one or more introduced birds; recent record of 2–3 on Curaçao, apparently wild vagrants. Large dark vulture, a bit smaller than Turkey Vulture with a dark head (juvenile Turkey Vulture also has a dark head) and short tail. In flight shows silvery wingtips.

Odontophoridae: New World Quail

3. Crested Bobwhite (*Colinus cristatus*), p. 108
L 8 in (20 cm). Resident breeding bird on Aruba and Curaçao. Small, stocky, short-tailed quail with pointed crest; mottled brown overall. Typically seen walking through shrubby areas in small groups.

Heliornithidae: Sungrebe

4. Sungrebe (*Heliornis fulica*), p. 143
L 11-12 in (28-30.5 cm). Very rare vagrant with single record from Bonaire. Generally quite distinctive with black-and-white striped head, white underparts, olive upperparts. Female has reddish bill, male's bill is pale with dark along upper ridge of upper mandible.

immature

1

2

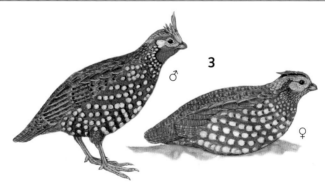

♂

3

♀

4

male/
nonbreeding female

breeding

♀

PLATE 16
Rails, Gallinules, Jacanas

Rallidae: Rails, Gallinules

1. Sora (*Porzana carolina*), p. 140
L 8–10 in (20–25 cm). Regular winter visitor, quite common around wetlands in some years. A small chickenlike waterbird. Most likely seen creeping along the edges of a marshy wetland or out onto floating vegetation. Note short, cocked tail and long. yellow legs and feet. Brown above, streaked and speckled with white; adult with gray face and upper breast; barring on flanks; yellow bill. Immatures similar but more nondescript, with tan replacing gray on face and upper breast, and with darker bill.

2. Purple Gallinule (*Porphyrio martinicus*), p. 140
L 11–13 in (28–33 cm). Uncommon and irregular visitor. Small, stocky waterbird. Greenish back and wings; deep bluish-purple head, neck, and underparts. White undertail. Yellow legs, red eye, red bill with yellow tip, and pale blue frontal shield. Immature much more nondescript, with bluish-purple colors replaced by browns; also note dusky bill.

3. Common Gallinule (*Gallinula galeata*), p. 141
L 13 in (33 cm). Common resident at permanent or semipermanent water bodies on Aruba and Curaçao. Medium-sized, stocky waterbird. All dark. with darker head and neck; white horizontal line along flank; white patch at rear. Startling red bill and frontal shield. Yellowish legs and toes.

4. Carribean Coot (*Fulica caribaea*), p. 141
L 13–15 in (33–38 cm). Common resident at permanent or semipermanent water bodies on all three islands. Medium-sized waterbird. Dark gray overall, with neck and head darker. Silvery-white bill with reddish dot near tip. White frontal shield, lacking any dark reddish tones (but see species account, p.142, to compare with American Coot).

Jacanidae: Jacanas

5. Wattled Jacana (*Jacana jacana*), p. 152
L 9.5 in (24 cm). Very rare visitor. Very long-legged and long-toed waterbird with black body and chestnut back and wings. In flight, note greenish-yellow flight feathers. Bill yellow, with red frontal shield and reddish wattle.

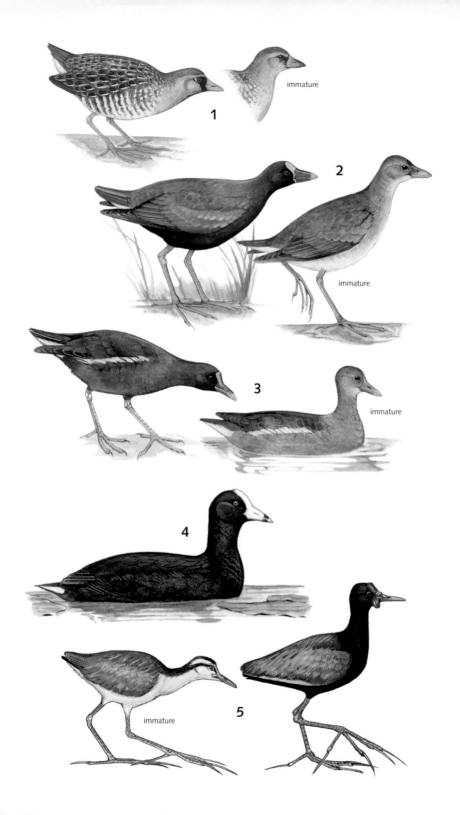

1

immature

2

immature

3

immature

4

5

immature

PLATE 17
Oystercatchers, Stilts, Avocets, Thick-knees

Haematopodidae: Oystercatchers

1. American Oystercatcher (*Haematopus palliatus*), p. 145
L 16–18 in (41–46 cm). Uncommon resident breeding bird on all three islands. Large, stocky shorebird with brown back, black head and neck, and white underside. Also note red bill, pinkish legs, and yellow eye with red eye-ring.

Recurvirostridae: Stilts, Avocets

2. Black-necked Stilt (*Himantopus mexicanus*), p. 144
L 13–16 in (33–41 cm). Common resident breeding bird. Large, striking black-and-white shorebird with very long, pink legs. Commonly occurs in flocks in mangrove wetlands and salinas.

3. American Avocet (*Recurvirostra americana*), p. 145
L 16–20 in (41–51 cm). Exceptionally rare vagrant. Large shorebird with long gray legs, long upcurved bill. Striking black and white pattern on body. Head and neck buffy-orange in breeding male, pale gray in female and immature.

Burhinidae: Thick-knees

4. Double-striped Thick-knee (*Burhinus bistriatus*), p. 144
L 18–20 in (46–50 cm). Exceptionally rare vagrant. Very unlike other shorebirds, with long, thick yellowish legs, stout dark bill, very large yellow eye, and with white eyebrow bordered with black to give it a scowling look.

breeding
adult

nonbreeding
adult

1

2

3

4

PLATE 18
Plovers, Lapwings

Charadriidae: Plovers, Lapwings

1. Southern Lapwing (*Vanellus chilensis*), p. 147
L 13–15 in (33–38 cm). Increasing and now resident breeding bird at multiple locations in Aruba and Curaçao and regular on Bonaire. Large, crested shorebird with grayish-brown upperparts, white belly, and black breast band extending to forehead.

2. Black-bellied Plover (*Pluvialis squatarola*), p. 146
L 10–13 in (25–33 cm). Regular winter visitor in small numbers, occasional birds seen throughout the year. Stocky with short, relatively thick bill. Most likely seen in nonbreeding or juvenile plumage, when this species shows black "armpits" and white tail. In breeding plumage, note that black stops at lower belly, with white undertail.

3. American Golden-Plover (*Pluvialis dominica*), p. 146
L 9–11 in (23–28 cm). Uncommon migrant stopover visitor in late fall and early winter. Stocky with short, relatively thick bill. Most likely seen in nonbreeding or juvenile plumage, when this species lacks the black "armpits" and white tail of similar Black-bellied Plover. In breeding plumage, note black undertail; undertail white in Black-bellied Plover.

1

nonbreeding
adult

2

nonbreeding
adult

nonbreeding
adult

breeding
adult

nonbreeding
adult

3

nonbreeding
adult

breeding
adult

PLATE 19
Plovers, Lapwings

Charadriidae: Plovers, Lapwings *continued*

1. Collared Plover (*Charadrius collaris*), p. 148

L 5.5–6 in (14–15 cm). Rare to uncommon migrant visitor from South America. Small shorebird with muddy brown upperparts and white underparts with dark breast band. Dark area between bill and eyes, white stripe behind eye. Slender black bill; light-colored legs.

2. Piping Plover (*Charadrius melodus*), p. 150

L 7.25 in (18 cm). Very rare vagrant. Small shorebird very similar to more common Snowy Plover. Note that this species always shows yellow legs.

3. Snowy Plover (*Charadrius nivosus*), p. 148

L 6.5 in (16.5 cm). Resident breeding bird on Bonaire and perhaps at some locations on Curaçao. Small shorebird with sandy upperparts. Dark patches at front of head, behind eye, and on side of breast in breeding plumage. Legs and bill dark.

4. Wilson's Plover (*Charadrius wilsonia*), p. 149

L 7–8 in (18–20 cm). Resident breeding bird on Bonaire and perhaps at some locations on Curaçao. Rare visitor on Aruba. Medium-sized shorebird with muddy brown upperparts, single dark breast band, and large, thick bill that is black in any plumage or age. Pinkish legs. Rusty cinnamon tones to plumage around head and neck.

5. Semipalmated Plover (*Charadrius semipalmatus*), p. 150

L 6.5–7.5 in (16.5–19 cm). Regular winter visitor in small numbers. Small shorebird with muddy brown upperparts; white underparts with single dark breast band. Orange-yellow legs; small dark-tipped bill usually with orangish-base, brightest in breeding plumage.

6. Killdeer (*Charadrius vociferous*), p. 151

L 9–11 in (23–28 cm). Regular winter visitor on all three islands and occasional breeder on Aruba. Medium-sized shorebird with mud-brown upperparts and two dark breast bands. Dark, slender bill; pale legs.

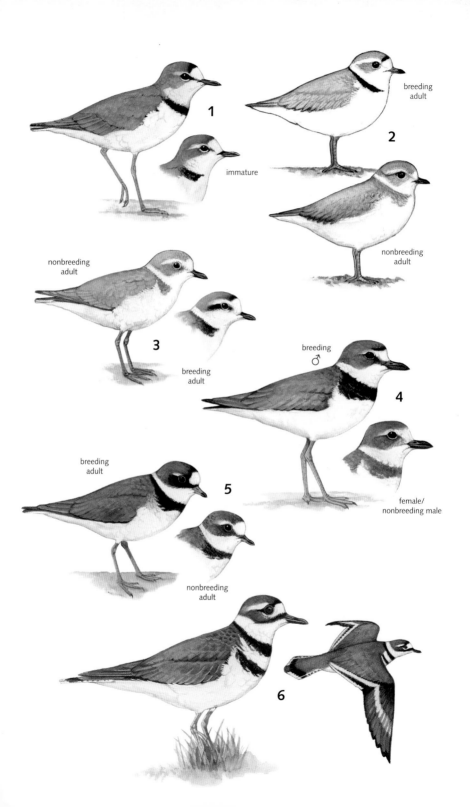

1

immature

2 breeding adult

nonbreeding adult

nonbreeding adult

3

breeding adult

breeding ♂

4

5 breeding adult

nonbreeding adult

female/ nonbreeding male

6

PLATE 20
Sandpipers, Allies

Scolopacidae: Sandpipers, Allies

1. Hudsonian Godwit (*Limosa haemastica*), p. 154
L 15 in (38 cm). Rare stop-over migrant, more likely during fall migration period. Large shorebird with very long, dark legs and very long, slightly upturned bill. In all plumages note black tail, white rump, and black underwing.

2. Whimbrel (*Numenius phaeopus*), p. 153
L 17 in (43 cm). Regular winter visitor in small numbers. Large shorebird with long, down-curved bill, brown upperparts, black stripes on head.

3. Upland Sandpiper (*Bartramia longicauda*), p. 153
L 11–12 in (28–30 cm). Rare stopover migrant. Medium-sized. A long-necked and small-headed shorebird. Eye appears proportionately large, set on pale face. Short, straight bill is pale with dark tip. Yellow legs.

4. Greater Yellowlegs (*Tringa melanoleuca*), p. 167
L 14 in (36 cm). Regular winter visitor. Medium-sized grayish shorebird. Long, bright yellow legs. Long, rather stout bill is usually slightly upturned. In flight, note white rump and dark tail.

5. Lesser Yellowlegs (*Tringa flavipes*), p. 168
L 10.5 in (27 cm). Regular winter visitor, sometimes quite abundant. Medium-sized grayish shorebird smaller than Greater Yellowlegs. Long, bright yellow legs. Long bill is thinner and shorter than that of Greater Yellowlegs and usually does not look upturned. In flight, note white rump and dark tail.

6. Solitary Sandpiper (*Tringa solitaria*), p. 166
L 8.5 in (22 cm). Regular winter visitor and stop-over migrant. Medium-sized grayish-brown shorebird. Sometimes confused with Lesser Yellowlegs, but note: Solitary is smaller; shows a white eye-ring; has shorter, greenish legs; and in flight shows a dark rump rather than the white rump on both yellowlegs.

7. Willet (*Tringa semipalmata*), p. 167
L 15 in (38 cm). Uncommon but regular winter visitor. Stocky, medium-sized grayish shorebird with thick bill and dark gray legs. In flight, note flashy black and white pattern on wings.

1 nonbreeding adult

breeding adult

nonbreeding adult

2

3

4

5

6

7 nonbreeding adult

breeding adult

nonbreeding adult

PLATE 21
Sandpipers, Allies

Scolopacidae: Sandpipers, Allies *continued*

1. Spotted Sandpiper (*Actitis macularius*), p. 166
L 7.5 in (19 cm). Regular winter visitor. Small shorebird with short pale legs. Behavior of constantly tipping back and forth is best key to identification. In all plumages note white "comma" mark in front of wing on sitting bird. In breeding plumage, underside is heavily spotted with black.

2. Ruddy Turnstone (*Arenaria interpres*), p. 154
L 8–10 in (20–25 cm). Regular winter visitor. Highly distinctive, small, stocky shorebird with short orangish legs and short, black bill. Face marked with unusual black and white markings; breeding plumaged birds very strikingly marked on wings and body with bright rust color. Only shorebird species that regularly begs for food around open-air restaurants and bars.

3. Red Knot (*Calidris canutus*), p. 155
L 10.5 in (27 cm). Uncommon but perhaps regular winter visitor on Bonaire. Stocky, rather squat shorebird. Legs vary from dark gray to dull yellow; note short, straight black bill. In nonbreeding plumage, nondescript flat gray upperparts, head, and breast. In breeding plumage, bright reddish head and underparts.

4. Sanderling (*Calidris alba*), p. 157
L 8 in (20 cm). Regular winter visitor, seen in small flocks along beaches. A slim shorebird seen running on beaches. In the ABCs, usually seen in pale, nonbreeding plumage, with light gray upperparts and very white underparts. Black leading edge to wing shows as black shoulder mark in folded wing.

5. Semipalmated Sandpiper (*Calidris pusilla*), p. 160
L 6.5 in (17 cm). Regular winter visitor, sometimes in flocks during migration. Small black-legged shorebird with stubby "cigar" of a bill, lacking any major droop at tip. In winter plumage very difficult to distinguish from rarer Western Sandpiper. In breeding plumage, lacks the rusty tones of Western Sandpiper.

6. Western Sandpiper (*Calidris mauri*), p. 161
L 6.5 in (17 cm). Rare winter visitor. Small black-legged shorebird with long bill, drooped at tip. Juveniles show bright rust-colored upper scapulars; pale around face and upper breast. In breeding plumage shows bright reddish in scapulars, face, and crown.

7. Least Sandpiper (*Calidris minutilla*), p. 158
L 6 in (15 cm). Regular winter visitor in small numbers. Smallest of the "peeps" and with the brownest tones. Also note yellow legs and fine short bill.

nonbreeding
adult

1

breeding
adult

nonbreeding
adult

2

breeding
adult

nonbreeding
adult

3

breeding
adult

nonbreeding
adult

4

breeding
adult

nonbreeding
adult

6

breeding
adult

nonbreeding
adult

5

breeding
adult

nonbreeding
adult

7

breeding
adult

PLATE 22
Sandpipers, Allies

Scolopacidae: Sandpipers, Allies *continued*

1. White-rumped Sandpiper (*Calidris fuscicollis*), p. 159
L 7.5 in (19 cm). Uncommon to rare stop-over migrant. One of the larger and stockier of the "peeps" (larger than Semipalmated and Western Sandpipers), and the only one that shows a white rump in flight. At rest the wingtips project beyond the tail, a feature shared only with the Baird's Sandpiper. Black legs.

2. Baird's Sandpiper (*Calidris bairdi*), p. 158
L 7.5 in (19 cm). Uncommon to rare stop-over migrant. Like the White-rumped Sandpiper, larger and longer-winged than Semipalmated and Western Sandpipers. Buffy breast, with scaly back. Wingtips extend beyond tail at rest. Has a dark rump (white rump on White-rumped Sandpiper). Black legs.

3. Pectoral Sandpiper (*Calidris melanotos*), p. 160
L 9 in (23 cm). Uncommon stop-over migrant. One of the largest and longest legged of the "peeps" and with very brownish tones. Combination of yellow legs and sharply demarcated breast streaking is distinctive.

4. Dunlin (*Calidris alpina*), p. 157
L 8.5 in (21 cm). Very rare nonbreeding vagrant. In breeding plumage, shows distinctive black belly patch that makes it virtually unmistakable. In nonbreeding and juvenile plumage, it is more nondescript, with gray-brown upperparts and white underparts; note rather long, black bill, down-curved at tip, and black legs.

5. Stilt Sandpiper (*Calidris himantopus*), p. 156
L 8.5 in (22 cm). Regular stop-over migrant and winter visitor. Uncommon on Aruba and Curaçao but can sometimes can occur in larger flocks on Bonaire. Slim, long-legged shorebird with long slightly drooped bill. Often confused with dowitchers but has shorter bill and, in flight, white rump rather than wedge of white on back, as in dowitchers. Distinctive feeding style suggests a pumping oil derrick tipping back and forth at both ends; note up-and-down sewing-machine motion of dowitchers. In breeding plumage, heavily barred on underparts, with chestnut cheeks. Yellow legs.

6. Buff-breasted Sandpiper (*Calidris subruficollis*), p. 159
L 8 in (20 cm). Rare stop-over migrant. Rather stout shorebird with yellow legs and black bill. Distinctively buff-colored underparts; buff extends up and surrounds dark eye.

nonbreeding adult

1

breeding adult

rump detail

nonbreeding adult

2

breeding adult

3

nonbreeding adult

4

breeding adult

breeding adult

nonbreeding adult

5

6

PLATE 23
Sandpipers, Allies

Scolopacidae: Sandpipers, Allies *continued*

1. Short-billed Dowitcher (*Limnodromus griseus*), p. 162
L 11 in (28 cm). Regular winter visitor. Stout, short-legged shorebird with very long straight bill. In flight, white triangle on back. Almost identical to Long-billed Dowitcher and best distinguished by voice.

2. Long-billed Dowitcher (*Limnodromus scolopaceus*), p. 163
L 11.5 in (29 cm). Rare, poorly documented, winter visitor. Stout, short-legged shorebird with very long straight bill. In flight, white triangle on back. Almost identical to Short-billed Dowitcher and best identified by voice.

3. Wilson's Snipe (*Gallinago delicata*), p. 163
L 10.5 in (27 cm). Regular and sometimes quite abundant winter visitor. Small, dark shorebird with very long bill, short yellow-green legs, black stripes on head, and light-colored stripes on back. Most often seen when flushed from edges of grassy pools.

4. Wilson's Phalarope (*Phalaropus tricolor*), p. 164
L 9 in (23 cm). Rare stop-over migrant. Distinctively shaped shorebird with tiny-looking head, black needle-like bill. In nonbreeding plumage, pale gray upperparts and white underparts and yellow legs. In flight, white rump and unmarked upperwings.

5. Red Phalarope (*Phalaropus fulicarius*), p. 165
L 8.5 in (22 cm). Rare vagrant. Small, distinctively shaped shorebird with small head, dumpy body. Typically seen floating and spinning on surface of water. In nonbreeding plumage, pale unmarked gray back, white underparts with black ear mark, heavy black bill. In flight, lacks white rump and shows bold white wingstripe. In breeding plumage, unmistakable, with brick-red underparts, white cheek, black cap, and yellowish bill with black tip.

6. Red-necked Phalarope (*Phalaropus lobatus*), p. 164
L 7.75 in (20 cm). Rare vagrant. Small, distinctively shaped shorebird with small head and dumpy body. Typically seen floating and spinning on surface of water. In nonbreeding plumage, note gray-streaked back, white underparts with black ear mark, fine black bill. In flight, lacks white rump and shows bold white wingstripe and dark streaked back. In breeding plumage, shows reddish neck.

1

nonbreeding adult

nonbreeding adult

detail of juvenile tertials

breeding adult

2

nonbreeding adult

detail of juvenile tertials

breeding adult

3

4

♂ breeding

♀ breeding

nonbreeding

nonbreeding adult

5

nonbreeding adult

nonbreeding adult

♀ breeding

6

nonbreeding adult

nonbreeding adult

♀ breeding

PLATE 24
Jaegars, Skuas

Stercorariidae: Jaegars, Skuas

1. Pomarine Jaegar (*Stercorarius pomarinus*), p. 171
L 18.5 in (47 cm); breeding adults longer. Rare and only occasionally seen from shore. Larger than Laughing Gull. A predatory seabird; rarely seen but usually occurs far offshore, harassing gulls and/or terns; almost falconlike in flight. White flash in outer wing. Largest of the three jaeger species, with thick-set body, blocky head, and heavy pink bill with black tip. In breeding plumage, tail feathers blunt, though this plumage not likely to be seen.

2. Parasitic Jaeger (*Stercorarius parasiticus*), p. 171
L 16.5 in (42 cm); breeding adults longer. Rare and only occasionally seen from shore. Slightly larger than Laughing Gull. A predatory seabird; rarely seen but usually occurs far offshore, harassing gulls and/or terns; almost falconlike in flight. White flash in outer wing. Appears slim, with a proportionately small, rounded head and thin bill; bill all dark in adults but light with dark tip in juveniles. In breeding plumage, central tail feathers pointed, though this plumage not likely to be seen.

3. Long-tailed Jaeger (*Stercorarius longicaudus*), p. 172
L 15 in (38 cm); breeding adults longer. Rare and only occasionally seen from shore. Smaller than Laughing Gull. A predatory seabird; rarely seen but usually occurs far offshore, harassing gulls and/or terns; almost falconlike in flight. White flash in outer wing is limited in adults and often difficult to see.

4. Great Skua (*Stercorarius skua*), p. 169
L 23 in (58 cm). Very rare well offshore. Large, brownish, predatory seabird. Rarely seen but usually occurs far offshore, harassing gulls and/or terns. Bold white wing patches. Tail broad and blunt. Much larger, bulkier, and broader-winged than jaegers, but Pomarine Jaeger, similar in size, could be mistaken at a distance. Great Skua typically shows more brown or reddish tones, with pale streaking on back and pale spotting on upperwing.

5. South Polar Skua (*Stercorarius maccormicki*), p. 170
L 21 in (53 cm). Very rare well offshore. Large brownish predatory seabird. Rarely seen but usually occurs far offshore harassing gulls and/or terns. Bold white wing patches. Tail broad and blunt. Much larger, bulkier and broader-winged than jaegers, but Pomarine Jaeger, similar in size, could be mistaken at a distance. South Polar Skua typically appears more uniformly dark (or sometimes uniformly light in "blonde" form) overall, with a contrasting pale nape and sometimes a prominent pale area at base of bill.

breeding adult

breeding adult pale morph

immature pale morph

immature pale morph

1

adult

breeding adult pale morph

immature pale morph

immature pale morph

2

breeding adult

breeding adult

immature

3

immature

4

5

PLATE 25
Gulls, Terns

Laridae: Gulls, Terns

1. Laughing Gull (*Leucophaeus atricilla*), p. 174
L 16.5 in (41 cm). The only gull commonly seen. A small gull with a dark gray back. In breeding season, note black hood with white eye crescent, reddish bill, black legs. In nonbreeding season, dark on nape forms a partial hood; also note black bill. Immature with black tail band, brownish tones in back and wings.

2. Franklin's Gull (*Leucophaeus pipixcan*), p. 175
L 14.5 in (37 cm). Only a few records. Adult has black wing tips separated from gray wing feathers by white band. Breeding adult has black hood, red bill. Nonbreeding adult shows partial dark hood, black bill. In immature, note that black tail band does not extend to outer edge of tail, unlike in Laughing Gull.

3. Black-headed Gull (*Chroicocephalus ridibundus*), p. 173
L 15 in (38 cm). Extremely rare. Light gray back and wings with large white wedge in outer wing. Note small red bill (at least at base). Reddish legs. In breeding plumage, adult has all-black hood that is reduced to black spot behind eye in nonbreeding plumage. Immature shows mottled browns and blacks on back and wings, with black trailing edge to wing and large area of white in wingtip.

4. Bonaparte's Gull (*Chroicocephalus philadelphia*), p. 173
L 12–14 in (30–36 cm). Very rare gull. Very small. Light gray back and wings with large white wedge in outer wing. Small, all-black bill. Pink legs. In breeding plumage, adult has all-black hood that is reduced to black spot behind eye in nonbreeding plumage. Immature shows mottled browns and blacks on back and wings, with black trailing edge to wing and large area of white in wingtip.

5. Ring-billed Gull (*Larus delawarensis*), p. 175
L 18–20 in (46–51 cm). Rare but regular visitor to the islands. Small gray-backed gull with black wing tips, slightly larger than common Laughing Gull. No hood. Adult has dark ring on yellow bill. Yellow legs. Immature has pink-based bill with dark tip; brown and black mottling in back, wings, and dark tail band.

6. Herring Gull (*Larus argentatus*), p. 176
L 23–26 in (58–66 cm). Rare but regular visitor. Has gray back and black wing-tips. No hood (though winter-plumaged adult may have streaking on nape and back of head). Large, yellow bill lacks black ring, though some immature in later stages can show extensive black near bill tip. Pink legs. Takes four years to reach adult plumage; in first two years, birds are brown overall, with a broad, dark tail band and an all-dark bill that takes on a pink base with age.

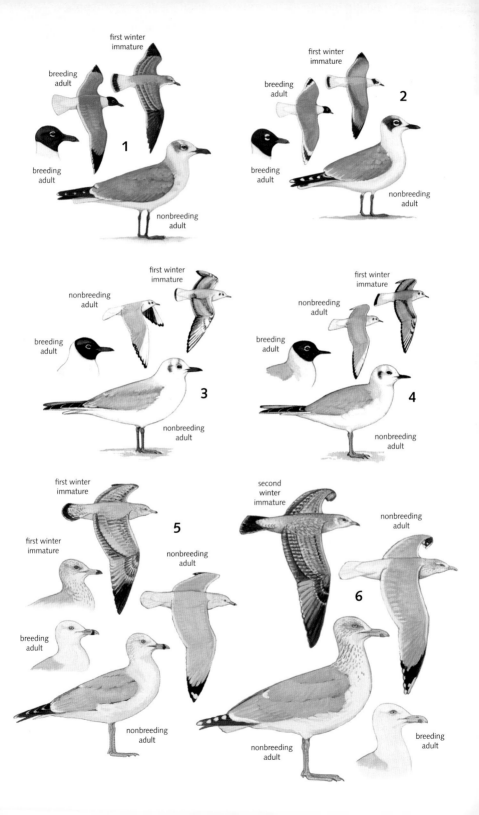

PLATE 26
Gulls, Terns

Laridae: Gulls, Terns *continued*

1. Lesser Black-backed Gull (*Larus fuscus*), p. 177
L 21 in (53 cm). Rare but regular visitor to the islands. Large gull with dark back and dark wingtips. In adult plumage, back and upperwing usually noticeably darker than on Herring Gull, but some forms are only slightly darker. On adult, note yellow legs and yellow bill. Adult in winter plumage shows heavily streaked nape and head, unlike Great Black-backed adult in winter plumage, which has limited streaking. Immatures are similar to immature Herring Gulls and can be difficult to identify, but note the darker outer wing and wing coverts and more contrasting rump of Lesser Black-backed Gull.

2. Great Black-backed Gull (*Larus marinus*), p. 178
L 30 in (76 cm). This rare visitor to the islands is larger than any other gull found there. Back is jet black; large bill. In adult plumage, note pink legs, unlike the yellow legs of Lesser Black-backed Gull. Immatures in first and second years are more mottled than Herring and Lesser Black-backed Gulls, with very checkered appearance on the back, very light head, pale rump, and large bill changing from all black to pink-based with age.

3. Brown Noddy (*Anous stolidus*), p. 179
L 15 in (38 cm). Breeds on Aruba's San Nicolas reef islands. Rare elsewhere. Medium-sized tern. Dark brown body with white crown. Double-rounded tail.

4. Black Noddy (*Anous minutus*), p. 179
L 13.5 in (34 cm). Breeds on Aruba's San Nicolas reef islands. Rare elsewhere. Medium-sized tern. Black body with demarcated white crown.

5. Sooty Tern (*Onychoprion fuscatus*), p. 180
L 16 in (41 cm). Breeds on Aruba's San Nicolas reef islands. Rare elsewhere. Medium-sized tropical tern, a little larger and stockier than Bridled Tern. Black on wings and back, black cap (with white on forehead extending back to top of eye but not beyond), slim black bill, and deeply forked black tail with narrow white edges. White does not extend as far around back of head as in Bridled Tern. Juvenile has a dark head, breast, and upperparts but white underwing.

6. Bridled Tern (*Onychoprion anaethetus*), p. 180
L 14–15 in (36–38 cm). Breeds on Aruba's San Nicolas reef islands. Rare elsewhere. Medium-sized tropical tern. Dark gray on wings and back, black cap, white forehead (with white extending back behind eye), slim black bill, and deeply forked gray tail with broad white edges. White extends around nape, almost forming a full collar.

nonbreeding
adult

1

first winter
immature

breeding
adult

nonbreeding
adult

2

first winter
immature

breeding
adult

3

4

immature

5

6

breeding
adult

immature

PLATE 27
Gulls, Terns

Laridae: Gulls, Terns *continued*

1. Black Tern (*Chlidonias niger*), p. 184

L 9–10 in (23–25 cm). Rare stopover migrant. Unmistakable in breeding plumage, when head and body are jet black and wings dark gray. In nonbreeding and immature plumage, note dark upperwing, black "helmet," dark shoulder hook extending onto white undersides, and all-gray underwing. In all plumages, note short, blunt tail and fine bill.

2. Least Tern (*Sternula antillarum*), p. 181

L 8–10 in (20–25 cm). Breeding colonies occur along coastal locations on all three islands from April through July. Smallest tern on the islands. Yellow legs, yellow bill with black tip, black cap, and white forehead in breeding season. In nonbreeding and immature plumage, bill is black; in place of black cap, black extends from eye across back of head.

3. Gull-billed Tern (*Gelochelidon nilotica*), p. 183

L 13–15 in (33–38 cm). Occasional rare migrant. Medium-sized, stocky tern with stout, black, gull-like bill; black legs; and very pale gray on back and upperwing. In nonbreeding plumage, black cap is replaced by a black patch behind eye.

4. Yellow-billed Tern (*Sternula superciliaris*), p. 182

L 8–10 in (20–25 cm). Exceptionally rare, at best; has never been unequivocally documented. Very similar to Least Tern but bill is slightly larger and has a thicker base; bill is all yellow in breeding plumage. In nonbreeding plumage, bill is mostly yellow, with dusky tip and dusky shading around nostrils. Immature very similar to immature Least Tern.

5. Large-billed Tern (*Phaetusa simplex*), p. 182

L 14.5 in (37 cm). A very rare visitor from the South American mainland, with only one record. Large tern. Quite unmistakable, with large, yellow bill, black cap, slate-colored mantle, bold white wing patches, black wingtips, short dark tail, and yellow legs.

6. Caspian Tern (*Hydroprogne caspia*), p. 183

L 19–23 in (48–58 cm). A rare visitor. Large tern; same size as Laughing Gull. Has large, red bill and black cap. Even in nonbreeding plumage, does not show white forehead of similar Royal Tern, though forehead is sometimes streaked with black rather than solid black.

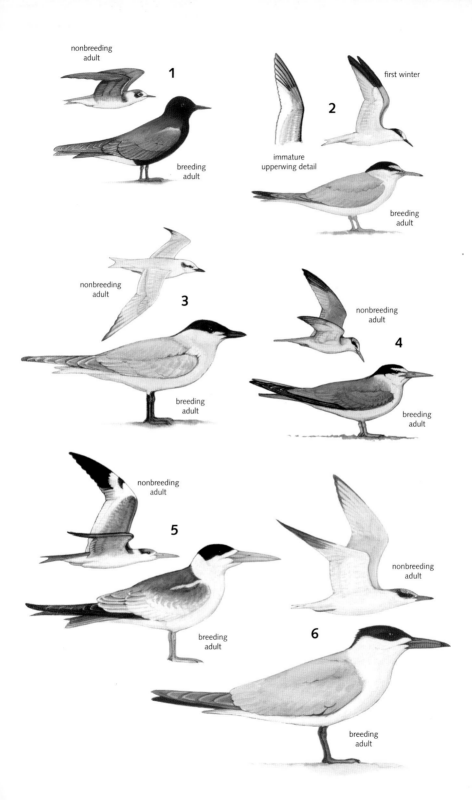

1
nonbreeding adult
breeding adult

2
first winter
immature upperwing detail
breeding adult

3
nonbreeding adult
breeding adult

4
nonbreeding adult
breeding adult

5
nonbreeding adult
breeding adult

6
nonbreeding adult
breeding adult

PLATE 28
Gulls, Terns

Laridae: Gulls, Terns *continued*

1. Roseate Tern (*Sterna dougalli*), p. 184
L 12.5 in (32 cm). Uncommon breeding bird. In flight, upperwings are very pale whitish, with limited darker gray-black on leading edge of outer wing. Has a very long tail; at rest, tail projects beyond wingtips. Proportionately shorter wings than similar Common Tern, so has faster, choppier wingbeats. Bill sometimes all dark; note that on Caribbean race lower two-thirds of bill is red (as on Common Tern) but still usually shows a more extensive dark tip. On the underwing, Roseate lacks the thin, dark trailing edge to the outermost primaries and the undersides appear translucent. In juvenile and first-year plumage, shows dark along front edge of inner wing but lacks any bars on trailing edge of wing.

2. Common Tern (*Sterna hirundo*), p. 185
L 12 in (30 cm). Occurs in small numbers, as a summer breeding species. Medium-sized tern. Black cap, red legs, and orange-red bill with black tip in breeding season. In nonbreeding plumage, bill black and cap is replaced by black patch. Note large wedge of black on wingtip of upperwing and thick black trailing edge to outer part of underwing in flight. Immature shows dark bars across leading and trailing edge of inner part of upperwing in flight.

3. Royal Tern (*Thalasseus maximus*), p. 186
L 18–21 in (46–53 cm). The most commonly seen tern in the islands, especially in winter. A large tern with large, bright orange bill. Except for a very brief period at the height of the breeding season, always shows a white forehead.

4. Sandwich Tern (*Thalasseus sandvicensis acuflavida*), p. 187
L 16 in (41 cm). Common to uncommon resident, breeding in some locations. A little smaller than Royal Tern. Has a slim black bill with a yellow tip and black legs. Immature lacks strongly contrasting dark bars on inner wing, as seen in other species, and has a dark bill.

5. Sandwich "Cayenne" Tern (*Thalasseus sandvicensis eurygnathus*), p. 187
L 16 in (41 cm). Common to uncommon resident, breeding in some locations. Like Sandwich Tern, but with all-yellow bill. Intermediates between the two forms are seen quite regularly.

6. Black Skimmer (*Rynchops niger*), p. 188
L 18 in (46 cm). Uncommon but increasingly regular migrant visitor from South America. Large, distinctive, ternlike seabird. Large black and bright orange bill with longer lower mandible.

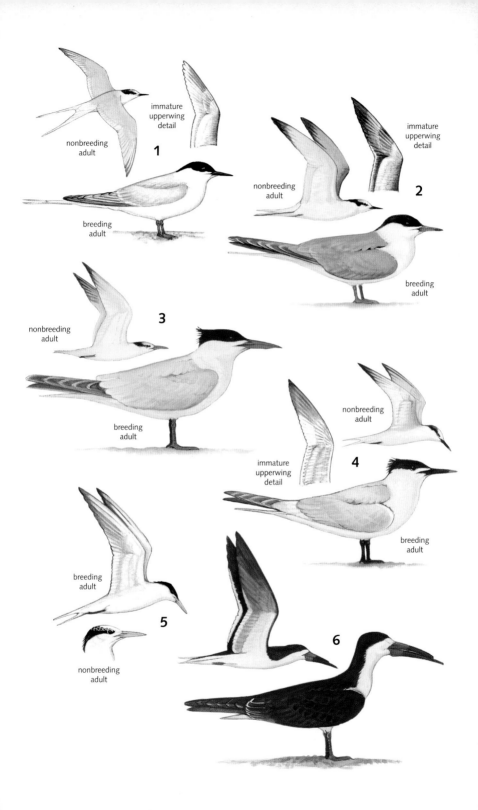

1
nonbreeding adult
immature upperwing detail
breeding adult

2
nonbreeding adult
immature upperwing detail
breeding adult

3
nonbreeding adult
breeding adult

4
nonbreeding adult
immature upperwing detail
breeding adult

5
breeding adult
nonbreeding adult

6

PLATE 29
Pigeons, Doves

Columbidae: Pigeons, Doves

1. Rock Pigeon (*Columba livia*), p. 189
12.5 in (32 cm). Common in urban and suburban areas. Medium-sized, stocky dove, with a variety of color morphs. This is the pigeon most commonly seen in cities.

2. Scaly-naped Pigeon (*Patagioenas squamosa*), p. 190
L 14–16 in (36–41 cm). Found only on Curaçao and Bonaire. Large dark pigeon with scaly nape. Note bare patch of skin around eyes (reddish in males, yellowish in females).

3. Bare-eyed Pigeon (*Patagioenas corensis*), p. 190
L 12–14 in (32–37 cm). Common and widespread on all three islands. Large pale pigeon with bold white wing patches in flight. At close range, note distinctive black "goggles" around eyes and pale bill.

4. White-tipped Dove (*Leptotila verreauxi*), p. 192
L 10 in (26 cm). Uncommon but widespread, especially in less developed areas. Medium-sized dove. Plain, unmarked dove; short, square tail with white edges.

5. Eared Dove (*Zenaida auriculata*), p. 193
L 9.5 in (24 cm). Abundant, particularly near suburban areas and resorts. Medium-sized dove. Square tail with cinnamon corners. Note black line extending from behind eye.

6. Common Ground-Dove (*Columbina passerina*), p. 191
L 6.5 in (17 cm). Common and widespread. Tiny dove. Rufous wings; reddish or yellowish at base of bill. Scaly appearance behind crown and on neck and upper breast.

7. Ruddy Ground-Dove (*Columbina talpacoti*), p. 192
L 6.5 in (17 cm). Very rare with only a single record. Tiny dove. Rufous back; gray head; black spots on wings.

PLATE 30
Cuckoos

Cuculidae: Cuckoos

1. Greater Ani (*Crotophaga major*), p. 194
L 19 in (48 cm). Rare vagrant to Aruba and Curaçao. A much larger version of the Groove-billed Ani, with pale eyes and larger, more keeled bill. Groove-billed Ani has dark eyes and a smaller, less ridged bill. Male Carib Grackle is all black and has pale eyes but is much smaller and has a slimmer bill. Male Great-tailed Grackle is large with pale eyes but has a slim bill that lacks a high ridge.

2. Smooth-billed Ani (*Crotophaga ani*), p. 194
L 14.5 in (37 cm). Exceptionally rare, with a single recent record from Aruba. Very difficult to separate from Groove-billed Ani (especially immature birds) but with a higher ridged bill that has no grooves or only weak grooves on part of bill closest to face. Most definitive way to distinguish Smooth-billed Ani from Groove-billed Ani is by call. Smooth-billed Ani makes a whining, rising *qu-weeeik*, while Groove-billed Ani makes a two-syllabled, descending *TEE-who*.

3. Groove-billed Ani (*Crotophaga sulcirostris*), p. 195
L 13.5 in (34 cm). Note thick bill, round-headed appearance, and long, loose tail. Grooves on bill not visible unless seen at close range. Greater Ani which has been documented on Curaçao and Aruba has white (not dark) eyes and larger keeled bill.

4. Yellow-billed Cuckoo (*Coccyzus americanus*), p. 195
L 12 in (30 cm). Regular fall stop-over migrant and winter visitor, occasionally in large numbers. Note long tail and long, slightly curved. mostly yellow bill (black on upper part of upper mandible and bill tip). White underparts. Rufous in wings. Large white spots in tail.

5. Mangrove Cuckoo (*Coccyzus minor*), p. 196
L 12 in (30 cm). Uncommon visitor. Note long tail and long, slightly curved, bill with black upper mandible and yellow lower mandible. Buffy underparts; dark stripe through eye; lacks rufous in wings; large white spots in tail. Although not re-corded on the islands, very similar Dark-billed Cuckoo shows an entirely dark bill.

6. Gray-capped Cuckoo (*Coccyzus lansbergi*), p.196X
L 10.5 in (27 cm). Only a single previous record, so unlikely. A richly colored cuckoo, with rufous-buff underparts, rufous back and wings, gray hood, and dark tail with small white spots in tail.

7. Guira Cuckoo (*Guira guira*), p. 193
L 14.5 in (36 cm). Only a single record from 1954, but if seen, unmistakable.

undertail

PLATE 31
Barn Owls, Owls, Oilbird

Tytonidae: Barn Owls

1. Barn Owl (*Tyto alba*), p. 197
L 16 in (41 cm). Rarely seen and never during daylight hours. Occurs only on Curaçao and Bonaire. Long legs, heart-shaped face, pale breast.

Strigidae: Owls

2. Burrowing Owl (*Athene cunicularia*), p. 197
L 9 in (23 cm). Occurs only on Aruba. Long legs, yellow eyes, stubby tail. Often seen during day, at mouth of underground burrow.

Steatornithidae: Oilbird

3. Oilbird (*Steatornis caripensis*), p. 198
L 16–19 in (41–48 cm). Only a single previous record, so unlikely. Large, rufous-brown bird with large, dark eyes (shine red when illuminated at night), hooked pale bill. Also note white spots in wing, head, and tail edges.

PLATE 32
Nightjars, Allies

Caprimulgidae: Nightjars, Allies

1. White-tailed Nightjar (*Hydropsalis cayennensis*), p. 200

L 8.5 in (22 cm). The only common resident nightjar; apparently more common on Curaçao and Bonaire than Aruba. Most likely seen after dark, on or near roads where headlights may pick up reflected, red eyeshine. If seen clearly, note long tail, short wings (unlike nighthawks), and reddish collar. Males have white in wings and tail but difficult to see under low light conditions. Note that these birds sometimes forage at night under street lamps, where fieldmarks are more easily seen.

2. Common Nighthawk (*Chordeiles minor*), p. 199

L 10 in (25 cm). Rare to uncommon stop-over migrant; status of nighthawks still poorly known. Most likely seen in flight; swift flight and long pointed wings could suggest a falcon. Nighthawks fly much more erratically than falcons, however, with sudden changes of direction and quick tips of the body. Common Nighthawk is the largest and longest-winged of the three nighthawks that have been recorded from the islands. Bold white wing patches are farther from wing tip than in Lesser Nighthawk. Wings extend beyond tail when bird is at rest. (Nighthawk species are very difficult to distinguish from each other, especially when in flight.)

3. Antillean Nighthawk (*Chordeiles gundlachii*), p. 200

L 8.5 in (22 cm). Rare stop-over migrant; status of nighthawks still poorly known. Smaller than Common Nighthawk, with shorter wings and tail. Wing patch may appear smaller than that of Common Nighthawk. Wingtips do not extend beyond tail when bird is at rest.

4. Lesser Nighthawk (*Chordeiles acutipennis*), p. 198

L 9 in (23 cm). Rare to uncommon stop-over migrant; status of nighthawks still poorly known. Smaller than Common Nighthawk, with shorter wings and tail. Wing patch closer to tip of wing than in Common Nighthawk. Wingtips do not extend beyond tail when bird is at rest. Usually shows more obvious spotting on inner flight feathers than do Common and Antillean Nighthawks, and a more uniformly colored upperwing.

5. Chuck-will's-widow (*Antrostomus carolinensis*), p. 201

L 12 in (30 cm). Rare bird, with only four previous records. Most likely seen roosting on ground or flushed from ground rather than in flight. Compared to resident White-tailed Nightjar, larger, browner, with shorter tail, and lacks rufous collar. At rest, wings much shorter than tail, unlike nighthawks.

PLATE 33
Swifts, Hummingbirds

Apodidae: Swifts

1. Chimney Swift (*Chaetura pelagica*), p. 202
L 4.75–5.5 in (12–14 cm). Thought to be the swift most likely to occur on the islands. Small all-gray bird always seen in flight, when it looks like a "flying cigar." Their short, stiff wings are flapped so quickly they appear as a blur.

2. Black Swift (*Cypseloides niger*), p. 202
L 7.25 in (18.4 cm). Single sight record for Curaçao, undocumented by photos or video. All-dark, large swift with proportionately long, curved wings that appear very broad where they connect with the body; also note broad tail.

Trochilidae: Hummingbirds

3. Ruby-topaz Hummingbird (*Chrysolampis mosquitus*), p. 204
L 3.5 in (9 cm). Relatively common. Male dark with ruby red crown and nape, orange-yellow throat and upper breast, and rufous tail with dark tips. Larger than Blue-tailed Emerald, with longer tail; more crested appearance; slightly decurved bill; tail rounded rather than forked, and either entirely rufous or with just a rufous base, with dark terminal band and large white tips to outer tails feathers (females and immatures can have very reduced amounts of rufous). Females can look almost sooty-brown. Immatures typically show more more green on back and and are pale underneath, sometimes with a dark line down center of throat.

4. Blue-tailed Emerald (*Chlorostilbon mellisugus*), p. 204
L 3 in (7.5 cm). One of two permanent resident and relatively common hummingbird species of the islands, and the one most commonly seen on the grounds of resorts and hotels. Smaller than Ruby-topaz Hummingbird, with a relatively short, straight bill and relatively short, slightly forked tail. Male has dark green body and dark blue tail, but in weak light can look all dark, especially the tail. Females and immatures are green-backed and light underneath, but look for forked, bluish tail with white tips on outer tail feathers.

5. Rufous-breasted Hermit (*Glaucis hirsutus*), p. 203
L 4.2 in (10.7 cm). Only one record. With long, down-curved bill; rufous underparts; long, rounded tail with chestnut sides and black terminal band.

6. White-necked Jacobin (*Florisuga mellivora*), p. 203
L 4.5 in (11 cm). Rare hummingbird, with only two records for the islands. Large hummingbird. Male has a dark blue head with all-dark bill. Lower breast and belly white. Tail mostly white. Female scaly on breast.

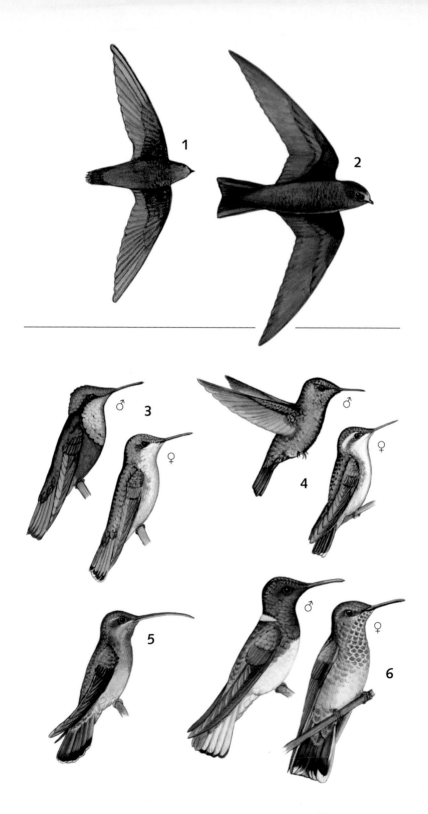

PLATE 34
Kingfishers, Woodpeckers

Alcedinidae: Kingfishers

1. Ringed Kingfisher (*Megaceryle torquata*), p. 205
L 15.5 in (40 cm). Only three records for the islands. Dark blue above and almost entirely dark rufous below, with large bill.

2. Belted Kingfisher (*Megaceryle alcyon*), p. 205
L 12 in (30 cm). Relatively common winter visitor around mangrove wetlands and other bodies of water. Blue-gray upperparts separated from white underparts by blue breast band (male has narrow rufous band as well); also note white spotting in wings.

3. American Pygmy-Kingfisher (*Chloroceryle aenea*), p. 206
L 5–5.5 in (12.7–14 cm). Exceptionally rare, with only a single record from Bonaire. Tiny kingfisher with a dark green head; on male note rust collar, throat, breast, and flanks. On female and immature, the throat and neck collar are whitish tinged with rust. Female has green chestband. White belly extends down to tail.

4. Amazon Kingfisher (*Chloroceryle amazona*), p. 206
L 11 in (28 cm). Only three records for the islands for this species. Dark green upperparts with no white spotting in wings. Female largely white underneath, male with rufous breast band.

Picidae: Woodpeckers

5. Yellow-bellied Sapsucker (*Sphyrapicus varius*), p. 207
L 8.5 in (22 cm). Less than 20 documented occurrences; a rare but relatively regular winter migrant to the islands. The only woodpecker ever documented on the islands, most in immature plumage. Note long white patch on wing.

PLATE 35
Old World Parrots, New World Parrots

Psittaculidae: Old World Parrots

1. Rose-ringed Parakeet (*Psittacula krameri*), p. 210
L 16 in (41 cm). Introduced population now resident in some parts of Curaçao. Note long thin tail, reddish bill, and black feathers in flight. Adult male has black collar by second year.

Psittacidae: New World Parrots

2. Yellow-crowned Parrot (*Amazona ochrocephala*), p. 211
L 14 in (36 cm). Escaped cage birds seen occasionally on Curaçao. Note short tail, rapid wingbeats in flight, red patch in inner wing, and yellow restricted to forehead. Extent of head markings can be very difficult to discern in flight.

3. Yellow-shouldered Parrot (*Amazona barbadensis*), p. 212
L 13 in (33 cm). Historically found on all three islands; now only on Bonaire except for possible birds released from captivity. Short, square tail; yellow on face; yellow and red on wing; yellow thighs. Other Amazon species have been released and possibly have bred on Curaçao and perhaps Aruba and Bonaire. These formerly captive birds are most likely seen in urban or suburban areas.

4. Orange-winged Parrot (*Amazona amazonica*), p. 213
L 13 in (33 cm). Escaped cage birds occasionally seen on Curaçao and Aruba. Note short tail, rapid wingbeats in flight, red patch in inner wing, and yellow in tail tip. Yellow on crown and cheek separated by greenish-blue line extending from bill through eye. Extent of head markings can be very difficult to discern in flight.

5. Red-lored Parrot (*Amazona autumnalis*), p. 211
L 13 in (33 cm). Escaped caged birds seen occasionally on Curaçao; they have bred there and might now be resident. Note short tail, rapid wingbeats in flight, red patch in inner wing, and red on forehead. Extent of head markings can be very difficult to discern in flight.

PLATE 36
New World Parrots

Psittacidae: New World Parrots *continued*

1. Chestnut-fronted Macaw (*Ara severus*), p. 214

L 19 in (48 cm). Small introduced population in suburbs of Willemstad, Curaçao; escaped birds spotted a number of times on Aruba. Note large size, long tail, white face, green body, and reddish underwing and undertail.

2. Blue-crowned Parakeet (*Thectocereus acuticaudatus*), p. 214

L 14.5 in (37 cm). Escaped caged birds seen in suburban areas of Curaçao, where it might be resident. Note long tail. All green, with white patch around eye, pale pinkish upper mandible, darker lower mandible. In flight, orangish on trailing edge of underwing and in tail. Light blue forehead only visible at close range.

3. Scarlet-fronted Parakeet (*Psittacara wagleri*), p. 215

L 13.5 in (34 cm). Likely to be seen only on Curaçao where a feral population is established. A large, green parakeet with red foreheard (reduced in young birds), pale bill, white eye-ring (reduced in young birds), and red thighs (lacking in immatures).

4. Green-rumped Parrotlet (*Forpus passerines*), p. 213

L 4.7 in (12 cm). A small population once established on Curaçao from released captive birds, but no birds have been seen there for many years. Not found on Aruba or Bonaire. Very small, entirely green; also note extremely short tail and pale bill.

PLATE 37
New World Parrots

Psittacidae: New World Parrots *continued*

1. Brown-throated Parakeet (*Eupsittula pertinax*), p. 213

L 10 in (25 cm). Resident breeding bird; each of the three islands with a distinct subspecies. Long tail. Flash of blue in outer part of upperwing and in tail tip; yellow line in wing lining visible only in flight. Aruba subspecies with very brown face and throat and small, bright yellow patch surrounding eye. Bonaire subspecies with bright yellow extending from crown through face to throat. Curaçao subspecies intermediate, with green crown but extensive yellow face. There have been occasional releases of a subspecies from one island onto another, so it is possible to see intermediate birds, presumed hybrids.

Aruba

Bonaire

Curaçao

Curaçao

1

PLATE 38
Flycatchers

Tyrannidae: Flycatchers

1. Caribbean Elaenia (*Elaenia martinica*), p. 215
L 6 in (15 cm). Quite common on Bonaire, uncommon on Curaçao, rare on Aruba. Compared to very similar and much more common Northern Scrub-Flycatcher, larger, with wider-based and larger bill that has obvious pink base (not all dark). Also has more obvious crest and, sometimes, white in crest.

2. Lesser Elaenia (*Elaenia chiriquensis*), p. 217
L 5.5 in (14 cm). Very rare, with only two confirmed records from specimens from Curaçao and Bonaire. Typically has less of a crest than does Caribbean Elaenia, and with less white in crest. Extremely difficult to positively identify, especially if not vocalizing.

3. Small-billed Elaenia (*Elaenia parvirostris*), p. 216
L 5.75 in (14.5 cm). Very rare austral migrant, with only a single record from 1906; occurs regularly in Venezuela so might be expected to occur occasionally on the islands. Rounded head with no crest. Shows narrow white crown stripe. More obvious white eye-ring than in Caribbean and Lesser Elaenia. Pale gray throat and breast lack the yellow tones of Northern Scrub-Flycatcher and Caribbean and Lesser Elaenia.

4. Northern Scrub-Flycatcher (*Sublegatus arenarum*), p. 217
L 5.5 in (14 cm). Perhaps the most common permanent resident flycatcher across the three islands (though Brown-crested Flycatchers and Gray Kingbirds are sometimes common as well). Smaller than Caribbean Elaenia. Has small, all-black bill; all elaenias have a longer bill and show a prominent pale base on lower mandible. Narrow white supraloral contrasts with olive-brown upperparts. Pale gray throat and breast contrast with yellow belly (*Elaenia* are paler overall and lack contrast between breast and belly; and supraloral is usually not as prominent).

crest
raised

1

2

3

4

PLATE 39
Flycatchers

Tyrannidae: Flycatchers *continued*

1. Eastern Wood-Pewee (*Contopus virens*), p. 218
L 7 in (16.5 cm). Rare winter migrant, with only four records. Grayish-olive above, pale below, with dark wash on breast and sides. Whitish wingbars; eyering faint or absent. Virtually indistinguishable from Western Wood-Pewee, which could theoretically occur.

2. Olive-sided Flycatcher (*Contopus cooperi*), p. 218
L 7.5 in (19 cm). Very rare winter migrant visitor, with only two records. Proportionately large head, short tail. Pattern of contrasting dark and pale underside gives appearance of wearing a vest.

3. Brown-crested Flycatcher (*Myiarchus tyrannulus*), p. 220
L 8.75 in (22 cm). Common permanent resident flycatcher; typically rather retiring, often hidden in shrubs and trees. Large flycatcher with rufous fringes to primaries, dull white fringes to other flight feathers. White wing bars. Gray throat, face, and breast contrast with pale yellow underparts. Tail rufous. All-black bill. Several other *Myiarchus* flycatchers occur in northern South America and are theoretically possible, so any birds that seem different from what is described here should be documented with photographs and video, and any vocalizations recorded.

4. Vermillion Flycatcher (*Pyrocephalus rubinus*), p. 219
L 5 in (13 cm). An extreme rarity, with only a single record of an immature male, from 1957. On adult male, bright red crown and underparts contrast with black mask and upperparts. Female and immature show dusky, streaked undersides with pinkish wash on flanks, lower belly, and vent.

5. Streaked Flycatcher (*Myiodynastes maculatus*), p. 222
L 8.5 in (22 cm). Rare, with only a single record from Bonaire. Heavy streaking on breast; reddish rump and tail; white eyebrow; half of lower mandible pink.

PLATE 40
Flycatchers

Tyrannidae: Flycatchers *continued*

1. **Pied Water-Tyrant** (*Fluvicola pica*), p. 219
   L 5 in (13 cm). Small flycatcher with only a single record from the ABCs. Males are strikingly patterned, with white underparts extending onto face and crown. Black nape and back separated from black wings by white stripe. Black tail with white tip and white rump. All-black bill and legs. Females and immatures similar, but black replaced with gray.

2. **Cattle Tyrant** (*Machetornis rixosa*), p. 220
   L 7.5 in (19 cm). Recently established as a likely breeding species at one location on Aruba; South American populations generally expanding so may become more common—and end up on Bonaire and Curaçao. Bright yellow underparts, olive-gray upperparts, red eyes, long dark legs. Unlike other flycatchers, spends most of its time on or near the ground; runs along on the ground.

3. **Tropical Kingbird** (*Tyrannus melancholicus*), p. 222
   L 8.5 in (22 cm). Rare to uncommon but regularly seen on some locations on Curaçao. Dark gray head contrasts with olive back; yellow breast shows olive green wash; also note bright yellow belly. Tail slightly notched.

4. **Gray Kingbird** (*Tyrannus dominicensis*), p. 224
   L 8.5 in (22 cm). Widespread resident. Upperparts entirely gray; underparts white.

5. **Eastern Kingbird** (*Tyrannus tyrannus*), p. 223
   L 8 in (20 cm). Rare on Aruba and Bonaire; no records from Curaçao. Black upperparts contrast strongly with white underparts. White band at tip of tail.

6. **Fork-tailed Flycatcher** (*Tyrannus savana*), p. 224
   L 14 in (36 cm). Regular austral migrant visitor from South America in fall. Extremely long, forked tail and dark cap make this species unmistakable, but note that immatures or molting adults can have very short tails and cause confusion with other species.

PLATE 41
Swallows, Martins

Hirundinidae: Swallows, Martins

1. **Purple Martin** (*Progne subis*), p. 229
   L 7.3 in (18.5 cm). Rare stopover migrant but status of martins on the islands is still somewhat uncertain because of difficulties in identification of females and immatures. Male glossy purple-black all over. Female and immature Purple Martin with grayish forehead, pale collar, gray underparts. Females difficult to distinguish from female Caribbean and Cuban Martins.

2. **Cuban Martin** (*Progne cryptoleuca*), p. 230
   L 7.3 in (18.5 cm). Exceptionally rare migrant, though very difficult to distinguish from Purple Martin. Males are dark blue, as in Purple Martin, but in the hand show a narrow, usually hidden, strip of white in center of belly. Tail is somewhat more deeply forked than in Purple Martin; but with current state of knowledge, not reliably separable from Purple Martin in the field. Females and immatures are very similar to Caribbean Martin.

3. **Caribbean Martin** (*Progne dominicensis*), p. 231
   L 7.3 in (18.5 cm). Apparently regular stop-over migrant in small numbers. Male glossy purple-black with contrasting strip of white extending from middle of breast to undertail. Females difficult to distinguish from female Purple and Cuban Martins.

4. **Brown-chested Martin** (*Progne tapera*), p. 232
   L 7 in (18 cm). Four records from Aruba; unrecorded from Bonaire and Curaçao. Like a giant Bank Swallow; has brown upperparts and white underparts with brown breast band.

5. **Bank Swallow** (*Riparia riparia*), p. 233
   L 4.75–5.5 in (12–14 cm). Regular stop-over migrant in small numbers. Small swallow with all-brown back and single, brown breast band.

6. **Southern Rough-winged Swallow** (*Stelgidopteryx ruficollis*), p. 229
   L 5–5.75 in (13–15 cm). Relatively rare, with a handful of sight records from all three islands; not known to be documented by published or publically available photos or video or specimens. Brown back, with whitish rump, orangish throat, and dusky wash across upper breast. Note that Northern Rough-winged Swallow is also possible; some sight records might have confused the two species. Northern Rough-winged Swallow does not show strongly contrasting light rump and adult lacks orangish throat; juvenile sometimes has buffy throat.

PLATE 42
Swallows, Martins

Hirundinidae: Swallows, Martins *continued*

1. White-winged Swallow (*Tachycineta albiventer*), p. 232
L 5.2 in (13.2 cm). Very rare, with only a single record on the islands, although common on the South American mainland. White rump; large white wing patches.

2. Chilean Swallow (*Tachycineta meyeni*), p. 233
L 5 in (13 cm). Very rare, with only a single record. White rump. Lacks distinct, narrow white streak from base of bill to eye that is found in similar but unrecorded White-rumped Swallow.

3. Cliff Swallow (*Petrochelidon pyrrhonota*), p. 234
L 5.5 in (14 cm). A regular but generally uncommon migrant visitor. Buffy rump with square tail; white forehead in northern populations (dark rusty in Mexican populations). Rusty throat with black mark across upper breast and narrow, buffy collar. Immature has dark cheeks.

4. Cave Swallow (*Petrochelidon fulva*), p. 235
L 5.5 in (14 cm). Very rare, with one specimen record from Curaçao and one photographed on Bonaire; possible, given the species' propensity for appearing outside its normal, known range. Typically, shows a pale rusty throat that is paler or same color as rump. Immature shows pale cheeks. Very similar Cliff Swallow shows black mark on upper chest.

5. Barn Swallow (*Hirundo rustica*), p. 234
L 5.75–7.75 in (15–20 cm). Only commonly occurring swallow with deeply forked tail. Dark blue upperparts. Chestnut throat and forehead.

PLATE 43
Old World Flycatchers, Thrushes

Muscicapidae: Old World Flycatchers

1. Northern Wheatear (*Oenanthe oenanthe*), p. 235
L 5.5–6 in (14–15 cm). Very rare, with only two records. Distinctive long-legged, short-tailed bird often seen on or near the ground, slowly dipping tail up and down. In flight, note white rump, black upside-down T on tail. Possible sightings of this species should be documented with photographs and/or video.

Turdidae: Thrushes

2. Veery (*Catharus fuscescens*), p. 236
L 6.5–7.5 in (17–19 cm). Rare to uncommon migrant visitor. Note uniformly reddish upperparts. Spotting restricted to upper chest.

3. Gray-cheeked Thrush (*Catharus minimus*), p. 237
L 6.5–8 in (17–20 cm). Rare to uncommon migrant visitor. Dull gray-brown upperparts. Grayish cheeks. Faint eye-ring.

4. Swainson's Thrush (*Catharus ustulatus*), p. 237
L 6.5–7.5 in (17–19 cm). Rare to uncommon migrant visitor. Dull gray-brown upperparts. Buffy cheeks. Bold, light eye-ring.

5. Wood Thrush (*Hylocichla mustelina*), p. 238
L 8 in (20 cm). Very rare, with only a single record. Rufous head contrasts with more brownish back. Bold spotting on breast and lower breast, more so than on other thrushes.

PLATE 44
Mockingbirds, Thrashers, Waxwings, Starlings

Mimidae: Mockingbirds, Thrashers

1. Tropical Mockingbird (*Mimis gilvus*), p. 239
L 10 in (25 cm). One of the most common and widespread birds on the islands; often quite tame and seen around hotels, resorts, and homes, as well as throughout much of the countryside. Seen singing from the top of bushes, cactus, and rooftops. Also often seen foraging on the ground, moving tail from side to side. Pale gray overall, long tail with white in corners. Narrow white wingbars.

2. Brown Thrasher (*Toxostoma rufum*), p. 239
L 11.5 in (29 cm). Very rare, with only a single record. Rufous upperparts. Striped breast; white wingbars; long tail.

3. Pearly-eyed Thrasher (*Margarops fuscatus*), p. 238
L 11 in (28 cm). Found only on Bonaire (two exceptional records for Curaçao). Pearly-white iris; large pinkish bill. Brown upperparts; white underparts streaked with brown. White tip to outer tail feathers.

Bombycillidae: Waxwings

4. Cedar Waxwing (*Bombycilla cedrorum*), p. 240
L 6.5–8 in (17–20 cm). Very rare, with only single record. Note the crest, black mask, and yellow band on tail tip.

Sturnidae: Starlings

5. European Starling (*Sturnus vulgaris*), p. 240
L 7.5–8.5 in (19–22 cm). Very rare, with only three sightings. In breeding plumage, black with rusty spotting; yellow bill. In nonbreeding plumage, note white spotting and dark bill. Also note very short tail.

1

2

3

4

breeding

nonbreeding

5

immature

PLATE 45
Vireos, Bananaquits

Vireonidae: Vireos

1. Yellow-throated Vireo (*Vireo flavifrons*), p. 225
L 5.5 in (14 cm). Very rare, with only one record. Yellow throat, yellow spectacles, and white wingbars.

2. Red-eyed Vireo (*Vireo olivaceus*), p. 226
L 6 in (15 cm). Rare migrant visitor. Gray crown with dark lower border and distinct line through eye. White eyebrow stripe.

3. Black-whiskered Vireo (*Vireo altiloquus*), p. 227
L 6.25 in (16 cm). Breeding resident (populations possibly migratory?). Most regular and widespread on Curaçao and Bonaire but small numbers found in recent years at Spanish Lagoon on Aruba. Note "whisker," a thin, dark lateral throat stripe. Heavy bill. Some birds may lack obvious whisker marks, making identification challenging.

4. Philadelphia Vireo (*Vireo philadelphicus*), p. 226
L 5.25 in (13 cm). Rare winter visitor. A small songbird; a bit larger than a Bananaquit but generally much more sluggish (as are all vireos). Rather nondescript, with greenish upperparts, gray cap, and white eyebrow with dark markings between eye and bill. Yellow on underside can be variable but often extends from throat to undertail. Bill rather short and thick as compared to warblers.

Coerebidae: Bananaquits

5. Bananaquit (*Coereba flaveola*), p. 266
L 4.5 in (11 cm). The common bird of yard and garden. Often comes to outdoor tables for juices, syrup, and sugar. Jet black upperparts, yellow underparts, yellowish rump, white eyebrow, and thin, slightly down-curved bill.

adult
(Aruba and
Curaçao)

5

adult
(Bonaire)

immature

PLATE 46
New World Warblers

Parulidae: New World Warblers

1. Tennessee Warbler (*Oreothlypis peregrina*), p. 246
L 5 in (13 cm). There are only a handful of previous records for this species, though it is a relatively common migrant winterer in Venezuela. Note the thin bill, grayish cap, and greenish back and wings; wings lack obvious wingbars. Dark, narrow eye line with pale stripe over eye. White undertail coverts. Beware molting and young Yellow "Golden" Warblers, which can appear very greenish and pale.

2. Northern Parula (*Setophaga americana*), p. 254
L 4.5 in (11 cm). Uncommon but regular northern migrant winterer. Note the broken, white eye-ring, two white wingbars, yellow throat and upper breast, and white lower breast and belly.

3. Yellow "Golden" Warbler (*Setophaga petechia rufopileata*), p. 258
Yellow "Northern" Warbler (*Setophaga petechia aestivia* Group), p. 258
L 4.5–5 in (11–13 cm). The common resident warbler species. Males bright yellow, with reddish cap and red streaks on breast. Inner web of tail feathers yellow, unlike similar rare migrant warbler visitors. Females and immatures, especially when moulting, can be very confusing, as they are often either very drab, even grayish overall, or with patches of yellow on head and breast. These birds typically lack an obvious eye line and strong wingbars; eye appears dark on unmarked face; they also tend to have gray or yellowish underparts, with no dark streaking. It is very easy to confuse these birds with a number of the rare migrant visitor warbler species, so use caution and photograph or videotape any one you are not sure of for later study. On migrant Yellow "Northern" Warblers from North America, the males lack the red cap; note that they occur regularly as winterers in northern South America and certainly must occur on the islands, though they have not been documented with certainty.

4. Chestnut-sided Warbler (*Setophaga pensylvanica*), p. 259
L 4.5–5 in (11–13 cm). Rare northern migrant visitor, with a handful of sightings. Adult in breeding plumage has blood-red flanks and distinct facial pattern, with lemon-yellow cap. In immature plumage, shows lime green upperparts with two wingbars, white eye-ring, and clean white undersides.

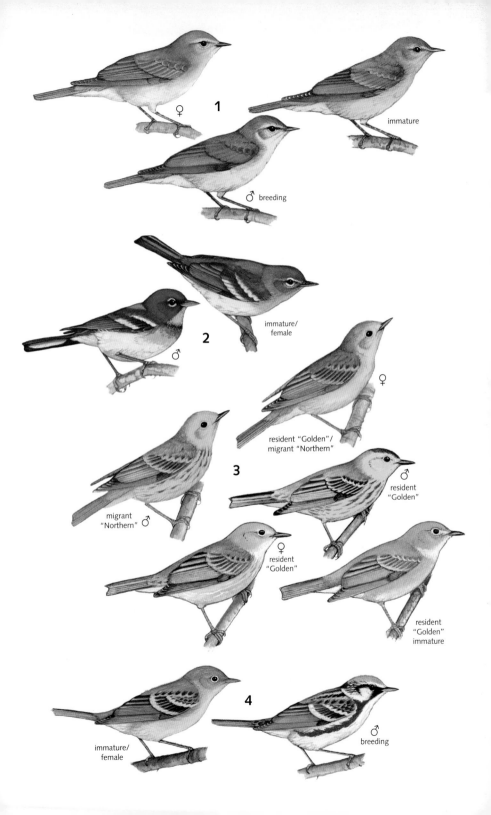

♀ **1**

immature

♂ breeding

immature/
female

♂ **2**

resident "Golden" /
migrant "Northern"

♀

3

migrant
"Northern" ♂

♂
resident
"Golden"

♀
resident
"Golden"

resident
"Golden"
immature

4

immature/
female

♂
breeding

PLATE 47
New World Warblers

Parulidae: New World Warblers *continued*

1. Magnolia Warbler (*Setophaga magnolia*), p. 255
L 4.5–5 in (11–13 cm). Rare northern migrant visitor, with a handful of sightings. Note yellow undersides; yellowish rump, wingbars, and streaking on breast or belly; white squares in tail.

2. Cape May Warbler (*Setophaga tigrina*), p. 252
L 5 in (13 cm). Uncommon northern migrant visitor. Breeding male has distinctive rufous cheeks; also note yellow rump. Female has yellowish patch on side of neck; shows streaking below but unstreaked back.

3. Black-throated Blue Warbler (*Setophaga caerulescens*), p. 262
L 5 in (13 cm). Uncommon northern migrant visitor. Breeding male shows distinctive blue, black, and white pattern. Female has small, white "handkerchief" at base of primaries, white eyebrow, and white arc below eye.

4. Yellow-rumped Warbler (*Setophaga coronata*), p. 263
L 5-6 in (13-15 cm). Rare northern migrant visitor, with a handful of sightings. Brown or gray upperparts, bright yellow rump, streaked breast. Yellow patch on sides of breast.

5. Black-throated Green Warbler (*Setophaga virens*), p. 265
L 4.5–5 in (11–13 cm). Rare northern migrant visitor, with a handful of sightings. Upperparts green, unstreaked; sides of head bright yellow; indistinct olive patch on cheek; yellow wash across upper breast in immature and female.

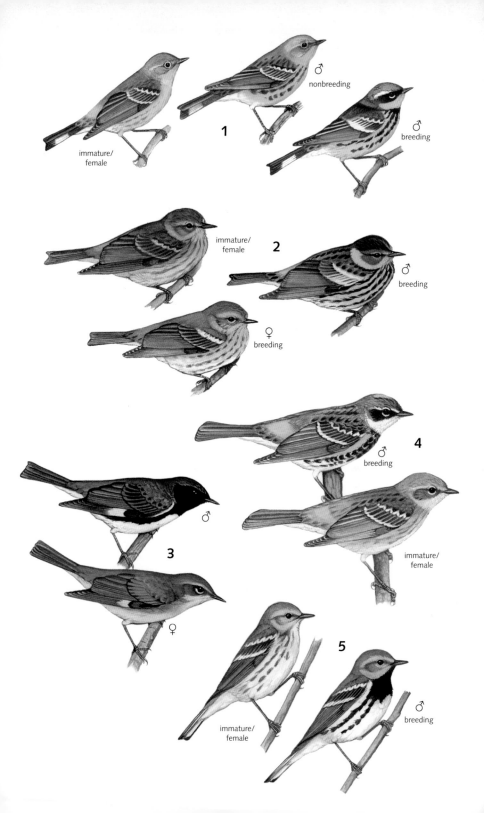

1

immature/
female

♂
nonbreeding

♂
breeding

2

immature/
female

♂
breeding

♀
breeding

4

♂
breeding

immature/
female

3

♂

♀

5

immature/
female

♂
breeding

PLATE 48
New World Warblers

Parulidae: New World Warblers *continued*

1. Blackburnian Warbler (*Setophaga fusca*), p. 257
L 5 in (13 cm). Rare northern stop-over migrant, with a handful of sightings. Breeding male is bright orange and has black arrowhead shape behind eye. All other plumages have same pattern, but duller.

2. Prairie Warbler (*Setophaga discolor*), p. 264
L 5 in (13 cm). Rare northern migrant visitor, with a handful of sightings. Dark semicircle around eye, below which is a yellow crescent. Constantly bobs tail.

3. Palm Warbler (*Setophaga palmarum*), p. 262
L 5.5 in (14 cm). Rare northern migrant visitor, with a handful of sightings. Breeding adult has chestnut cap. All other plumages show yellow rump and tail coverts. Lacks wingbars. Frequently bobs tail.

4. Bay-breasted Warbler (*Setophaga castanea*), p. 256
L 5.5 in (14 cm). Rare northern migrant visitor, with a handful of sightings. Breeding male has chestnut head, throat, and flanks; black mask; and buffy patch on side of neck. Breeding female has paler chestnut flanks and pale patch on neck. Nonbreeding adults usually show some traces of chestnut on flanks. Immature female shows streaked back but lacks streaking on upper breast and flanks and has buffy undertail coverts.

5. Blackpoll Warbler (*Setophaga striata*), p. 260
L 5.5 in (14 cm). One of the most common migrant visitors on the islands, especially during fall migration; small numbers can be seen throughout winter and early spring. All adult plumages show dark lateral stripes on throat; streaks on breast and flanks; white underparts; and white undertail coverts. Breeding male unmistakable, with black cap and white cheeks; note that he lacks white stripes on crown. Nonbreeding adult and immature lack black cap, show grayish neck; legs (or at least feet) usually yellowish; has white rather than buffy undertail coverts; streaked back (usually more obviously streaked than in Bay-breasted Warbler).

1 ♂ breeding

breeding

immature/female

2 ♂ breeding

immature/female

3 nonbreeding

breeding

4 ♂ nonbreeding

♀ breeding

♂ breeding

immature/female

5 ♂ breeding

♀ breeding

immature

PLATE 49
New World Warblers

Parulidae: New World Warblers *continued*

1. Cerulean Warbler (*Setophaga cerulea*), p. 253
L 4.5 in (12 cm). Rare northern stop-over migrant, with a handful of sightings. Male striking, with sky blue upperparts; two white wingbars; dark face contrasting with white throat and dark chest bar; streaking on sides extends to flanks. Female and immature greenish-blue above; two white wingbars; light eyebrow; blurry streaking on sides of breast.

2. Black-and-white Warbler (*Mniotilta varia*), p. 244
L 5.25 in (13 cm). One of the more frequently seen winter migrant visitors. Often seen in mangroves. Strong black-and-white streaking below and above, including on head.

3. Prothonotary Warbler (*Protonotaria citrea*), p. 245
L 5.5 in (14 cm). Regular but uncommon winter migrant visitor, virtually always seen in mangroves. Yellow head and breast. Unmarked blue-gray wings. Solid white undertail coverts.

4. American Redstart (*Setophaga ruticilla*), p. 251
L 5.25 in (13 cm). Regular winter visitor in small numbers. Small songbird, similar in size to Bananaquit. Adult male has jet black upperside, head, and throat, with bright orange patches in wing, sides of breast, and tail. Females and first-year males have yellow instead of orange and show greenish-gray back and wings, and gray head. Regularly fans tail out as it hops around.

5. Ovenbird (*Seiurus aurocapilla*), p. 241
L 6 in (15 cm). Uncommon winter migrant visitor; but note that it is shy and perhaps occurs more regularly than previously thought. Usually seen skulking low in mangroves or thick brush, like Northern Waterthrush, but note orange crown bordered by black and white eye-ring (no eyebrow as seen in both waterthrush species).

1

♂ breeding

immature/ female

2

♂

♀

3

♂

♀

4

♂

♀

first year ♂

5

PLATE 50
New World Warblers

Parulidae: New World Warblers *continued*

1. **Worm-eating Warbler** (*Helmitheros vermivorum*), p. 241
   L 5.25 in (13 cm). Extremely rare, with only two records for the islands. Very distinctive, with black head stripes, olive back, and tannish undersides.

2. **Northern Waterthrush** (*Parkesia noveboracensis*), p. 243
   L 6 in (15 cm). Perhaps the most abundant winter resident and migrant warbler. Most commonly seen in mangroves or brushy areas near water, where often its harsh *chink* call note is first sign of its presence. Unmarked brown above, streaked below with prominent pale eyebrow. While walking on ground, teeters in a very distinctive fashion.

3. **Louisiana Waterthrush** (*Parkesia motacilla*), p. 242
   L 6 in (15 cm). Rare northern migrant visitor, with a handful of sightings. Difficult to differentiate from Northern Waterthrush, but note that it generally lacks buffy tones to eyebrow and underside (except flanks); eyebrow widens behind eye (narrows in Northern); throat is unstreaked; bill slightly longer and heavier; and legs tend to look pinker.

4. **Golden-winged Warbler** (*Vermivora chrysoptera*), p. 243
   L 4.75 in (12 cm). Exceptionally rare vagrant. Small songbird (similar in size to Bananaquit) with gray upperparts and gold wing patch; whitish underparts. Shows a striking head pattern: dark mask and bib and gold crown contrast with white eyebrow and white moustache stripe.

5. **Blue-winged Warbler** (*Vermivora pinus*), p. 244
   L 4.75 in (12 cm). Exceptionally rare vagrant. Small (similar in size to Bananaquit), brightly colored songbird with yellow head and underparts; black line through eye; short, bluish-gray tail and wings, with two white wingbars.

PLATE 51
New World Warblers

Parulidae: New World Warblers *continued*

1. Kentucky Warbler (*Geothlypis formosa*), p. 248
L 5.25 in (13 cm). Rare northern stop-over migrant, with a handful of sightings. Olive-green upperparts and yellow underparts. Incomplete yellow spectacles surrounded by black are definitive.

2. Connecticut Warbler (*Oporornis agilis*), p. 246
L 5.75 in (15 cm). Rare northern stop-over migrant, with a handful of sightings. Note conspicuous, complete, white eye-ring. Undertail coverts extend more than halfway to end of tail. Pale yellow underparts. Tends to walk rather than hop.

3. Mourning Warbler (*Geothlypis philadelphia*), p. 247
L 5.25 in (13 cm). Rare northern stop-over migrant, with a handful of sightings. Gray hood and lack of eye-ring is distinctive in adults. Immature can have partial white eye-ring but usually shows yellowish throat; brighter yellow underparts; and undertail coverts that are not as long as in Connecticut Warbler.

4. Common Yellowthroat (*Geothlypis trichas*), p. 249
L 5 in (13 cm). Rare northern migrant visitor, with only a handful of sightings on Bonaire and Curaçao; sightings have increased in recent years on Aruba, largely at Bubali wetlands. Male has strong black "robber mask" edged with white above; yellow throat and upper breast. Female drab brownish-olive above, with yellow throat and undertail coverts, and tan or buffy belly. Immature male can resemble female-type, but with a steely gray mask initially lacking the white upper border.

5. Hooded Warbler (*Setophaga citrina*), p. 250
L 5.25 in (13 cm). Uncommon to rare, but regular, northern migrant visitor, often seen in mangroves. Upperparts olive-green, with no wingbars. Male very distinctive, with black hood and bright yellow face. Female lacks black hood but shows extensive yellow face. Both show characteristic white in tail, and birds often slightly fan tail as they hop around, making white obvious.

6. Canada Warbler (*Cardellina canadensis*), p. 265
L 5.25 in (13 cm). Rare northern stop-over migrant, with only a single record. Unmarked grayish-blue upperparts, yellow underparts, yellow in front of eye, white eye-ring, and white undertail coverts. Male has obvious dark "necklace," female and immature with fainter "necklace."

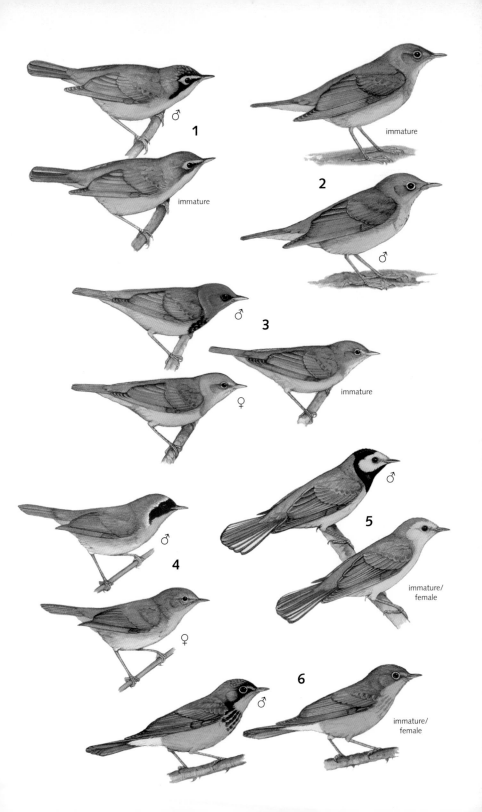

PLATE 52
Tanagers

Thraupidae: Tanagers

1. Swallow Tanager (*Tersina viridis*), p. 267
L 5.5 in (14 cm). Very rare, with only a single record for the islands. Male unlike anything else on the islands. In both sexes, note barring on flanks and upright posture. With close view, note reddish eyes.

2. Red-legged Honeycreeper (*Cyanerpes cyaneus*), p. 267
L 4.5 in (11 cm). Very rare, with only one record for the islands. Breeding male has bright red legs and pale blue crown. Female shows reddish legs and faint streaking on breast.

3. Saffron Finch (*Sicalis flaveola*), p. 268
L 5 in (13 cm). Breeding resident (presumably from escaped caged birds) on all three islands, though local on Aruba (suburbs of Orenjestad). Male is bright yellow overall, with orange wash on head, thick bill. Female similar but duller. Immature is streaked above, whitish below, with yellow band stretching from breast to back of neck, and yellow undertail.

1

♂

♀

2

underwing

♂

♀

breeding

♂

nonbreeding

3

♂

immature

PLATE 53
Tanagers

Thraupidae: Tanagers *continued*

1. Blue-back Grassquit (*Volatinia jacarina*), p. 269

L 4 in (10 cm). Very rare, with only a handful of records for the islands. Male all dark, with small white mark at bend of wing and white on underside of flight feathers, visible only in flight. Bill relatively slender and pointed. Immature males show brown edging to blackish feathers. Females with light brown unmarked upperparts and brown streaking on breast.

2. Lined Seedeater (*Sporophila lineola*), p. 269

L 4.5 in (11 cm). Very rare vagrant. Small, ground-loving bird; just a bit larger than a Black-faced Grassquit but with a stubbier, less pointed bill. Males have striking coloration, with black above and white below, white crown stripe, white cheek, and black throat connected by a line to the nape; also note small, white "handkerchief" on folded black wing, and black bill. Females are dull brown above with brownish-yellow throat, chest, and flanks contrasting with whitish belly and undertail coverts; they show a pale, or mostly pale, bill.

3. Black-faced Grassquit (*Tiaris bicolor*), p. 270

L 4 in (10 cm). One of the most common birds on the islands. Small, ground-dwelling bird that occurs in open and semi-open areas, including around homes and on the grounds of resorts. Note broad-based, short bill and short tail. Male has greenish back and black face and underparts. Female and immatures are brownish-green overall.

PLATE 54
Grosbeaks, Buntings

Cardinalidae: Grosbeaks, Buntings

1. Summer Tanager (*Piranga rubra*), p. 273

L 7 in (18 cm). Only a handful of records on the islands, though not uncommon on mainland Venezuela. Male entirely red; female and immature yellowish. Large, pale bill.

2. Scarlet Tanager (*Piranga olivacea*), p. 274

L 6.5 in (17 cm). Regular migrant visitor, especially in spring. Male with scarlet body and black wings and tail. Female and immature are greenish-yellow, with black wings and tail.

3. Western Tanager (*Piranga ludoviciana*), p. 274

L 6.5 in (17 cm). Very rare, with only a single record for the islands. Male unmistakable. In female and immature, note the white wingbar and yellow shoulder bar.

4. Dickcissel (*Spiza americana*), p. 278

L 6.25 in (17 cm). Rare to uncommon stopover migrant. Look for chestnut shoulders and fairly large bill. Adult male in breeding plumage has black V on chest; yellow on breast and over eye. Immature birds streaked underneath and with black line framing throat.

1
♀
♂

2
♂
nonbreeding
♂
breeding

3
♂
immature
♀
♂

4
♂
breeding
♂
immature

PLATE 55
Grosbeaks, Buntings

Cardinalidae: Grosbeaks, Buntings *continued*

1. Rose-breasted Grosbeak (*Pheucticus ludovicianus*), p. 275
L 8 in (20 cm). Uncommon migrant visitor. Male is black and white with bright red patch on breast. Female and immature have thick, pale bill; two strong, white wing bars; white eyebrow; and brown cheek patch.

2. Black-headed Grosbeak (*Pheucticus melanocephalus*), p. 276
L 8.25 in (21 cm). Very rare, with only a single record for the islands. Male has streaked, dark back; white wing bars; and white "shoulders." Female and immature very difficult to distinguish from immature Rose-breasted Grosbeak.

3. Blue Grosbeak (*Passerina caerulea*), p. 277
L 6.75 in (17 cm). Very rare, with only a single record for the islands. Male distinctive, dark blue (can look black in poor light) with two rust-colored wing-bars. Female and immature similar to female and immature Indigo Bunting, but note rust-colored wingbars, thicker bill, lack of streaking on breast, and larger size. Possible sightings should be documented with photos and/or video.

4. Indigo Bunting (*Passerina cyanea*), p. 277
L 5.5 in (14 cm). Uncommon but regular migrant visitor. Male in breeding plumage unmistakable, entirely bright blue. Male in nonbreeding plumage and first year male brownish-grayish, mottled with blue. On female and immature, undersides faintly streaked; also note indistinct buffy wing bars.

1

♂

♀

♂
immature

2

♂

♀

3

♂

♀

4

♂
breeding

♂
nonbreeding

♀

PLATE 56
Old World Sparrows, Weavers

Passeridae: Old World Sparrows

1. House Sparrow (*Passer domesticus*), p. 289
L 6 in (15 cm). Introduced species now resident and common on all three islands. Male easily recognized by head adorned with chestnut, gray, and black. Female and immature rather drab, with tan crown, light eyebrow stripe extending behind eye, single white wing bar, and plain unstreaked undersides.

Ploceidae: Weavers

2. Village Weaver (*Ploceus cucullatus*), p. 290
L 7 in (17 cm). Introduced species found only in a localized area of suburban Curaçao. Male yellow with black hood, large bill, red eye. Female similar but lacking black hood.

3. Baya Weaver (*Ploceus philippinus*), p. 291
L 7 in (17 cm). Introduced species found only in a localized area of suburban Curaçao. Male with yellow head and breast, black face and throat, thick bill. Female has dull tan upperparts and light underparts, with fine streaking on upper breast and sides.

PLATE 57
Sparrows, Crows

Emberizidae: Sparrows

1. Grasshopper Sparrow (*Ammodramus savannarum*), p. 270
L 4.5 in (11 cm). Rare breeding resident found only on Curaçao and Bonaire, in open areas with grass and scattered low shrubs. Buffy, unstreaked breast; pale median crown stripe; short tail; flat-headed profile.

2. Rufous-collared Sparrow (*Zonotrichia capensis*), p. 271
L 5 in (13 cm). Common on Curaçao. On Aruba, apparently once common in parts of Arikok National Park, but now seemingly rare there. Gray-and-black striped head with slight crest. Rufous collar, black mark on side of upper breast, two white wingbars, unstreaked breast. Immature birds have brown-and-tan striped heads, streaked breast, and white wingbars, but lack rufous collar.

3. White-throated Sparrow (*Zonotrichia albicollis*), p. 272
L 6–7 in (15–18 cm). Very rare, with only a single record for the islands. Striped crown. White throat.

Corvidae: Crows

4. House Crow (*Corvus splendens*), p. 228
L 16 in (40 cm). Very rare ship-assisted non-native species from Asia that has appeared on Curaçao. Readily identifiable as there is nothing similar on the islands. Birders from North America and Europe would certainly recognize this species as some type of crow. A large, dark bird with gray nape and black cap and throat. Large, black, stout bill and strong, black legs.

1

2

immature

3

4

PLATE 58
Blackbirds, Orioles

Icteridae: Blackbirds, Orioles

1. Bobolink (*Dolichonyx oryzivorus*), p. 279
L 6–8 in (16–20 cm). A spring and fall migrant throughout the islands; formerly more regular. Male in breeding plumage very striking, with black underside and face, golden nape, and white rump and white shoulder bar. In early spring. black feathers can be edged with brown. Females, nonbreeding adults, and immatures show brown-streaked upperparts, streaking on flanks and (sometimes) across upper breast, and black stripes on head and behind eye. Pale between eye and bill; bill has pink base.

2. Red-breasted Meadowlark (*Sturnella militaris*), p. 281
L 7.5 in (19 cm). Rare, with one record of a female-plumaged bird from Aruba, but common in parts of Venezuela. Male in breeding plumage should be unmistakable, with its bright throat and breast. Females and immatures are brown, with pale eyebrow and crown stripe, and usually show rosy tinge to upper breast, streaking on upper breast and flanks. Note barred tail.

3. Eastern Meadowlark (*Sturnella magna*), p. 280
L 9 in (23 cm). Rare, with only a single record on the islands, though the species is relatively common on Venezuelan mainland. Very distinctive, ground-dwelling blackbird, with bright yellow undersides and bold black V on chest. Streaked-brown upperparts; on very short tail, note white sides.

4. Great-tailed Grackle (*Quiscalus mexicanus*), p. 281
15.5 in (40 cm). Newly arrived species to Aruba and Curaçao; small numbers presently but might establish breeding populations. Like a much larger and longer-tailed Carib Grackle, but beware size variation in Carib Grackles.

5. Carib Grackle (*Quiscalus lugubris*), p. 283
L 9.75–10.5 in (24.6–27.5 cm). Although a relatively recent arrival to the islands, now common in many areas, including around resorts. Male is glossy, purplish-black overall; female is rust-colored. Yellow iris. Wedge-shaped tail. Makes loud screeching calls.

6. Shiny Cowbird (*Molothrus bonariensis*), p. 285
L 9 in (23 cm). A recent arrival to the islands; now well-established on Aruba and Curaçao; current status on Bonaire is uncertain. Smaller than Carib Grackle, with a shorter, squared tail and thicker-based bill. Male black, with glossy purple sheen (in good light) to head and body. Eyes dark. Female drab gray-brown; also has dark eye.

♂
breeding

1

nonbreeding
adult

2

♂

♀

3

♀

4

♂

5

♂

♀

6

♂

♀

PLATE 59
Blackbirds, Orioles

Icteridae: Blackbirds, Orioles *continued*

1. Oriole Blackbird (*Gymnomystax mexicanus*), p. 284
L 12 in (30.5 cm). There is only single, undocumented report from Aruba. Large, Troupial-sized blackbird with large, pointed bill, bright yellow underside and head; black back, wings, and tail. Yellow shoulder on wing; black between eye and base of bill; and black line extends down from bill base to form whisker mark. Immature has black cap.

2. Yellow-hooded Blackbird (*Chrysomus icterocephalus*), p. 284
L 7 in (18 cm). Rare, with a handful of records; common in parts of Venezuela and Colombia. Male unmistakable, with bright yellow hood. In female, note yellowish eyebrow, yellow throat and upper breast, brownish belly, dark bill.

3. Troupial (*Icterus icterus*), p. 286
L 9–9.5 in (23–24 cm). Relatively common breeding resident on all three islands. Unmistakable, with bright orange body and black hood and bib. Bold white wingstripe. Yellow iris. Also note large, light blue fleshy patch behind eye.

4. Baltimore Oriole (*Icterus galbula*), p. 288
L 7.5 in (19 cm). Rare winter migrant visitor, with a handful of records. Smaller than Troupial and male lacks that species' blue fleshy patch behind eye. Male bright orange with black hood and back. Unlike Troupial, has orange in outer tail feathers. In female and immature, note white wingbars, orangish upper breast.

5. Yellow Oriole (*Icterus nigrogularis*), p. 287
L 8 in (20.5 cm). Resident breeding bird on all three islands, but much more common on Curaçao and Bonaire than on Aruba. Bright yellow overall, with black bib and tail; shows black from eye to base of bill. Narrow white edging on wing and single white wingbar.

6. Orchard Oriole (*Icterus spurius*), p. 286
L 6.5 in (16 cm). Rare migrant winter visitor, with only a single record; wintering birds occur to northern South America, so future sightings are not unlikely. Male quite unmistakable, with chestnut lower breast, belly, and rump contrasting with black hood, back, and tail; black wings show white wingbar and edgings. Immature male yellowish-green, with black throat and two white wingbars. Female yellowish-green overall, with two white wingbars.

1

2 ♂

♀

3

4 ♀

♂

immature

5 immature

♂

6 ♂

♂ immature

♀

Bird Conservation

The major conservation challenges on Aruba, Bonaire, and Curaçao fall within a few broad categories: invasive species; habitat loss and degradation; unsustainable resource use; pollution, toxins, and waste disposal; and climate change. Islands—especially small islands like Aruba, Bonaire, and Curaçao—pose special challenges to conservation, due to the simple fact that they have less land area. Less land area means reduced habitat for animals and plants and a higher risk of extinction. In fact, extinction rates documented for island flora and fauna are much higher than those on continents. More than 88% of the bird species that have gone extinct since the year 1500 lived only on islands (Butchart et al. 2006); and about half of all animal extinctions since then have been of island-restricted species. As dire as this sounds, these estimates might be low, as many island-restricted species have not been recognized or identified.

Aruba, Bonaire, and Curaçao have witnessed their share of extinctions. Yellow-shouldered Parrots and Scaly-naped Pigeons once occurred on Aruba but have been gone for more than 50 years (they do occur elsewhere). White-tailed Hawks nested on Aruba at one time but now appear to be absent from the island. Indeed each of the three islands has a number of bird species that have declined in numbers or whose population status is uncertain.

We have not been able to find any official government lists of endangered, threatened, or special concern bird species for any of the islands. What we present here is a proposed list for each island based on our experience, judgment, and anecdotes from colleagues, recognizing, of course, that other people may hold different opinions. Absent from the list are some migrants that rarely occur or do so with no regularity or in small numbers. Among these are Piping Plover, Olive-sided Flycatcher, and Cerulean Warbler. Our hope is that this list will encourage more discussion within each island's own governmental and non-governmental conservation agencies and organizations about the priorities for bird species conservation, monitoring, and action.

Aruba

Island Extinct:

> Yellow-shouldered Parrot
> Scaly-naped Pigeon

Island Endangered:

> Black-capped Petrel
> White-tailed Hawk
> Brown-throated Parakeet (Aruban form)
> Rufous-collared Sparrow (Aruban form)

Island Threatened:

Crested Bobwhite (local form)
Crested Caracara
Burrowing Owl (Aruban form)
White-tailed Nightjar (local form)
Caribbean Elaenia
Black-whiskered Vireo

Special Concern:

White-cheeked Pintail
Brown Booby
Brown Pelican (local breeding population)
Magnificent Frigatebird
American Kestrel (local form)
Caribbean Coot
Semipalmated Sandpiper
American Oystercatcher
Sandwich "Cayenne" Tern
Least Tern
Black Noddy
Brown Noddy
Reddish Egret
Yellow Oriole

Bonaire

Island Endangered:

Barn Owl (apparent local form)
Yellow-shouldered Parrot
Grasshopper Sparrow (local form)

Island Threatened:

White-tailed Hawk
Crested Caracara
Pearly-eyed Thrasher

Special Concern:

White-cheeked Pintail
Brown Booby
American Flamingo
Caribbean Coot
Snowy Plover

Wilson's Plover
American Oystercatcher
Red Knot
Sandwich "Cayenne" Tern
Royal Tern
Least Tern
Reddish Egret
Scaly-naped Pigeon
Brown-throated Parakeet (Bonaire form)
White-tailed Nightjar
Caribbean Elaenia
Black-whiskered Vireo
Yellow Oriole

Curaçao

Island Endangered:

Barn Owl (local form)
Scaly-naped Pigeon
Grasshopper Sparrow (local form)

Island Threatened:

White-tailed Hawk
Crested Caracara

Special Concern:

White-cheeked Pintail
Brown Booby
Magnificent Frigatebird
American Flamingo
Caribbean Coot
Snowy Plover
Wilson's Plover
American Oystercatcher
Semipalmated Sandpiper
Sandwich "Cayenne" Tern
Least Tern
Reddish Egret
Brown-throated Parakeet (Curaçao form)
White-tailed Nightjar
Caribbean Elaenia
Black-whiskered Vireo

Major Threats

Certainly since the arrival, in the 1500s, of European explorers to the islands, there have been a variety of important, changing pressures on its avifauna. Early European explorers left behind goats, cows, donkeys, sheep, and horses with the idea that the animals would reproduce and the explorers could return and use them. The ships also carried rats that sometimes came ashore. Thus began 500 years of changes wrought by invasive species; rats alone would have deleterious effects on bird nests and young. Although likely an exaggeration, an English explorer—some say pirate—named John Hawkins stopped on Curaçao in 1565 and claimed that there were about 100,000 cattle on the island (Rupert 2012). A report in the 1620s by a Spanish priest estimated 10,000 cattle, 14,000 sheep, 6,000 horses and mules, and 200 goats on Curaçao (Rupert 2012). In 1672 it was reported that the meat from 1,200 goats raised on Aruba was being stored in the Dutch West India Company warehouse on Aruba, and that Aruba was often referred to as "goat island" during that period (Hartog 2003). On all three islands, the vegetation consumed by grazing animals surely resulted in a change to bird populations. During the first few hundred years of European colonization, there were also periods in which a lot of wood was taken from the islands, mostly so-called "dye-wood" (Rupert 2012). It is unknown how widespread wood harvesting might have been, but it is interesting to note that a mayor of the mainland South American city of Coro reported in August, 1632, that 18 vessels had come to Bonaire to take on wood (Rupert 2012), and in 1633 the Venezuelan governor reported that 15 vessels had come to Bonaire and Curaçao in June alone to take on wood (Rupert 2012).

Goats and other livestock were left on the islands by early explorers beginning in the 1500s. As populations of these animals soared, they had major impacts on vegetation.

Although the domestic animals raised on the islands and revenue from the exportation of wood and salt would have supported more people, the human population on the islands remained small from the 1500s to 1800 (an 1804 census found only 1,155 people on Aruba [Hartog 1961]). In the late 1700s and early 1800s, especially when slave trading on Curaçao began to take off, slightly larger human populations were found on all three islands. As a result, larger areas of land were cleared of native vegetation for crops and charcoal production or were more heavily grazed by goats, donkeys, and horses. Yet here again, it is difficult to surmise the damage to native bird populations, as we lack an estimate of the amount of habitat converted and have no assessments of the bird life on the islands at that time. Certainly, hunting, probably of larger birds like doves and bobwhite and perhaps of parrots and parakeets, would have increased.

From the mid-1800s well into the early 1900s, people harvested divi-divi pods (also known as *watapana* on Aruba) for export to Europe, where the tannins in the seed pods were extracted for use in tanning hides (Hartog 1961). During each year, harvesting would peak in January and in June and July (Hartog 1961). According to Hartog (1961), as late as 1911, Aruba exported 668,000 pounds (303,000 kilograms) of the pods, Curaçao exported 1,437,414 pounds (652,000 kilograms), and Bonaire an astounding 2,429,494 pounds (1,102,000 kilograms). One cannot help but wonder how the removal of a major food source for many birds might have impacted their populations. Despite these changes, however, there is no documentation pointing to the disappearance of any bird species from the islands during this time.

While prospectors attempted to find gold or other valuable minerals on all three islands, it was only on Aruba that more intensive gold mining took place. After the discovery in 1824 of some gold nuggets in a dry stream valley not far from present-day Arikok National Park, the island experienced its own gold rush of sorts, as people scrambled out in search of more nuggets. Gold mining as an eventual industry on Aruba lasted until 1916, during which time the corporate structures that controlled the industry and the technology used for mining underwent changes. Eventually miners employed a more modern and technologically driven process that involved digging mine shafts at a number of locations around the island. Processing facilities were built at Bushiribana and Bushiri, the latter near present day Spanish Lagoon (Hartog 1961; Morin and Hutt 1998). The environmental impacts of gold mining on Aruba apparently have never been studied in detail, but the cyanide leaching process was used extensively at the Bushiri facility (Hartog 1961). This process can leave behind a variety of contaminants, including heavy metals, cyanide itself, and tailings that generate highly acidic runoff when exposed to air and water. Along the edge of the lagoons at Spanish Lagoon, below the remnants of the former Bushiri gold processing facility, piles of fine tailings devoid of vegetation continue to be a prominent feature (Morin and Hutt 1998). One would like to know if there are any lingering impacts from tailings here or at any other former gold mines or processing locations.

Beginning in the late 1800s, phosphate mining was carried out on Curaçao, Klein Curaçao, and Aruba. On Aruba, this continued until the early 1900s, and even later on Curaçao (Hartog 1961, 1968; Steinstra 1991). The most extensive landscape changes from phosphate mining probably occurred on Klein Curaçao, while relatively small open-pit mines were dug on Aruba and Curaçao (Hartog 1961, 1968). A fairly extensive underground phosphate mine was built in the Sero Colorado section of Aruba. These phosphate deposits are all from ancient seabird nesting sites. On Curaçao, some waste rock piles and water-filled phosphate pits remain, though the water in the pits is likely of such high salinity that it limits the aquatic life found there, and thus probably fails to attract birds.

There was limited manganese mining on Curaçao, in what is now Christoffel National Park. Here you can still find several heaps of stone rubble taken from the mines, all free of vegetation, but there is no obvious evidence of pollutants or acid drainage. Today, some operations on Aruba and Curaçao mine limestone, sand, and gravel for construction on the islands and—at least on Curaçao—for export. On Aruba, several such operations lie close to the borders of Arikok National Park and have thus been of concern to some conservationists. Bonaire has also had mining operations for limestone, sand, and gravel that have gone on for many years. There was a moratorium placed on this so-called "coral mining" in 2010, but enforcement was a problem and the moratorium proved to be controversial, as these materials must be imported if not mined locally. Sand mining is a major problem at many beaches, where its takes many years to rebuild the volume of sand. Under normal conditions, these beaches are sometimes used by nesting sea turtles and by feeding shorebirds.

From the late 1800s until the first half of the twentieth century, new industries and a generally growing human population brought new pressures to bear on the plant and animal communities of Aruba, Bonaire, and Curaçao. Aloe plantations

Aloe production was once a major industry and a lot of habitat was converted to aloe plantations, particularly on Aruba. Today it is only a small segment of the economy.

essentially required the complete removal of native vegetation. These plantations reached their peak in the late 1800s and again in the 1950s (Hartog 1961); people have claimed that as much as 30% of the land area of Aruba was under some form of aloe cultivation at the height of the industry. Regardless of exact numbers, it is clearly the case that many bird habitats were lost due to aloe cultivation. Voous (1983), writing of Aruba, noted that "Semi-deserts, opuntia and cultivated aloe fields have extended considerably at the expense of the original thorny woodland. Fields of aloe with scattered asymmetrical dividivi trees now form a kind of artificial savanna in which birdlife is scarce."

On Bonaire, the production of solar salt (salt produced by evaporating seawater) on the very low, flat southern end of the island has been ongoing for hundreds of years. The salt beds, or salinas, on the northern end of the island (in what is now in Washington-Slagbaai National Park), often have large populations of brine shrimp and larvae of brine flies. These are fed upon by American Flamingos, Lesser Yellowlegs, Stilt Sandpipers, Snowy Plovers, and other shorebirds. Salinas and bays with lower levels of salinity can host small fish that are a food source for herons, egrets, and other species. Salt was probably once gathered at the southern end of Bonaire—and at similar areas on Curaçao and Aruba—by indigenous peoples, even before the arrival of Europeans. Europeans certainly began to export salt from Bonaire in the seventeenth century, if not before (Hartog 1978). For several hundred years—up until slavery was abolished in 1863—salt was harvested by slaves toiling under the broiling sun (Hartog 1978). In the early 1800s, several salt ponds were created in order to control water levels for increased salt production (Hartog 1978). At times, the saltworks on Bonaire produced a significant amount of salt—an amazing 29.6 million pounds (13,426,334 kilograms) in 1837—but production varied greatly from year to year, depending both on demand and production capacities. Thus the industry did not grow, and nor did the area of salinas on the southern end of the island (Hartog 1978). In the late 1800s, when the government sold virtually the entire island to private interests, someone developed salt ponds on the northern end of the island, at Goto and Slagbaai, although these never reached the output of the saltworks at the Pekelmeer, on the southern end of the island (Hartog 1978). Presumably the ponds and salinas in these areas would have supported shorebirds and waterbirds of various species, although, historically, flamingos, sandpipers, and plovers were likely popular game birds for residents and may have been reduced in numbers.

In 1966 salt production on Bonaire was transformed into a technologically advanced industrial operation, with the creation of a facility with a series of shallow impoundments in which water with increasing salinity is moved from pond to pond until solar evaporation leaves a highly refined salt. This facility, now owned by the global Cargill Corporation, produces about 400,000 tons of salt per year (Barendson et al. 2006). Impoundments within and near the complex are frequented by a variety of sandpipers and plovers, herons and egrets, terns, and the most famous avian inhabitant—the American Flamingo. Since 1969, when an agreement was reached between the government of Bonaire and the Antilles International Salt Company, the owner at the time, 120 acres (55 hectares) within the complex have been designated as a flamingo sanctuary (Voous 1983). The water levels of the sanctuary are managed by Cargill, and the area is kept off limits to the public to prevent disturbance. However, it appears that there has been a long-term decline in the numbers of breeding American Flamingos within the sanctuary, with just 500 pairs breeding there in recent years (Slijkerman et al. 2013). As reference, there were 1,300 pairs reported nesting in 1996 (Wells and Debrot 2008) and 2,300 nesting pairs from 1969 to 1970 (Voous 1983).

Beginning in the 1920s, the arrival of the oil industry brought major changes to Aruba and Curaçao. With it came a hefty increase in revenues, followed by a massive increase on both islands in the number of people. The facilities on Aruba and Curaçao received thousands of oil tanker visits as oil was shuttled from Venezuela's Maracaibo Basin to the islands and then moved from there to larger tankers for transport around the world. Major oil refining plants were built and continued to expand on Curaçao and Aruba, and this rapidly booming oil industry required more workers than were available on the islands. The number of people on Curaçao increased from 33,000 in 1916 to 134,250 by 1965; on Aruba the number jumped from 9,000 in 1926 to 59,858 by 1965 (Hartog 1968). In turn, this lead to a boom in new businesses and the construction of housing and roads. It is probably no coincidence that the Yellow-shouldered Parrot and Scaly-naped Pigeon disappeared from Aruba during this period, although these species may also have been targeted by hunters.

Volatility in the world oil market and modernization of the industry on Aruba and Curaçao resulted in constant changes in the number of people employed. Indeed, after 1950 the total number employed decreased significantly on Aruba and Curaçao (Hartog 1978). The extent to which pollutants attributable to the oil industry have affected birds on and near the islands has not been carefully studied. The potential impacts would have come from air pollution from oil refinery operations and from oil and other contaminants spilled directly into the water or on land (Wells et al. 2008). Certainly, in the initial decades of oil industry growth on Aruba and Curaçao, environmental awareness and regulations would have been minimal or nonexistent. On Curaçao, a number of areas have heavy petroleum contamination from operations that were unregulated or poorly regulated, especially at Schottegat, Bullenbaai, and Spaanse Water (Woodard 2008); and many samples of marine sediment show high levels of contamination from metals (Pors and Nagelkerken 2008). Unfortunately, clean-up costs have been estimated to be as high as one billion dollars; Shell, the previous owner of the facilities, has not accepted any responsibility and claims that the government of Curaçao agreed to absolve the company of any future cleanup costs (Woodard 2008). The Aruba oil refinery, under ownership of Valero Corporation, was closed in 2012, but a recent agreement with the Venezuela government appears to mean the refinery may reopen.

A survey of Curaçao's coastline found oil and tar residues at various shore communities, many years after known oil spill events, with highly reduced marine invertebrate populations on the most contaminated sites (Nagelkerken and Debrot 1995). We are not aware of any more recent research on the long-term effects of petroleum on the ecosystems of Curaçao, although research following oil spills in other parts of the world has shown that populations of birds, fish, other wildlife, and plants are often impacted for many years after the initial spill and

Unfortunately, oil pollution sometimes impacts wildlife on the islands, as demonstrated by these oiled American Flamingos on Curaçao.

cleanup (Wells et al. 1995; Henkel et al. 2012; Mearns et al. 2013; Burns et al. 2014). It is important to note that shorebirds like Ruddy Turnstones, American Oystercatcher, and Whimbrel are among the bird species that often probe into the coral rubble beaches where tar and oil contamination often linger the longest.

Problems with oil pollution continue, unfortunately. On Curaçao, oil from land sources spilled into the water around Bullenbaai and Jan Kok, in 2012 and 2013. And on Bonaire, a fire in 2010 at an oil facility near Gotomeer contaminated an area nearby and perhaps explains why fewer flamingos were found in Gotomeer in recent years (Slijkerman et al. 2013). Air pollution from the refinery operations on Curaçao are a continuing source of health concern to people living downwind of the plume of sulfur dioxide emitted by the operations, measured as recently as February, 2015 (Beumer et al. 2011; Curaçao Chronicle, Feb 2, 2014; Curaçao Chronicle, Sept 22, 2014). We do not know of any studies on the possible effects of air pollutants on the terrestrial and marine ecosystems of Curaçao and Aruba.

A potential threat to marine conservation arose in 2015, when surveys for off-shore oil began off Aruba; a plan has also been initiated for oil surveys in Curaçao's offshore waters (Curaçao Chronicle, Jan 16, 2015). It goes without saying that the governments of the islands should procede slowly and thoughtfully, given the potential impacts to the environment and considering the revenue generated by tourism. Further, meaningful progress in cleaning up petrochemical pollution should be a priority on all three islands, and many wisely advocate for instituting the highest level of environmental safety standards for the oil industry (Beumer et al. 2011).

The people and the governments of the islands have become so dependent on the oil industry that occasional downturns in the price of oil or oil production have serious consequences. Government leaders have therefore recognized the need for diversification of the economic sector and have looked to tourism as one of the important revenue generating industries for the islands.

The tourism industry effectively started on Curaçao in the 1920s, when cruise ships began coming to the island to obtain cheap fuel from the oil refinery (Hartog 1968). Numbers increased fairly modestly, with 11 visiting tourist vessels in 1927, 37 in 1934, and 90 in 1964 (Hartog 1968). Despite the increase in the number of cruise ship visitors, hotels were not required for passengers' stay, so terrestrial habitats remained largely unaffected by tourism. Beginning in the 1920s, a number of small hotels were established on Aruba and Curaçao (Bongers 2009), mostly to accommodate business travelers and a small number of tourists. But in the 1950s, all three islands saw hotel-based tourism (Hartog 1968; Hartog 2003; Bongers 2009) begin to take off, with consequent effects on the environment (Thielman 2013).

On Aruba, the island with the largest tourism sector, the number of hotel rooms increased from 64 in 1956 to 1,303 in 1976, and then to an astounding 7,483 in 2007 (Bongers 2009). This increased the number of stay-over visitors on Aruba from 981 in 1956 to 732,392 in 2005 (Bongers 2009) and 903,934 in 2012 (Central Bureau of Statistics Aruba 2015). Beginning in the 1980s, a number of large hotels were built on Aruba, and the demand for hotel employees and workers in other sectors of the tourist market jumped upward. The population on Aruba had remained fairly static in the 1960s and 1970s (58,189 in 1972) but by 1995 the number stood at about 81,000. As of 2014, the total population was at 107,394 (Central Bureau of Statistics Aruba 2015).

Tourism has now become one of the most important sectors of the economy in Curaçao as well, with the number of stay-over visitors increasing from 226,248 in 2002 to 440,063 in 2013; on top of this, in 2013 over 580,000 one-day visitors disembarked from cruise ships. The number of hotel rooms in large hotels increased from 2,860 in 2003 to 4,367 in 2009; and as of 2009, there were 772 small hotels and guesthouses (Halcrow 2010) on the island. Curaçao has seen ups and downs in population numbers in modern times, with the rise and fall of the oil industry, going from 125,000 people in 1960 to a peak of about 153,000 in the mid-1980s, then declining to about 127,000 in the early 2000s (ter Bals 2014) before the current increase to 150,563 in 2011.

Bonaire is now highly reliant on tourism to sustain its economy, although unlike Aruba and Curaçao, its main focus is ecotourism, especially scuba diving. Bonaire's number of stay-over tourists increased from 50,000 in 2001 to slightly more than 70,000 in 2007 and 2008, reaching a peak in 2013 of 165,898 visitors. The number of hotel units (or similar accommodations) on Bonaire was at 2,139 in 2011. The number of cruise ships stopping at Bonaire increased from 64 in 2001 to 143 in 2010, with the number of cruise ship passengers totaling over 230,000. As you might expect, the resident population on the island, many of whom work in the tourism industry, went from about 8,000 in 1972 to 10,000 in 1992, and then skyrocketed to 17,400 as of 2013.

The hotel-based tourism industry on all three islands, along with the increase in the resident population required to service that industry, means that the need for water, electricity, sewage treatment and disposal, food and other resources is vastly greater than before the advent of a large tourism sector. In turn, this has affected the wildlife on the islands. The most obvious change is the amount of habitat lost to development for hotels, golf courses, restaurants, stores, housing, and the like. Curaçao is estimated to have lost approximately 30% of its natural habitat to housing and industry (Debrot and Wells 2008). It stands to reason that similar trends have taken place on Aruba and Bonaire. Unquestionably, habitat loss means there is less land available to support most native bird populations. Research has documented and mapped many of the unique habitat associations, important bird areas, wilderness areas, large ecologically intact landscapes, and other areas of high conservation value on all three islands (Debrot and de Freitas 1991; de Freitas et al. 2005; Wells and Wells 2006; Debrot and Wells 2008; del Nevo 2008; Nijman et al. 2008; Wells and Debrot 2008; Debrot et al. 2009; Smith et al. 2012; Geelhoed et al. 2013; CARMABI 2014; Vermeij 2014). Conservationists are in agreement that such areas should be placed within permanent protection as swiftly as possible in order to maintain healthy ecosystems where populations of native species of birds, other animals, and plants can thrive.

Large hotels and a growing numbers of residents on the islands have increased the demand for water and sewage disposal facilities, difficult challenges for small islands, particularly on Aruba. Beginning in the 1970s, the sewage treatment facilities on Aruba began discharging the treated water into the salinas at Bubali, transforming the intermittent wetlands there into permanent wetlands, and resulting in a change in the composition of aquatic plant species. The wetlands at Bubali have become one of the best birding locations on Aruba. The area has hosted many first bird records on the island, as well as many rare birds. For example, when the wetlands were first established in the 1970s, four species of ibis and a Wood Stork all made an appearance on the island for the first time (Voous 1983). Neotropic Cormorants, Pied-billed Grebes, Least Grebes, and Caribbean Coots appeared there and eventually began nesting. Bubali continues to be a year-round magnet for many bird species. Another Aruba location, the Tierra del Sol golf course and ponds, and, to a lesser extent, the grounds at the Divi Links, have played host to many new and rare birds, including Southern Lapwing, Cattle Tyrant (not a wetland bird but this species prefers grassy golf course habitats), Buff-breasted Sandpiper, and a variety of uncommon or rare northern duck species. On Curaçao, the sewage treatment facility at Klein Hofje, one of the only permanent water habitats on the island, has hosted Least Grebes, Caribbean Coot, and Black-bellied Whistling Ducks, among other species. The importance of wetland habitats for birdlife highlights the importance of ominous news in 2015, in which it was revealed that—due to insufficient capacity for disposing of wastewater—large quantities of wastewater and other liquid waste had been illegally disposed for years near Hato (Curaçao Chronicle, Feb 3, 2015).

The good news is that sewage treatment facilities and the water conservation measures in place at most hotels on all three islands have been instrumental in lowering the amount of damaging nitrogen and other contaminants that leach out into the surrounding marine waters and cause major problems for those ecosystems (Bak et al. 2005; Van Der Lely et al. 2013). Lac Bay on Bonaire is a unique ecosystem highly susceptible to overload from nitrogen, as well as from excess

Tourism is one of the largest contributors to the economies of all three islands, and the number of hotels and other tourism related infrastructure continues to grow, as does the resulting challenges to maintaining healthy ecosystems.

sediments washing into the bay, and recommendations have been developed to decrease these problems (Wentink and Wulfsun 2011; Wosten 2013). Proposals for a new tourist development in what is still a fairly pristine place, the Oostpunt Eastpoint area of Curaçao, with its important inland bays and nearby coral reefs, are of major concern because of these water issues, among others (Vermeij 2014).

Waste disposal is another particularly difficult problem on small islands for the obvious reason that there is limited space for landfilling (Polido et al. 2014). Although a lot of waste was disposed at sea prior to the 1970s, Aruba, Bonaire, and Curaçao have also had large unlined landfills for decades that undoubtedly leach contaminants into the ocean. The landfill on Aruba (and at times, those on Curaçao and Bonaire) has often burned and released unknown quantities of pollutants into the air. On Curaçao, the Malpais landfill, on the verge of reaching its capacity, is just upstream from the Malpais Important Bird Area. Unfortunately, illegal dumping of household waste, construction materials, and industrial wastes is common on Aruba—there are at least 40 illegal landfills according to Aruba Birdlife Conservation, including some large ones in abandoned quarries—and on Curaçao and, to a lesser extent, Bonaire. The governments of all three islands are aware of such problems and the need for increased recycling and the adoption of modern technologies and sustainability practices. They are in various stages of implementing solutions (Barendsen et al. 2006, Government of Aruba 2009, Government of Curaçao 2014). Some bird species like Laughing Gulls, Cattle Egrets, and small numbers of Crested Caracas forage in and near landfills on the islands. The impacts to birds from contaminants and from lost and degraded habitat as a result of waste management problems has not been studied on the islands.

The question of how to balance the increasing demands of a growing tourism sector with the capacity of each island to maintain healthy ecosystems and human communities is so critical that it merits extensive study (Debrot and de Freitas 1991; Ponson 1997; Cole and Razak 2009; Halcrow Limited 2010; Thielman 2013; Van Der Liely et al. 2013; Polida et al. 2014; Ridderstaat et al. 2014). Modern conservation science shows that in order to maintain all native species, ecosystem services, and the ability of animals and plants to adapt to climate change and other stressors, requires the protection of at least 50% of the landscape (Noss et al. 2012; IBCSP 2013, Locke 2013, Wells et al. 2013; Wells et al. 2014; Carlson et al. 2015). Along with setting aside terrestrial and marine habitats for protection, much work is still needed to deal with managing energy needs, waste disposal, water consumption, and the removal of pollutants and toxins from the environment on all three islands. Fortunately, technological advances are being made annually to help deal with these issues (LIGTT 2014).

With increasing numbers of tourists and an increase in imports to the islands, the introduction of harmful non-native species to the islands has also increased (van Buurt 2005; Hulsman et al. 2008; van Buurt and Debrot 2012; Smith et al. 2014). Historically, the introduced species that have perhaps had the biggest long-term impact on birds were probably goats and donkeys, whose grazing habits have diminished plant populations and caused changes to their species composition. Rats and, more recently, cats have likely been significant predators of bird eggs and young, especially of ground-nesting seabirds on places such as Klein Curaçao and Klein Bonaire. Recent management efforts have focused on removing cats from those two islands to increase the likelihood of success for birds that breed there (van Buurt and Debrot 2012).

Since the inadvertent introduction of the Boa Constrictor to Aruba, the species has exploded in numbers and is thought to be driving many native bird species toward potential extinction on the island. This one was captured at Fontein in Arikok National Park.

Recently, perhaps the most worrisome biological invasion is the Boa Constrictor's arrival to Aruba, where it was first discovered in the wild in 1999. The assumption is that the first individual was either an escaped pet or was intentionally released, a pregnant female or multiple individuals (Quick et al. 2005; van Buurt and Debrot 2012). In either case, the snakes have bred prolifically and now number in the thousands across most of the island, including within Arikok National Park (Quick et al. 2005; van Buurt and Debrot 2012). A study on Aruba of prey revealed that, of the prey items that could be identified, 40% were birds, including Brown-throated Parakeets (Quick et al. 2005). Observers have seen boas eating Yellow Orioles and other birds (Greg Peterson, Aruba BirdLife). Between 2003 and 2013, we and others noted a major decline in the numbers of some bird species on Aruba, notably Brown-throated Parakeets and Rufous-collared Sparrows. Although not enough research has been done to show that the decline is caused exclusively by predation from boas, we suspect that Boa Constrictors are the major contributor to recent bird declines. Along with measures to control boa numbers, it may be necessary to develop ways to maintain nesting sites that are free of boas or, in the case of parakeets, to provide artificial nest boxes placed in locations from which the snakes are excluded through various methods.

There is no doubt that introduced species are causing other ecological problems on land and in the surrounding marine waters of the islands, but attention has been focused on this issue only relatively recently (van Buurt and Debrot 2012). Problems caused by the lionfish on reef communities is now well known, and efforts continue to decrease the populations of this fish around Bonaire and Curaçao (de Leon et al. 2013), and, more recently, in Aruba (Aruba Ports Authority Newsletter, Sept 2014). Potential impacts from introduced insects are poorly documented, but there is increasing concern that Africanized honeybees and perhaps other introduced hymenoptera could prevent cavity-nesting birds like Yellow-shouldered Parrots and Brown-throated Parakeets from being able to use many otherwise suitable nesting sites.

In recent decades, a number of bird species that were either introduced or whose colonization was perhaps ship assisted, have established populations on the islands. House Sparrows commonly occur on all three islands and have spread to most of the occupied parts of all three islands over the last 20 years. It is unclear whether House Sparrows compete with native bird species like Rufous-collared Sparrows. Saffron Finches have been introduced on all three islands and are fairly widespread on Curaçao and Bonaire. Since first spotted on the islands in the 1980s and 1990s, Carib Grackles and Shiny Cowbirds are now fairly abundant on Aruba and Curaçao and increasingly abundant on Bonaire. Great-tailed Grackles now occur in small numbers on Curaçao and Aruba, with nesting documented on Curaçao. One of the oddest avian introductions was that of Troupials to Bonaire; they are now common throughout the island. Little is known about the interactions of introduced birds with native birds and other wildlife on the islands, but it is important to prevent the introduction of new non-native animals and plants to the islands, as history has proved the devastating consequences that such introductions can have on native fauna and flora (Smith et al. 2014).

Organizations, Initiatives, and Strategies

Fortunately, conservation organizations, educators, researchers, and advocates continue to play an active role on Aruba, Bonaire, and Curaçao to help protect wildlife and flora. Established in 2005, the Dutch Caribbean Nature Alliance (DCNA) is a forward-thinking conservation organization that works across these and other Dutch Caribbean islands (Saba, St. Eustatius, and St. Maarten). The DCNA was first established to deal with the long-term challenges faced by parks (both terrestrial and marine) and other protected areas, which, because they are managed by nonprofit organizations whose funding is not guaranteed by the government, face constant budget shortfalls. In addition, there have historically been many barriers preventing island-based organizations from seeking international funding. DCNA established and administers a trust fund to assist these parks. Specifically, the long-term goals of DCNA, as provided on their website, (www.dcanature.org) are:

- Securing long-term financing for the parks and establishing the trust fund;
- Providing professional training for park staff and developing their operating procedures, management, and strategic planning;
- Creating a shared pool of information and expertise about biodiversity, protected areas and conservation management in the Dutch Caribbean;
- Formulating an international fundraising and communications strategy.

Through DCNA there is now an integrated approach across the Dutch Caribbean in biodiversity monitoring, information sharing, education, communication, fund-raising, and other critical aspects of conservation. The numerous successful projects initiated by DCNA include:

- Development and implementation of a standardized bird monitoring protocol across the islands;
- Development and implementation of a bat monitoring and tagging study that has documented movements of bats within the islands and back and forth with mainland Venezuela;
- Establishment of an online repository for biological and ecological reports and data on the islands;
- Establishment of an educational website about biodiversity and conservation of the islands;
- Publication of a monthly online newsletter about new research, monitoring, education, and other conservation efforts underway on the islands.

For those interested in learning more about conservation in the Dutch Caribbean, the DCNA is a great resource and one that merits our support. A number of international organizations have been involved in conservation efforts in the islands, including the Institute for Marine Resources and Ecosystem Studies at Wageningen UR; US National Park Service; RARE; Vogelbescheming Nederland; World Wildlife Fund; Toledo Zoo; National Audubon Society; and many others. Finally, it is important to note the research and conservation organization CARMABI that, although based in Curaçao, has a rich history of also carrying out work on Bonaire and Aruba. We describe CARMABI in the section under Curaçao.

Aruba

A significant triumph for conservation was the formal establishment in 2002 of Arikok National Park, an 8,400-acre (3,400-hectare) protected area that encompasses about 17% of the total land area of the island. Even so, Important Bird Areas were identified on Aruba in 2008, most of these without formal protection (del Nevo 2008). Other important areas for birds on the islands are becoming better known but also lack protections. Aruba Birdlife Conservation has been working in recent years to gain full protection for 16 areas identified as important habitat for birds and other wildlife. Unlike Bonaire and Curaçao, Aruba does not currently have any marine parks.

Aruba has a number of active conservation and environmental organizations. The Arikok National Park Foundation, the nonprofit that manages Arikok National Park, is one of the most important. The park has a new visitor's center and road infrastructure completed in 2009 from an investment of over $6 million from the European Development Corporation. It also has many ongoing conservation research and monitoring initiatives and education projects that encourage kids and adults from the island population to learn more about nature and ecology. These include a research program on the Aruban Rattlesnake (Reinert et al. 2002); research on the introduced Boa Constrictor (Quick et al. 2005); and a bird monitoring program that functions in cooperation with the tern research and conservation initiative on the islands off San Nicolas, led by Adrian del Nevo (del Nevo 2008, 2009). The US National Park Service is partnering with Arikok National Park to help plan for the future of the park, focusing initially on the development of documents that will identify the park's key resources, values, management challenges, and planning needs. This work will help Arikok National Park staff prioritize their activities over the coming decades.

Another important organization, Aruba Birdlife Conservation, was started in 2010 under the leadership of Greg Peterson (who also has a pivotal role with the Arikok National Park Foundation). Aruba Birdlife Conservation has taken an active role in promoting bird conservation efforts and education on Aruba. These include a public awareness campaign about the devastating impacts of Boa Constrictors on Aruba's birds, a campaign to make the Burrowing Owl a national symbol of Aruba, and, perhaps most significantly, a campaign to designate many of the island's important areas for birds as protected areas. Another important initiative led by Aruba Birdlife Conservation was the Aruba National Bird Count, quite similar to the Great Backyard Bird Count in the US and Canada. Aruba's Central Bureau of Statistics was even enlisted to complete a report on the 2011 Aruba National Bird Count (Derix et al. 2013).

There are other environmental nonprofit organizations active on Aruba. These include organizations that are doing important work on the marine systems of the

Aruba Birdlife Conservation is actively engaged in a variety of important bird conservation and education efforts on Aruba and has a vibrant local membership.

island: Aruba Marine Park Foundation, Aruba Marine Mammal Foundation, Reef Care Aruba, and Turtugaruba Foundation (sea turtle conservation). These groups are working on exciting initiatives, such as looking at the possibility of developing a marine park in Aruba, establishing whale watching tours, monitoring sea turtle nest sites, and controlling non-native lionfish on Aruba's reefs.

In 2008, the non-governmental organization WildAruba invited the IUCN Conservation Breeding Specialist Group to lead a workshop to consider the top conservation policy priorities for Aruba. Sixty-five stakeholders came together over several days, and a final report with recommendations was completed (Barendson et al. 2008). The top recommendations from that workshop included:

1. Institute an Environmental Management Authority (EMA) including a Biodiversity Office;
2. Establish a Marine Park (implement protected marine areas leading to this);
3. Reinforce a moratorium on commercial development (hotels, malls, restaurants, etc.);
4. Create a permanent education committee of key stakeholders involved in Aruba's natural history and heritage;
5. Gather information on species/habitat and distribution;
6. Change/modify human behavior to create a more environmental-minded population.

Along with these broad conservation recommendations, we recommend a number of bird conservation actions for Aruba:

1. Assess the status of Brown-throated Parakeet populations, distribution, and reproductive success on the island and begin experimenting with provisioning of predator-proof artificial nesting boxes for the species;

2. Continue efforts to eradicate or decrease the numbers of boa constrictors on the island;

3. Assess the status of Rufous-collared Sparrow populations, distribution, and reproductive success on the island;

4. Increase the amount of habitat across the island under permanent protection for birds and other wildlife, beginning with the areas proposed by Aruba Birdlife Conservation and also including establishment of a marine protected area;

5. Assess the status of Burrowing Owl, Crested Bobwhite, White-tailed Nightjar, Crested Caracara, American Kestrel, and Yellow Oriole on the island;

6. Complete a study on the feasibility of reintroducing Scaly-naped Pigeon and Yellow-shouldered Parrot to the island;

7. Assess the impact of cat and rat predation on birds on the island and take steps to address it;

8. Continue efforts to decrease free-ranging goat and donkey populations on the island;

9. Formally protect the San Nicolas tern islands and make them part of Aruba's National Park system and continue to support monitoring and research at tern nesting islands at San Nicolas and Orenjestad and fund a seasonal tern and seabird warden;

10. Assess status of Least Tern populations, distribution, and reproductive success on the island;

11. Assess status of mangroves and seasonal wetland salinas habitats and their use by birds; develop policies to ensure the maintenance and protection of such habitats;

12. Create a nature-based tourism plan that capitalizes on Aruba's spectacular birds, animals, and plants.

Bonaire

Bonaire has demonstrated an awareness of the importance of maintaining healthy ecosystems for tourism, especially dive tourism. And Bonaire was at the forefront of land conservation efforts in the region when it created Washington Slagbaai National Park (in 1969) and Bonaire National Marine Park (in 1979). Washington Slagbaai National Park, located on the northwestern part of Bonaire, has 14,000 acres (5,600 hectares), encompassing approximately 20% of the island. The Bonaire

National Marine Park is 6,700 acres (2,700 hectares) in size. Six Important Bird Areas were identified on Bonaire in 2008 (Wells and Debrot 2008), though many of the areas are without strong, formal protection (Wells and Debrot 2008; Geelhoed et al. 2013). Other important areas for birds on the island are becoming better known but also remain without protections (Smith et al. 2012; Geelhoed et al. 2013).

Much exciting conservation work is underway on Bonaire, including research projects, conservation, and education focused on Yellow-shouldered Parrot conservation (O'Callahan 2009, Williams 2009). Other laudable work includes efforts to limit problems from introduced goats and donkeys (Debrot et al. 2011; van Buurt and Debrot 2012); construction of new sewage treatment facilities to lessen impacts from pollutants on the surrounding marine environment; identification of areas that need greater protection (Debrot et al. 2011; Smith et al. 2012); implementation of bird monitoring programs in Washington Slagbaai National Park and other areas (Simal and Frank Rivera-Milán 2010, Simal et al. 2010a, Simal et al. 2010b; Debrot et al. 2011; Rivera- Milán et al. 2012; Geelhoed et al. 2013); programs to remove or reduce feral cat and rat populations and to develop predator-free tern nesting areas (van Buurt and Debrot 2012; IMARES Wageningen UR 2014); and work to increase the health of the Lac Bay ecosystem (Debrot et al. 2010, 2012; Wosten 2013).

One of the most important nonprofit organizations is the Bonaire National Park Foundation (STINAPA). STINAPA is responsible for managing the protected areas of Bonaire (both Washington Slagbaai National Park and the Bonaire National Marine Park). It also has a robust environmental education program and engages in monitoring and research in addition to its conservation and enforcement activities. An important nonprofit bird conservation organization on

Echo, a parrot conservation organization on Bonaire, carries out research, education, and applied conservation programs on Bonaire to ensure the survival of Yellow-shouldered Parrots, Brown-throated Parakeets, and other native birds and wildlife.

Bonaire called ECHO was started by parrot conservation biologist Sam Williams in 2010. ECHO has made amazing progress in reducing the poaching of Yellow-shouldered Parrots and Brown-throated Parakeets and restoring habitat and key tree and shrub species needed by the birds. Adolph Debrot has been a leader in many research and conservation efforts on Bonaire for decades, initially through CARMABI and now through the Institute for Marine Resources and Ecosystem Studies at Wageningen UR, where there are a host of research and conservation initiatives underway.

Sea Turtle Conservation Bonaire, an organization started in 1991, carries out turtle nesting surveys, satellite tracking, education, advocacy, and research. The Council on International Educational Exchange Research Station Bonaire offers courses in ecology and biology and carries out research and education activities.

The 2006 Bonaire National Marine Park Management Plan identified a number of priority conservation issues and actions related to the entire island, including:

- Need for an integrated approach to coastal zone management that acknowledges the intimate link between activities on the land and the health of marine ecosystems surrounding the island;
- Need for comprehensive ongoing outreach to all stakeholder groups;
- Promotion and implementation of sustainable development concepts in economic planning, urban, and land use planning and construction;
- Reducing impacts from wastewater and sewage runoff;
- Reducing poaching of conch and sea turtles (we would add to that Yellow-shouldered Parrots and Brown-headed Parakeets);
- Managing artisanal fishing for long-term sustainability.

Along with these broad conservation recommendations, we would recommend a number of specific bird conservation actions for Bonaire:

1. Continue efforts to maintain and increase Yellow-shouldered Parrot populations;

2. Assess status of Brown-throated Parakeets on the island and increase efforts to ensure the maintenance of populations of the species on the island;

3. Increase amount of habitat across the island under permanent protection for birds and other wildlife;

4. Assess the status of Reddish Egret, Snowy Plover, Wilson's Plover, American Oystercatcher, White-tailed Hawk, Crested Caracara, Barn Owl, White-tailed Nightjar, Pearly-eyed Thrasher, Scaly-naped Pigeon, Black-whiskered Vireo, and Grasshopper Sparrow on the island;

5. Assess the impact of cat and rat predation on birds on the island and take steps to address it;

6. Continue efforts to decrease free-ranging goat and donkey populations on the island;

7. Continue to support monitoring and protection of tern nesting areas and fund a seasonal tern and seabird warden;

8. Develop an oil spill clean-up contingency plan;

9. Create a nature-based tourism plan that capitalizes on Bonaire's spectacular birds, animals, and plants.

Curaçao

Curaçao began conserving land in the 1960s, when its government purchased several former plantations in the northwestern part of the island. In 1978, Christoffel National Park was officially established, and in 1994 the government added the 494 acre (200 hectare) Shete Boka Park, followed in 1999 by the 272 acre (110 hectare) Daaibooi Beach natural area. Together, these parks encompass about 6% of the land area of Curaçao. Important Bird Areas were identified on Curaçao in 2008 (Debrot and Wells 2008), but many of these are without formal protection (Debrot and Wells 2008). Other important natural areas on the island also remain without protection (Debrot and DeFreitas 1991; CARMABI 2014; Vermeij 2014). A marine park known as the Curaçao Underwater Park, covering 2,559 acres (1,036 hectares), was established along the southeastern coast in 1983, but it has no regulations and is thus considered a so-called "paper park."

Curaçao's parks are managed by the nonprofit organization CARMABI, established in 1955 originally to focus on marine biological research. It now encompasses park management, education, and terrestrial as well as marine research.

All three islands have sustainability measures underway to increase use of cleaner energy-production sources, lessen pollution impacts, and to increase recycling.

CARMABI also has an impressive facility for visiting researchers and graduate students that supports 70–100 visiting researchers annually. CARMABI's education programs reach 12,000 school children each year.

Curaçao has an active conservation, education, and environmental research community, although there are no organizations that focus specifically on birds. Some of the organizations that have been active on Curaçao include Defensa Ambiental (Environmental Defense), Amigu di Tera (Friends of the Earth), Reef Care, Kids for Corals, Uniek Curaçao (focusing on awareness and guided tours), Korsou Limpi i Bunita (cleanups), Stichting Dierenbescherming (animal cruelty and welfare), Curaçao Nature Conservation, Foundation for a Clean Environment on Curaçao (Stichting SMOC), and the Curaçao Sea Aquarium.

The Government of Curaçao prepared a report in 2014 on sustainable development successes and challenges (Government of Curaçao 2014). The following conservation recommendations are taken from that report.

1. Develop a public awareness campaign to bring attention to the waste management challenges and potential future problems the island could face without major changes in waste management planning on Curaçao;

2. Develop a national policy and well-enforced regulations to vastly reduce solid waste stream, effectively manage hazardous and toxic waste, and incorporate mandatory recycling programs;

3. Develop a national policy and well-enforced regulations to manage sewage, wastewater, and runoff from hotels and housing developments;

4. Develop a comprehensive coastal zone management strategy that includes concepts related to cumulative impacts from land use;

5. Develop a long-term strategic environmental assessment and land-use plan for Curaçao that integrates the need for protection of land and marine ecosystems on the island that provide essential ecological services to human communities and includes mandatory environmental impact assessment for any new tourism, commercial, or residential development project;

6. Develop an annual environmental indicators monitoring system for Curaçao and provide an easy-to-understand summary to the public on an annual basis;

7. Increase capacity in non-governmental and governmental institutions focused on biodiversity conservation, ecosystem health, and sustainable development;

8. Develop a national water management strategy that considers catchments, groundwater resources, and surface impoundments;

9. Develop a national plan to understand risks from ship transport of toxic and hazardous materials (including petroleum products), include management of those risks as well as how to effectively deal with spills.

Along with these broad conservation recommendations we recommend a number of specific bird conservation priority actions for Curaçao:

1. Increase amount of habitat across the island under permanent protection (only about 6% is currently within parks) for birds and other wildlife starting with the Oostpunt area and Klein Curaçao;

2. Develop a more robust, protected marine park system;

3. Assess status of mangroves and seasonal wetland salina habitats and use by birds; develop policies to ensure maintenance and protection of such habitats;

4. Create a broad, nature-based tourism plan that capitalizes on Curaçao's spectacular birds, animals, and plants. Consider tapping into such a program to create a steady, sustainable funding source to ensure conservation, perhaps through a small tourist user fee that is dedicated and guaranteed to be used for nature conservation;

5. Develop monitoring and protection of tern nesting islands to rehabilitate tern breeding populations including at Jan Thiel and Klein Curaçao and other locations, and consider supporting a seasonal tern and seabird warden;

6. Assess status of Least Tern populations, distribution and reproductive success on the island;

7. Assess the status of Grasshopper Sparrow populations, distribution, and reproductive success on the island and increase efforts to ensure the maintenance of populations of the species on the island;

8. Assess status of Brown-throated Parakeets on the island and increase efforts to ensure the maintenance of populations of the species on the island;

9. Assess the status of Reddish Egret, Snowy Plover, Wilson's Plover, American Oystercatcher, White-tailed Hawk, Crested Caracara, Barn Owl, White-tailed Nightjar, Scaly-naped Pigeon, and Black-whiskered Vireo on the island;

10. Institute a broad-based annual bird population monitoring system for the island including a citizen-science based initiative like the Great Backyard Bird Count;

11. Assess the impact of cat and rat predation on birds on the island and take steps to address it;

12. Continue efforts to decrease free-ranging goat and donkey populations on the island;

13. Develop an oil spill contingency plan for the island.

Species Conservation

It is important to draw attention to several priority bird species—or species groups—and to summarize their conservation status and what is being done to protect them.

Yellow-shouldered Parrot

BirdLife International designates the Yellow-shouldered Parrot as the most globally threatened of the breeding species that occur on the islands (it no longer occurs on Aruba). The global assessment for the species places the total population size at between 2,500 and 9,999 individuals in 6 geographically disjunct populations. Three of these populations are on mainland Venezuela, two are on the Venezuelan islands of Isla Margarita and La Blanquilla, and the sixth is on Bonaire. In 2012, the number of Yellow-shouldered Parrots on Bonaire was estimated at between 650 and 800 (Birdlife International 2014), which makes it a crucial population for the long-term prospects of the species. It was known as a breeding species on Aruba until 1947. There is some evidence that it also once occurred on Curaçao (BirdLife International 2014). Escaped caged birds have been seen on Curaçao in modern times but there is no known established breeding population there. In recent decades, the main threat to the population of Yellow-shouldered Parrots on Bonaire has been from the poaching of nestlings to sell into the caged bird trade. The level of poaching is thought to have been reduced because of a major public awareness campaign and a new law that gave amnesty to all owners of Yellow-shouldered Parrots, so long as

Washington-Slagbaai National Park on Bonaire is one of the most important places in the world for Yellow-shouldered Parrots and American Flamingos.

they registered and banded the birds. However, monitoring of nests over the last 5–6 years has revealed that about 30% of nestlings that make it to fledging age are still taken by poachers (Species Management Plan 2012). It is thought that many of these birds end up as captives in Curaçao and perhaps make their way into the international illegal parrot trade. Unfortunately, there is limited law enforcement capability to stop poachers, even within Washington-Slagbaai National Park.

Even if poaching can be eliminated, the long-term survival of Yellow-shouldered Parrots requires large blocks of intact and healthy natural habitat. The vegetation of Bonaire has been greatly altered over centuries from overgrazing and overharvest of trees (Wells and Debrot 2011). This has caused a decrease in tree diversity and in the number of large, old trees that can provide nesting cavities. Reduced tree diversity means that there are fewer fruit and seed-bearing trees on which Yellow-shouldered Parrots can feed and that food is available during fewer periods in the course of a year. During dry periods, Yellow-shouldered Parrots are known to move into urban backyards near Kralendijk to feed on fruits and seeds in trees that are maintained through gardening and watering. A plan to lower goat densities in Washington-Slagbaai National Park is underway as of 2014 (OLB/STINAPA 2014). It is hoped that this will begin to allow the natural regeneration of a broader diversity of plants in the park.

The ECHO foundation was established in 2010 on Bonaire by parrot researcher Sam Williams to monitor Yellow-shouldered Parrot populations. A management plan developed in 2010 provides many useful, specific recommendations to ensure the long-term survival of a native population of Yellow-shouldered Parrots on Bonaire. Interestingly, the plan suggests the possibility of re-establishing a breeding population of Yellow-shouldered Parrots on Aruba.

American Flamingo

The American Flamingo is currently officially listed by Birdlife International as of Least Concern (i.e., not endangered or threatened) at the global level. It is one of the most visible and important species in the islands, especially on Bonaire, which harbors the only current nesting colony among the three islands. American Flamingos are quite vulnerable because they have relatively few nesting colony locations throughout their range; their food supplies are sensitive to changes in salinity and temperature and are often ephemeral; they take up to six years to reach breeding age; they don't breed every year; and when they do breed, they lay only a single egg at a time.

The breeding colony on Bonaire is very important to the well-being of the regional population of American Flamingos, which includes birds that move among the islands and back and forth to the Venezuelan mainland. The birds breed on Bonaire in a 120-acre (55-ha) section within the industrial solar salt-making complex on the southern end of the island. This area has been set aside as a flamingo sanctuary since 1969, when an agreement was reached between the Bonaire government and the then-owner of the salt works, the Antilles International Salt Company (Voous 1983). The salt operation is now owned by the Cargill

Corporation, which is responsible for managing water levels in the sanctuary and keeping the area off limits to the public in order to prevent disturbance. Unfortunately, it appears that there has been a long-term decline in the numbers of breeding American Flamingos within the sanctuary, with only about 500 breeding there in recent years (Slijkerman et al. 2013).

American Flamingos feed in bays, salinas, and other wetlands on Bonaire and Curaçao, sometimes in the dozens or hundreds. Small numbers have become regular on Aruba in several locations in recent years as well.

To ensure the long-term health of American Flamingos on the islands, there is a need for renewed research and monitoring to better understand their current movements, demography, health of their foraging habitats and food supply, and the factors contributing to the apparent declines in breeding numbers on Bonaire.

Caribbean Coot and the Importance of Water

Until recently, the Caribbean Coot was considered Near Threatened under the IUCN BirdLife International Redlist criteria. Today, BirdLife International considers it a form of the more widespread American Coot. Whatever its taxonomic status, it is clear that the form is localized and uncommon in much of the Caribbean basin. Now that sewage and water treatment facilities on Aruba, Bonaire, and Curaçao provide water sources that maintain permanent wetlands, the number of Caribbean Coots has greatly increased. Formerly, the birds showed up in larger numbers in wet years, when more and larger wetlands were formed in places like Muizenberg and Malpais on Curaçao. Here, the birds would breed

Caribbean Coots are commonly found in freshwater wetlands on all three islands.

when there was sufficient water and, if the wetlands remained long enough, significant numbers of the species would build up. When these temporary wetlands eventually dried up, the birds would disappear, though it is unclear whether they died or migrated somewhere else. It may be that coots and other waterbirds move regularly between Venezuelan wetlands and those on the islands.

Many of the wetlands on the islands that are used by Caribbean Coot and other waterbirds are neither formally protected nor actively managed. Many suffer from the effects of pollution, sedimentation, and disturbance. We are not aware of any formal management plans for wetland habitats or wetland-dependent birds on Aruba, Bonaire, or Curaçao. The issue is also complicated by the fact that many wetlands, salinas, ponds, and streams on the islands experience dramatically changing water levels depending on seasonal variations and changes in precipitation from year to year. A few areas have more or less continual small flows of water, and these areas probably became known quite early in the human occupation of the islands as they afforded the only sources of freshwater. With the advent of agriculture and livestock husbandry, such water sources began to serve another important function. So, it is perhaps not surprising that each island has a location known as a *Fontein*, or fountain, where water flows from limestone escapements. Bonaire's Fontein has a long history of hosting a fruit tree plantation and other crops as well as a great variety of birds. Similarly, each island has a number of places named *pos* or *put*, places with waterholes or wells, or sites where they used to exist. On Bonaire, for example, Dos Pos, Pos Mangel, and Put Bronswinkel have freshwater wells or springs that are great for birds.

Another type of wetland found on all islands is a small pond formed when a stream is dammed. These generally small impoundments can grow quite large when there is an extended period of rain and can hold water sometimes for months or even years depending on precipitation. On Curaçao, a large earthen dam at Malpais has created a freshwater pond that teems with Caribbean Coots, Pied-billed and Least grebes, Blue-winged Teal, White-cheeked Pintails, and other birds when there are frequent rains; this pond dries up completely under extended periods without rain. Muizenberg, also on Curaçao, sometimes forms into an even larger freshwater pond behind its concrete dam, at which time it is often filled with American Flamingos, Caribbean Coots, grebes, wading birds, shorebirds, and other species. On Bonaire, a now cracked concrete dam at Onima once formed a freshwater impoundment regularly used by birds.

These ponds and impoundments have often been modified—and in many cases now have their water levels controlled—by input of treated wastewater from sewage operations and/or water from island desalinization plants that make freshwater from seawater for human use. This effectively turns these wetlands into permanent water bodies that allow populations of fish, amphibians, and invertebrates to grow in number. As a result, there has been greater opportunity for expansion of some introduced species as well as more habitat for a variety of native birds. One of the most well-known of these wetlands is at Bubali, on Aruba. The area had always been a site of seasonal shallow salinas but in the 1970s it began to receive input from wastewater treatment facilities. Unusual numbers of wading birds and wetland-dependent birds began using the site soon after, resulting in many new bird records for Aruba. Over the years, the site has changed in response to its

The Reddish Egret is often sought after by birders. Never common, it is more abundant on Bonaire, where it is often seen in the Pekelmeer near the salt industry operations there.

management and use, and as sedimentation and vegetation filled in certain areas. Nonetheless, it has remained a relatively large permanent wetland that hosts a great many wetland-dependent birds throughout the year. The salinas or ponds beside the Tierra del Sol and Divi Links golf courses on Aruba are another example of natural salinas whose water levels are now maintained by inputs from treated wastewater and which host great numbers of wetland birds. The Blue Bay Golf Course on Curaçao has at least one small pond that attracts birds, and the small ponds at the sewage treatment plant at Klein Hofje on Curaçao are also great for birds. In recent years, the sewage treatment facility on the outskirts of Kralendijk, Bonaire, has become host to numbers of waterbirds.

Inland bays and salinas that have permanent or intermittent connection to the ocean are very important wetland habitats. The Pekelmeer, on Bonaire, is a series of interconnected salinas that have long been managed for solar salt production (evaporating salt water from shallow basins) but that also provide large expanses of habitat for nesting American Flamingos, terns, and plovers, as well as feeding grounds for herons and egrets, and for wintering shorebirds. Also on Bonaire, the Gottomeir and Slagbaai are among the larger salty salinas and bays that are especially important. Curaçao has many such wetlands. Some, like the large Schottegat Bay at Willemstad, once hosted breeding colonies of terns, but are now highly degraded from years of pollution from nearby oil industry and industrial shipping facilities. Smaller salinas on Curaçao like Salina Sint Michiel and Salina Sint Marie can host non-breeding American Flamingos along with various egrets, herons, and shorebirds. The Jan Thiel Lagoon on Curaçao once hosted large numbers of nesting terns but now they arrive only intermittently, largely, it seems, because of regular and repeated human disturbance in the area. Mangrove-dominated lagoons like those at Lac Bay on Bonaire, St. Jorisbaai and Spaanse Water on Curaçao, and Spanish Lagoon on Aruba, are highly important natural coastal wetlands that serve as fish nurseries and harbor great numbers and diversity of birds.

Snowy Plover

The Snowy Plover is listed by BirdLife International as Near Threatened. The largest nesting population in the islands is thought to be on Bonaire. Significant nesting populations may also occur on Curaçao but we are not aware of any surveys for the species there. Aruba may have a few nesting pairs but high levels of human use of most preferred beach habitats have probably reduced opportunities for successful breeding. We surveyed a large proportion of potential Snowy Plover habitat on Bonaire in July of 2001 and documented at least 46 adults, a population size that approaches 2% of the estimated breeding population in eastern North America, Mexico, and the Caribbean (Wells and Childs Wells 2006). These numbers, and the fact that these are the southernmost known breeding locations for this eastern form, make these populations regionally significant. There is a need for further research to document levels of breeding success on Bonaire and for surveys on Curaçao to better understand whether significant populations are breeding there. Wilson's Plover nests in similar habitats (we documented 24–26 adults on Bonaire in 2001) and research into Snowy Plover would likely shed light on both species.

Beach- and shoreline-nesting plovers around the world often suffer from human disturbance on beaches as well as from increased predation of nests and young from predators associated with humans (cats, rats, etc.). Human use of areas with significant nesting populations of Snowy Plovers (and Wilson's Plovers) should be restricted during the breeding season to maximize nest success for these species.

A major effort was made in the late 1990s and early 2000s to rid Klein Curaçao of feral cats (van Buurt and Debrot 2012). This would decrease cat predation on eggs and young of any nesting plovers or terns that might be there. Apparently the introduced rat population increased following cat eradication, so the overall impact on bird productivity is unknown. Controlling cat and rat populations near important nesting areas like Klein Bonaire and within Washington-Slagbaai National Park on Bonaire could help increase productivity for Snowy Plovers and Wilson's Plovers as well as for nesting terns. A new project was initiated in 2014 on Bonaire to engineer a new low island in the saltworks area of the Pekelmeer that was free of predators. Tern decoys were used to attract Least Terns and Common Terns to the island resulting in at least 90 Least Tern nests and two Common Tern nests. Such areas may also eventually be used by Snowy Plovers (IMARES Wageningen UR 2014).

Reddish Egret

The Reddish Egret, of concern to conservationists throughout its range, is confined to the Caribbean basin, from the US south through the Caribbean coast of Mexico and Central America and through the Caribbean to Aruba, Bonaire, and

Curaçao. It reaches highest numbers on Bonaire (although numbers are still small—typically less than 10 or so seen in an average day of birding on the island) and typically occurs only in ones and twos on Curaçao and Aruba. Historically, nesting has been confirmed on Bonaire in mangroves at Lac Bay and in button-bush shrubs at the Pekelmeer but not on Curaçao or Aruba (Voous 1983). The species is a specialist of shallow salinas and marine lagoons and mangroves, and it is thus highly dependent on the health of such ecosystems. Sedimentation and high nitrogen loads have been identified as among the problems negatively affecting some of the preferred habitats for Reddish Egret on Bonaire and Curaçao (Debrot et al. 2010). Lac Bay, which has been recognized as a RAMSAR site and an Important Bird Area (Wells and Debrot 2008; Geelhoed et al. 2013), has been the focus of management efforts and research for decades (Debrot et al. 2010). It continues to experience problems resulting from sediments from eroded, overgrazed lands in the watershed and from increased use for recreation (Debrot et al. 2010).

Terns, Noddies, and Other Seabirds

There are important tern nesting areas on all three islands. It appears from recent surveys, however, that many of the most important tern nesting locations on Curaçao are no longer active. Some of the sites there seem to have been abandoned because of increased human recreational activity, especially during nesting, which could cause mass abandonment. Other historical sites, particularly around the industrial Shottegat Harbor area, have suffered from pollution, including from oil and tar.

Over the last decade or so there has been a concerted effort to propose measures—and implement them—to safeguard the tern-nesting islets off San Nicolas in Aruba. These tiny islands have been known to host a large number of nesting terns, even in the late 1800s, when they were first reported by Ernst Hartert (Voous 1983). In recent years, the San Nicolas Bay islets have hosted thousands of nesting Sandwich "Cayenne" Terns and sometimes hundreds of Roseate, Common, Sooty, and Bridled terns, in addition to Brown and Black noddies (del Nevo 2008). There has also been a concerted effort to educate residents that live near the islets of the importance of the islets to nesting terns and to point out that the success of the terns depends on keeping the islets free from human disturbance. In the past, staff from the nearby oil refinery watched the area and reported undocumented activity to the local police and Coast Guard to ensure that people didn't visit the islets, especially during the tern breeding season (del Nevo 2009; Holian 2012). While this informal arrangement has allowed terns to continue to nest on the islands, neither the islands nor the terns have any formal protection. Some plans have been considered for a new resort complex at nearby Rodger's Beach that could vastly increase the level of human disturbance to nesting terns if jet skiing and other water sports are allowed, as they are at other locations on Aruba. Rats have been a problem on some of the islets; they take tern eggs and young; and Laughing Gulls are often a major predator on eggs and young. Efforts to protect the terns have been largely successful on these islets as they continue to support healthy numbers of nesting terns. Efforts to safeguard the islets from human disturbance will always be required in order to maintain them as important tern nesting sites.

Smaller numbers of terns also nest on the small reef islets off Orenjestad. Most of these islets are covered with mangroves, with small, sandy openings or narrow sand fringes, though a few of the very smallest islets are largely free of vegetation. In some years, thousands of Sandwich "Cayenne" Terns nest on these islands, with smaller numbers of Common Terns (del Nevo 2008). Various herons and egrets as well as Brown Pelicans have historically nested on some of these islands, within the mangroves, but it is unclear if any nesting has been documented in recent years. Dozens of Magnificent Frigatebirds regularly roost at night on one small mangrove-covered island off Orenjestad. These islands may receive more human disturbance from recreational boaters, jet skiers, and the like, since they are relatively close to hotel complexes, and there is no regular patrolling of the islands to protect them from disturbance.

On Bonaire in 2014, a small islet was built within the Pekelmeer saltworks complex specifically as a nesting site for terns, free from industrial activity, making it more difficult for rats and cats to get to it. Least Tern decoys and sound recordings were deployed on the small islet in spring 2014, and by May there were at least 90 Least Tern nests and two Common Tern nests (IMARES Wageningen UR 2014).

We are not aware of any other specific tern conservation and protection measures underway at other locations, although efforts to eradicate feral cats and possibly rats on Klein Curaçao and Klein Bonaire could prove very helpful to terns

Terns and other seabirds rely on the marine environment around the islands for their food supply and on small islands free of disturbance and predators in order to lay their eggs and raise their young.

nesting on those islands. Least Terns nest along the coastlines of Aruba, Bonaire, and Curaçao in small colonies, and nesting efforts of those species could potentially be improved with control of cats and rats near nesting sites (Geelhoed et al. 2013). Maintaining the areas free from human disturbance by site wardens, signs, or fencing could also prove beneficial. There is a need for more detailed surveys of historical and present tern nesting sites on Curaçao.

In addition to doing research into the nesting sites of terns, it is important to also consider the places where they forage for fish, both during the breeding season and in the non-breeding season, when some species may remain in the general area. Other species of seabirds use the same waters surrounding the islands, though most tend to stay too far offshore to be readily observed from land. A few seabird species do occur regularly in near-shore waters. These include the various tern species (although the Bridled and Sooty terns and Black and Brown noddies are typically only seen near the nesting islets on Aruba) and Laughing Gulls, as well as Magnificent Frigatebird and Brown Booby.

The offshore avifauna of Aruba, Curaçao, and Bonaire has received scant attention since Ruud van Halewijn completed his seabird transects in the offshore waters of the area in the early 1970s. In recent years, there has been a resurgence in research on marine birds in the region (Luksenburg and Sangster 2013; Geelhoed et al. 2014). Impressive intensive efforts (415 boat-based surveys between April 2010 and November 2011) by researchers riding along on commercial sportfishing boats off Aruba resulted in a number of new or rare records for the waters around Aruba (Luksenburg and Sangster 2013). These included sightings of the globally endangered Black-capped Petrel only 1.9 mi (3 km) from the northern tip of Aruba (Luksenburg and Sangster 2013). New species reported in the region included Cory's Shearwater and South Polar Skua (Luksenburg and Sangster 2013). Aerial marine mammal surveys conducted in November 2013 in the waters around Aruba, Curaçao, and Bonaire added to our knowledge of the use of offshore waters by marine birds (Geelhoed et al. 2014).

The inshore and reef areas and offshore waters of Aruba, Bonaire, and Curaçao are incredibly important, as they support a rich diversity of birds, marine mammals, fish, and other marine species. Each island has a unique history with respect to management and protection of its inshore waters and reefs and developed different regulations concerning land-based processes and activities that impact those inshore waters: sedimentation, pollution, fishing, and other issues. Over the last decade, all of the islands have developed new sewage treatment facilities and solid waste management plans to reduce the amount of sewage and pollutants going into the surrounding waters and into the air. Aruba now has three large sewage treatment facilities so that very little untreated sewage makes its way into the sea. Similarly, a new sewage treatment facility is operating on Bonaire. Aruba has a solid waste facility that removes and recycles a large proportion of wastes, although smoke from the open landfill could still be seen as of 2016.

Annotated List of Species

X = documented on the island
? = possible record or uncertain status
IN = introduced

Anatidae	Ducks	A	B	C
Dendrocygna viduata	White-faced Whistling-Duck	X		X
Dendrocygna autumnalis	Black-bellied Whistling-Duck	X	X	X
Dendrocygna bicolor	Fulvous Whistling-Duck	X	X	
Anser albifrons	Greater White-fronted Goose	X		
Sarkidiornis melanotos	Comb Duck	X	X	X
Anas americana	American Wigeon	X	X	X
Anas platyrhynchos	Mallard		X	
Anas discors	Blue-winged Teal	X	X	X
Anas cyanoptera	Cinnamon Teal	X		
Anas clypeata	Northern Shoveler	X	X	X
Anas bahamensis	White-cheeked Pintail	X	X	X
Anas acuta	Northern Pintail	X	X	
Anas crecca	Green-winged Teal	X		X
Aythya collaris	Ring-necked Duck	X	X	X
Aythya affinis	Lesser Scaup	X	X	X
Bucephala albeola	Bufflehead			X
Nomonyx dominicus	Masked Duck		X	X

Odontophoridae	New World Quail			
Colinus cristatus	Crested Bobwhite	X		X

Podicipedidae	Grebes			
Podilymbus podiceps	Pied-billed Grebe	X	X	X
Tachybaptus dominicus	Least Grebe	X	X	X

Phoenicopteridae	Flamingos			
Phoenicopterus ruber	American Flamingo	X	X	X

Procellaridae	Petrels, Shearwaters			
Pterodroma hasitata	Black-capped Petrel	X		
Bulweria bulwerii	Bulwer's Petrel			X
Calonectris diomedea	Cory's Shearwater	X		
Ardenna gravis	Great Shearwater		X	
Puffinus lherminieri	Audubon's Shearwater	X	X	X

Hydrobatidae	Storm-Petrels	A	B	C
Oceanites oceanicus	Wilson's Storm-Petrel	X	X	X
Oceanodroma leucorhoa	Leach's Storm-Petrel		X	X

Phaethontidae	Tropicbirds			
Phaethon aethereus	Red-billed Tropicbird	X	X	X
Phaethon lepturus	White-tailed Tropicbird		X	X

Ciconiidae	Storks			
Mycteria americana	Wood Stork	X		

Fregatidae	Frigatebirds			
Fregata magnificens	Magnificent Frigatebird	X	X	X

Sulidae	Boobies			
Sula dactylatra	Masked Booby	X	X	X
Sula leucogaster	Brown Booby	X	X	X
Sula sula	Red-footed Booby	X	X	X

Phalacrocoracidae	Cormorants			
Phalacrocorax brasilianus	Neotropic Cormorant	X	X	X
Phalacrocorax auritus	Double-crested Cormorant		X	

Pelecanidae	Pelicans			
Pelecanus occidentalis	Brown Pelican	X	X	X

Ardeidae	Herons, Egrets, Bitterns			
Botaurus pinnatus	Pinnated Bittern	X		
Ixobrychus exilis	Least Bittern	X		X
Ardea herodias	Great Blue Heron	X	X	X
Ardea alba	Great Egret	X	X	X
Egretta garzetta	Little Egret	X		
Egretta thula	Snowy Egret	X	X	X
Egretta caerulea	Little Blue Heron	X	X	X
Egretta tricolor	Tricolored Heron	X	X	X
Egretta rufescens	Reddish Egret	X	X	X
Bubulcus ibis	Cattle Egret	X	X	X
Butorides virescens	Green Heron	X	X	X
Butorides striata	Striated Heron	X	X	X
Syrigma sibilatrix	Whistling Heron	X	X	
Nycticorax nycticorax	Black-crowned Night-Heron	X	X	X
Nyctanassa violacea	Yellow-crowned Night-Heron	X	X	X
Cochlearius cochlearius	Boat-billed Heron		X	

Threskiornithidae	Ibises, Spoonbills	A	B	C
Eudocimus albus	White Ibis	X		X
Eudocimus ruber	Scarlet Ibis	X	X	X
Plegadis falcinellus	Glossy Ibis	X	X	X
Plegadis chihi	White-faced Ibis	X		
Platalea ajaja	Roseate Spoobill	X	X	X

Cathartidae	Vultures			
Coragyps atratus	Black Vulture	X		
Cathartes aura	Turkey Vulture	X		

Pandionidae	Osprey			
Pandion haliaetus	Osprey	X	X	X

Accipitridae	Hawks, Kites			
Elanus leucurus	White-tailed Kite	X		
Elanoides forficatus	Swallow-tailed Kite	X	X	
Circus buffoni	Long-winged Harrier		X	X
Circus cyaneus	Northern Harrier			X
Geranoaetus albicaudatus	White-tailed Hawk		X	X

Rallidae	Rails, Gallinules			
Porzana carolina	Sora	X	X	X
Porphyrio martinicus	Purple Gallinule	X	X	X
Gallinula galeata	Common Gallinule	X	X	X
Fulica caribaea	Caribbean Coot	X	X	X
Fulica americana	American Coot	X	X	X

Heliornithidae	Sungrebe			
Heliornis fulica	Sungrebe		X	

Aramidae	Limpkin			
Aramus guarauna	Limpkin	X		

Burhinidae	Thick-knees			
Burhinus bistriatus	Double-striped Thick-knee			X

Recurvirostridae	Stilts, Avocets			
Himantopus mexicanus	Black-necked Stilt	X	X	X
Recurvirostra americana	American Avocet		X	

Haematopodidae	Oystercatchers			
Haematopus palliatus	American Oystercatcher	X	X	X

Charadriidae	Plovers, Lapwings	A	B	C
Pluvialis squatarola	Black-bellied Plover	X	X	X
Pluvialis dominica	American Golden-Plover	X	X	X
Vanellus chilensis	Southern Lapwing	X	X	X
Charadrius collaris	Collared Plover	X	X	X
Charadrius nivosus	Snowy Plover	X	X	X
Charadrius wilsonia	Wilson's Plover	X	X	X
Charadrius semipalmatus	Semipalmated Plover	X	X	X
Charadrius melodus	Piping Plover		X	
Charadrius vociferus	Killdeer	X	X	X

Jacanidae	Jacanas			
Jacana jacana	Wattled Jacana	X	X	X

Scolopacidae	Sandpipers, Allies			
Bartramia longicauda	Upland Sandpiper	X	X	X
Numenius phaeopus	Whimbrel	X	X	X
Limosa haemastica	Hudsonian Godwit	X	X	X
Arenaria interpres	Ruddy Turnstone	X	X	X
Calidris canutus	Red Knot	X	X	X
Calidris himantopus	Stilt Sandpiper	X	X	X
Calidris alba	Sanderling	X	X	X
Calidris alpina	Dunlin		X	X
Calidris bairdii	Baird's Sandpiper	X	X	X
Calidris minutilla	Least Sandpiper	X	X	X
Calidris fuscicollis	White-rumped Sandpiper	X	X	X
Calidris subruficollis	Buff-breasted Sandpiper	X	X	X
Calidris melanotos	Pectoral Sandpiper	X	X	X
Calidris pusilla	Semipalmated Sandpiper	X	X	X
Calidris mauri	Western Sandpiper	X	X	X
Limnodromus griseus	Short-billed Dowitcher	X	X	X
Limnodromus scolopaceus	Long-billed Dowitcher	?	?	?
Gallinago delicata	Wilson's Snipe	X	X	X
Phalaropus tricolor	Wilson's Phalarope		X	X
Phalaropus lobatus	Red-necked Phalarope		X	X
Phalaropus fulicarius	Red Phalarope		X	
Actitis macularius	Spotted Sandpiper	X	X	X
Tringa solitaria	Solitary Sandpiper	X	X	X
Tringa melanoleuca	Greater Yellowlegs	X	X	X
Tringa semipalmata	Willet	X	X	X
Tringa flavipes	Lesser Yellowlegs	X	X	X

Stercorariidae	Jaegars, Skuas	A	B	C
Stercorarius skua	Great Skua			X
Stercorarius maccormicki	South Polar Skua	X		
Stercorarius pomarinus	Pomarine Jaeger	X		
Stercorarius parasiticus	Parasitic Jaeger	X	X	X
Stercorarius longicaudus	Long-tailed Jaeger	X		

Laridae	Gulls, Terns			
Chroicocephalus philadelphia	Bonaparte's Gull		X	X
Chroicocephalus ridibundus	Black-headed Gull		X	
Leucophaeus atricilla	Laughing Gull	X	X	X
Leucophaeus pipixcan	Franklin's Gull	X		
Larus delawarensis	Ring-billed Gull	X	X	
Larus argentatus	Herring Gull	X	X	
Larus fuscus	Lesser Black-backed Gull	X		
Larus marinus	Great Black-backed Gull	X	X	
Anous stolidus	Brown Noddy	X	X	X
Anous minutus	Black Noddy	X	X	
Onychoprion fuscatus	Sooty Tern	X	X	X
Onychoprion anaethetus	Bridled Tern	X	X	X
Sternula antillarum	Least Tern	X	X	X
Sternula superciliaris	Yellow-billed Tern		?	
Phaetusa simplex	Large-billed Tern	X		
Gelochelidon nilotica	Gull-billed Tern	X	X	X
Hydroprogne caspia	Caspian Tern	X	X	X
Chlidonias niger	Black Tern	X	X	X
Sterna dougallii	Roseate Tern	X	X	X
Sterna hirundo	Common Tern	X	X	X
Thalasseus maximus	Royal Tern	X	X	X
Thalasseus sandvicensis acuflavida	Sandwich Tern	X	X	X
Thalasseus sandvicensis eurygnathus	Sandwich "Cayenne" Tern	X	X	X
Rynchops niger	Black Skimmer	X	X	X

Columbidae	Pigeons, Doves			
Columba livia	Rock Pigeon	IN	IN	IN
Patagioenas squamosa	Scaly-naped Pigeon		X	X
Patagioenas corensis	Bare-eyed Pigeon	X	X	X
Columbina passerina	Common Ground Dove	X	X	X
Columbina talpacoti	Ruddy Ground Dove		X	
Leptotila verreauxi	White-tipped Dove	X	X	X
Zenaida auriculata	Eared Dove	X	X	X

Cuculidae	Cuckoos	A	B	C
Guira guira	Guira Cuckoo			X
Crotophaga major	Greater Ani	X		X
Crotophaga ani	Smooth-billed Ani	X		
Crotophaga sulcirostris	Groove-billed Ani	X	X	X
Coccyzus americanus	Yellow-billed Cuckoo	X	X	X
Coccyzus minor	Mangrove Cuckoo	X	X	X
Coccyzus lansbergi	Gray-capped Cuckoo		X	

Tytonidae	Barn Owls	A	B	C
Tyto alba	Barn Owl		X	X

Strigidae	Owls	A	B	C
Athene cunicularia	Burrowing Owl	X	X	X

Steatornithidae	Oilbird	A	B	C
Steatornis caripensis	Oilbird	X		

Caprimulgidae	Nightjars, Allies	A	B	C
Chordeiles acutipennis	Lesser Nighthawk	?	?	?
Chordeiles minor	Common Nighthawk	X	X	X
Chordeiles gundlachii	Antillean Nighthawk			X
Hydropsalis cayennensis	White-tailed Nightjar	X	X	X
Antrostomus carolinensis	Chuck-will's-widow	X	X	X

Apodidae	Swifts	A	B	C
Cypseloides niger	Black Swift			?
Chaetura pelagica	Chimney Swift	X	X	X

Trochilidae	Hummingbirds	A	B	C
Florisuga mellivora	White-necked Jacobin	X		X
Glaucis hirsutus	Rufous-breasted Hermit			X
Chrysolampis mosquitus	Ruby-topaz Hummingbird	X	X	X
Chlorostilbon mellisugus	Blue-tailed Emerald	X	X	X

Alcedinidae	Kingfishers	A	B	C
Megaceryle torquata	Ringed Kingfisher	X		X
Megaceryle alcyon	Belted Kingfisher	X	X	X
Chloroceryle amazona	Amazon Kingfisher	X		X
Chloroceryle aenea	American Pygmy Kingfisher		X	

Picidae	Woodpeckers	A	B	C
Sphyrapicus varius	Yellow-bellied Sapsucker	X	X	X

Falconidae	Falcons, Caracaras	A	B	C
Caracara cheriway	Crested Caracara	X	X	X
Milvago chimachima	Yellow-headed Caracara		X	X
Falco sparverius	American Kestrel	X	X	X
Falco columbarius	Merlin	X	X	X
Falco peregrinus	Peregrine Falcon	X	X	X

Psittaculidae	Old World Parrots			
Psittacula krameri	Rose-ringed Parakeet			IN

Psittacidae	New World Parrots			
Amazona autumnalis	Red-lored Parrot			IN
Amazona ochrocephala	Yellow-crowned Parrot			IN
Amazona barbadensis	Yellow-shouldered Parrot		X	IN
Amazona amazonica	Orange-winged Parrot			IN
Forpus passerinus	Green-rumped Parrotlet			IN
Eupsittula pertinax	Brown-throated Parakeet	X	X	X
Ara severus	Chestnut-fronted Macaw			IN
Thectocercus acuticaudatus	Blue-crowned Parakeet			IN
Psittacara wagleri	Scarlet-fronted Parakeet			IN

Tyrannidae	Flycatchers			
Elaenia martinica	Caribbean Elaenia	X	X	X
Elaenia parvirostris	Small-billed Elaenia	X		
Elaenia chiriquensis	Lesser Elaenia		X	X
Sublegatus arenarum	Northern Scrub-Flycatcher	X	X	X
Contopus cooperi	Olive-sided Flycatcher		X	
Contopus virens	Eastern Wood-Pewee	X	X	
Pyrocephalus rubinus	Vermilion Flycatcher	X		
Fluvicola pica	Pied Water-Tyrant		X	
Machetornis rixosa	Cattle Tyrant	X		
Myiarchus tyrannulus	Brown-crested Flycatcher	X	X	X
Myiodynastes maculatus	Streaked Flycatcher		X	
Tyrannus melancholicus	Tropical Kingbird	X	X	X
Tyrannus tyrannus	Eastern Kingbird	X	X	
Tyrannus dominicensis	Gray Kingbird	X	X	X
Tyrannus savana	Fork-tailed Flycatcher	X	X	X

Vireonidae	Vireos			
Vireo flavifrons	Yellow-throated Vireo	X		X
Vireo philadelphicus	Philadelphia Vireo	X		X
Vireo olivaceus	Red-eyed Vireo	X	X	X
Vireo altiloquus	Black-whiskered Vireo	X	X	X

Corvidae	Crows	A	B	C
Corvus splendens	House Crow			X

Hirundinidae	Swallows, Martins			
Stelgidopteryx ruficollis	Southern Rough-winged Swallow	X	X	X
Progne subis	Purple Martin	X	X	X
Progne cryptoleuca	Cuban Martin			X
Progne dominicensis	Caribbean Martin	X	X	X
Progne tapera	Brown-chested Martin	X		
Tachycineta albiventer	White-winged Swallow			X
Tachycineta meyeni	Chilean Swallow			X
Riparia riparia	Bank Swallow	X	X	X
Hirundo rustica	Barn Swallow	X	X	X
Petrochelidon pyrrhonota	Cliff Swallow	X	X	X
Petrochelidon fulva	Cave Swallow		X	X

Muscicapidae	Old World Flycatchers			
Oenanthe oenanthe	Northern Wheatear		X	X

Turdidae	Thrushes			
Catharus fuscescens	Veery	X	X	X
Catharus minimus	Gray-cheeked Thrush		X	X
Catharus ustulatus	Swainson's Thrush	X	X	X
Hylocichla mustelina	Wood Thrush			X

Mimidae	Mockingbirds, Thrashers			
Margarops fuscatus	Pearly-eyed Thrasher		X	very rare
Toxostoma rufum	Brown Thrasher			X
Mimus gilvus	Tropical Mockingbird	X	X	X

Sturnidae	Starlings			
Sturnus vulgaris	European Starling	X	X	

Bombycillidae	Waxwings			
Bombycilla cedrorum	Cedar Waxwing	X		

Parulidae	New World Warblers			
Seiurus aurocapilla	Ovenbird	X	X	X
Helmitheros vermivorum	Worm-eating Warbler	X	X	
Parkesia motacilla	Louisiana Waterthrush	X	X	X
Parkesia noveboracensis	Northern Waterthrush	X	X	X
Vermivora chrysoptera	Golden-winged Warbler		X	

		A	B	C
Vermivora cyanoptera	Blue-winged Warbler	X		
Mniotilta varia	Black-and-white Warbler	X	X	X
Protonotaria citrea	Prothonotary Warbler	X	X	X
Leiothlypis peregrina	Tennessee Warbler	X	X	X
Oporornis agilis	Connecticut Warbler	X	X	X
Geothlypis philadelphia	Mourning Warbler	X		X
Geothlypis formosa	Kentucky Warbler	X	X	X
Geothlypis trichas	Common Yellowthroat	X	X	X
Setophaga citrina	Hooded Warbler	X	X	
Setophaga ruticilla	American Redstart	X	X	X
Setophaga tigrina	Cape May Warbler	X	X	X
Setophaga cerulea	Cerulean Warbler		X	
Setophaga americana	Northern Parula	X	X	X
Setophaga magnolia	Magnolia Warbler	X	X	X
Setophaga castanea	Bay-breasted Warbler	X	X	X
Setophaga fusca	Blackburnian Warbler	X	X	X
Setophaga petechia rufopileata	Yellow "Golden" Warbler	X	X	X
Setophaga petechia aestivia	Yellow "Northern" Warbler	?	?	?
Setophaga pensylvanica	Chestnut-sided Warbler	X	X	X
Setophaga striata	Blackpoll Warbler	X	X	X
Setophaga caerulescens	Black-throated Blue Warbler	X	X	
Setophaga palmarum	Palm Warbler	X	X	X
Setophaga coronata	Yellow-rumped Warbler	X	X	X
Setophaga discolor	Prairie Warbler	X		X
Setophaga virens	Black-throated Green Warbler	X	X	X
Cardellina canadensis	Canada Warbler		X	

Coerebidae — **Bananaquits**

		A	B	C
Coereba flaveola	Bananaquit	X	X	X

Thraupidae — **Tanagers**

		A	B	C
Tersina viridis	Swallow Tanager		X	
Cyanerpes cyaneus	Red-legged Honeycreeper		X	
Sicalis flaveola	Saffron Finch	IN	IN	IN
Volatinia jacarina	Blue-black Grassquit		X	X
Sporophila lineola	Lined Seedeater		X	
Tiaris bicolor	Black-faced Grassquit	X	X	X

Emberizidae — **Sparrows**

		A	B	C
Ammodramus savannarum	Grasshopper Sparrow		X	X
Zonotrichia capensis	Rufous-collared Sparrow	X		X
Zonotrichia albicollis	White-throated Sparrow	X		

Cardinalidae	Grosbeaks, Buntings	A	B	C
Piranga rubra	Summer Tanager	X	X	X
Piranga olivacea	Scarlet Tanager	X	X	X
Piranga ludoviciana	Western Tanager		X	
Pheucticus ludovicianus	Rose-breasted Grosbeak	X	X	X
Pheucticus melanocephalus	Black-headed Grosbeak			X
Passerina caerulea	Blue Grosbeak		X	
Passerina cyanea	Indigo Bunting	X	X	X
Spiza americana	Dickcissel	X	X	X

Icteridae	Blackbirds, Orioles			
Dolichonyx oryzivorus	Bobolink	X	X	X
Sturnella magna	Eastern Meadowlark		X	
Sturnella militaris	Red-breasted Meadowlark	X		X
Quiscalus mexicanus	Great-tailed Grackle	X		X
Quiscalus lugubris	Carib Grackle	X	X	X
Gymnomystax mexicanus	Oriole Blackbird	?		
Chrysomus icterocephalus	Yellow-hooded Blackbird	X	X	X
Molothrus bonariensis	Shiny Cowbird	X	X	X
Icterus spurius	Orchard Oriole			X
Icterus icterus	Troupial	X	IN	X
Icterus nigrogularis	Yellow Oriole	X	X	X
Icterus galbula	Baltimore Oriole	X	X	

Passeridae	Old World Sparrows			
Passer domesticus	House Sparrow	IN	IN	IN

Ploceidae	Weavers			
Ploceus cucullatus	Village Weaver			IN
Ploceus philippinus	Baya Weaver			IN

Works Cited

Alexander, C.S. 1961. The marine terraces of Aruba, Bonaire, and Curaçao, Netherlands Antilles. Annals of the Association of American Geographers 51:102–123.

Allen, R. P. 1956. The flamingos: their life history and survival. New York: National Audubon Society.

Angehr, G.R., and R. Dean. 2010. The birds of Panama: A field guide. San Jose: Zone Tropical Press, and Ithaca, NY: Cornell University Press.

Arends, R., and F. Boersma. 2000. Guide to Parke Nacional Arikok. Aruba: Arikok National Park Foundation.

Bairlein, F., D.R. Norris, R. Nagel, M. Bulte, C.C. Voigt, J.W. Fox, D.J. Hussell, and H. Schmaljohann. 2012. Cross-hemisphere migration of a 25 g songbird. Biology Letters. doi:10.1098/rsbl.2011.1223.

Bak, R.P.M., G. Nieuwland, and E.H. Meesters. 2005. Coral reef crisis in deep and shallow reefs: 30 years of constancy and change in reefs of Curaçao and Bonaire. Coral Reefs 24:475-479.

Barendsen, P., B. Boekhoudt, G. Boekhoudt, L. Carrillo, R. Derix, F. Franken, R.A. Odum, P. Portier, M. De Meyer, C., and D. MacRae. 2006. Bonaire National Marine Park Management Plan 2006. Bonaire: STINAPA.

Barendsen, P., B. Boekhoudt, G. Boekhoudt, L. Carrillo, R. Derix, F. Franken, R.A. Odum, P. Portier, M. Sweerts-de Veer, R. van der Wal, and O. Byers, eds. 2008. WildAruba conservation planning workshop final report. Apple Valley, MN: IUCN/SSC Conservation Breeding Specialist Group.

Beadle, D., and J.D. Rising. 2002. Sparrows of the United States and Canada: The photographic guide. London: Academic Press.

Beaman, M., and S. Madge. 1998. The Handbook of bird identification for Europe and the Western Palearctic. Princeton, NJ: Princeton University Press.

Beers C.E., J. de Freitas, and P. Ketner. 1997. Landscape ecological vegetation map of the island of Curaçao, Netherlands Antilles. Publications foundation for scientific research in the Caribbean region. No. 138. Amsterdam.

Bekker, J.P. 1996. The mammals of Aruba. The Netherlands: Vereniging voor Zoogdierkunde en Zoogdierbescherming.

Beumer, L., I. Van De Velde, and N. Verster. 2011. A sustainable future for Curaçao: Strategic options for ISLA and the ISLA site, Phase 1, part 1. Final report to the Ministry of Economic Development, Government of Curaçao. ECORYS Nederland BV, Rotterdam, Netherlands.

Beylevelt, K. 1995. Nature guide Netherlands Antilles & Aruba. Van Dorp Aruba, NV, Oranjestad, Aruba.

BirdLife International. 2014. Species factsheet: *Amazona barbadensis*. Downloaded from http://www.birdlife.org

Blomdahl, A., B. Breife, and N. Holmstrom. 2003. Flight identification of European Seabirds. London: Christopher Helm.

Bradley, P.E. and R.L.Norton. 2009. An inventory of breeding seabirds of the Caribbean. Gainesville: University Press of Florida.

Brewer, D. 2001. Wrens, dippers and thrashers. New Haven, CT: Yale University Press.

Buhrman-Deever, S.C., E.A. Hobson, and A.D. Hobson. 2008. Individual recognition and selective response to contact calls in foraging brown-throated conures, *Aratinga pertinax*. Animal Behavior 76:1715-1725.

Burns, C. M. B., J.A. Olin, S. Woltmann, P.C. Stouffer, S.S. Taylor. 2014. Effects of oil on terrestrial vertebrates: Predicting impacts of the Macondo blowout. BioScience 64:820-828.

Butchart, S. H. M., A.J. Stattersfield and T.M. Brooks. 2006. Going or gone: defining 'Possibly Extinct' species to give a truer picture of recent extinctions. Bull. Brit. Orn. Club. 126A: 7–24.

Bongers, E. 2009. Creating one happy island: The story of Aruba's tourism. Aruba: Evert Bongers.

Boyer, P. 1984. Birds of Bonaire. Bonaire: STINAPA Bonaire.

Byers, C., J. Curson, and U. Olsson. 1995. Sparrows and buntings: A guide to the sparrows and buntings of North America and the world. New York: Houghton Mifflin Company.

Carlson, M., J. Wells, and M. Jacobson. 2015. Balancing the relationship between protection and sustainable management in Canada's boreal forest. Conservation and Society.

CARMABI. 2014. Four wetlands protected on Curaçao under the RAMSAR Treaty. Carmabi, Curaçao.

Chantler, P. 2000. Swifts: A guide to the wwifts and treeswifts of the world, 2nd edition. New Haven, CT: Yale University Press.

Cleere, N. 2010. Nightjars, potoos, frogmouths, oilbird, and owlet-nightjars of the world. Princeton, NJ: Princeton University Press.

Cleere, N. 1998. Nightjars: A Guide to the nightjars, nighthawks, and their relatives. New Haven, CT: Yale University Press.

Clement, P. 1993. Finches & sparrows. Princeton, NJ: Princeton University Press,.

Coblentz, B.E. 1980. Goat problems in the national parks of the Netherlands Antilles. Report submitted to The Netherlands Antilles National Parks Foundation, Caribbean Marine Biological Institute, Curaçao.

Cole, S., and V. Razak. 2009. How far, and how fast? Population, culture, and carrying capacity in Aruba. Futures 41:414-425.

Cooper, G. 2011. Half a century of civil society participation in biodiversity conservation and protected areas management: a case study of Bonaire. Caribbean Natural Resources Institute (CANARI), Technical Report No. 397, Laventille, Trinidad.

Crossley, R. 2010. The Crossley ID Guide – Eastern Birds. Princeton, NJ: Princeton University Press.

Curson, J., D. Quinn, and D. Beadle. 1994. Warblers of the Americas: an identification guide. New York: Houghton Mifflin Company.

Davaasuren, N., and E.H.W.G. Meesters. 2012. Extent and health of mangroves in Lac Bay Bonaire using satellite data. IMARES Wageningen UR, Netherlands.

de Boer, B., E. Newton, and R. Restall. 2011. Birds of Aruba, Curaçao & Bonaire. Princeton, NJ: Princeton University Press.

de Boer, B.A. 1996. Our plants and trees, Curaçao, Bonaire, Aruba. Curaçao: Stichting Dierenbescherming.

De Buisonjé, P.H., 1974. Neogene and Quaternary geology of Aruba, Curaçao and Bonaire (Netherlands Antilles). Natuurwetenschappelijke Studiekring Voor Suriname en de Nederlandse Antillen, 78. Utrechet.

Debrot, A.O. 2006. Preliminary checklist of extant and fossil endemic taxa of the ABC-islands, Leeward Antilles. CARMABI, Curaçao.

Debrot, A.O., P.S. Bron, and R. de Leon. 2013. Marine debris in mangroves and on the seabed: largely neglected litter problems. Marine Pollution Bulletin 72:1.

Debrot, A.O., and J.A. de Freitas. 1991. Wilderness areas of exceptional conservation value in Curaçao, Netherlands Antilles. Nederlandse Commissie voor Internationale Natuurbescherming, Meded 26:1-25.

Debrot, A.O., J.A. de Freitas, A. Brouwer, and M. van Marwijk Kooy. 2001. The Curaçao barn owl: status and diet 1987-1989. Caribbean Journal of Science 37:185-193.

Debrot, A.O., M. de Graaf, R. Henkens, H.W.G. Meesters, and D.M.E. Slijkerman. 2011. A status report of nature policy development and implementation in the Dutch Caribbean over the last 10 years and recommendations towards the Nature Policy Plan 2012-2017. IMARES Wageningen UR, Netherlands.

Debrot, A., E. Meesters, D. Slijkerman. 2010. Assessment of Ramsar site Lac Bonaire – June 2010. IMARES Wageningen UR, Netherlands.

Debrot, A.O., A.B. Tiel, and J.E. Bradshaw. 1999. Beach debris in Curaçao. Marine Pollution Bulletin 38:795-801.

Debrot, A.O., J. van Rijn, P.S. Bron, and R. de Leon. 2013. A baseline assessment of beach debris and tar contamination in Bonaire, Southeastern Caribbean. Marine Pollution Bulletin 71:325-329.

Debrot, A.O., R. van Bemmelen, and J. Ligon. 2013. Bird communities of contrasting semi-natural habitats of Lac Bay, Bonaire, during the fall migration season, 2011. IMARES Wageningen UR, Netherlands.

Debrot, A.O., C. Wentink, and A. Wulfsen A, 2012. Baseline survey of anthropogenic pressures for the Lac Bay ecosystem, Bonaire. IMARES Report number C092/12.

Debrot, A.O., and J. Wells. 2008. Curaçao. In: D. Wege and V. Anadon V (eds), Important Bird Areas in the Caribbean: key areas for conservation: 143-149. BirdLife International, Cambridge, UK.

Debrot, A.O., R.H. Witte, M. Scheidat, and K. Lucke. 2011. A proposal towards a Dutch Caribbean Marine Mammal Sanctuary. IMARES Wageningen UR, Netherlands.

Debrot, A.O., C. Boogerd, D. van den Broeck. 2009. The Netherlands Antilles III. Curaçao and Bonaire. In: Bradley PE & Norton RL, eds. An inventory of breeding seabirds of the Caribbean. University Press of Florida.

Debrot, A.O., E. Meesters, R. de Leon, and D. Slijkerman. 2010. Lac Bonaire – Restoration Action Spear Points. September 2010. IMARES Report. No. C133/10.

Debrot, A.O., E. Meesters, and D. Slijkerman. 2010. Assessment of RAMSAR site Lac Bonaire – June 2010. IMARES Report. No. C066/10.

de Freitas, J.A., B.S.J. Nijhof, A.C. Rojer, and A.O. Debrot. 2005. Landscape ecological vegetation map of the island of Bonaire (Southern Caribbean). Amsterdam: Royal Netherlands Academy of Arts and Sciences.

de Freitas, J.A., and A.C. Rojer. 2013. New plant records for Bonaire and the Dutch Caribbean islands. Caribbean Journal of Science 47:114-117.

de León, R., K. Vane, P. Bertuol, V.C. Chamberland, F. Simal, E. Imms, and M.J. Vermeij. 2013. Effectiveness of lionfish removal efforts in the southern Caribbean. Endangered Species Research 22:175-182.

del Nevo, A. 2009. An assessment of nesting seabirds within San Nicolas Bay, Aruba, 2009. Report to Valero Aruba Refinery from Caribe Alaska N.V. prepared by Applied Ecological Solutions, Inc., Lawrence, KA.

del Nevo, A.. 2008. Aruba, in: Birdlife International (2008) Important bird areas in the Caribbean: key sites for conservation. Cambridge, UK: Birdlife International. (Birdlife Conservation Series no. 15).

De Meyer, K. Undated. Bonaire, Netherlands Antilles. Coastal Region and Small Island Papers 3, Environment and Development in Coastal Regions and Small Islands, UNESCO (http://www.unesco.org/csi/pub/papers/demayer.htm).

Derix, R., G. Peterson, and D. Marquez. 2013. The national bird count in 2011 in Aruba. Oranjestad, Aruba: Central Bureau of Statistics, Government of Aruba.

Dunn, J.L., and J. Alderfer. 2006. National Geographic field guide to the Birds of North America, 5th edition. Washington, DC: National Geographic Society,.

Dunn, J.L., and K. Garrett. 1997. A field guide to warblers of North America. New York: Houghton Mifflin Company.

Dutch Caribbean Nature Alliance. 2012. Annual Report 2012. Kralendijk, Bonaire: Dutch Caribbean Nature Alliance.

Dutch Caribbean Nature Alliance. 2013a. BioNews, Issue 6, June/July 2013. Dutch Caribbean Nature Alliance, Kralendijk, Bonaire (www.DCNAnature.org).

Dutch Caribbean Nature Alliance. 2013b. BioNews, Issue 7, August 2013. Dutch Caribbean Nature Alliance, Kralendijk, Bonaire (www.DCNAnature.org).

Engel, M., H. Brückner, V. Wennrich, A. Scheffers, D. Kelletat, A. Vött, F. Schäbitz, G. Daut, T. Willershäuser, and S. M. May. 2010. Coastal stratigraphies of eastern Bonaire (Netherlands Antilles): new insights into the palaeo-tsunami history of the southern Caribbean. Sedimentary Geology 231:14-30.

Ferguson-Lees, J., and D.A. Christie. 2001. Raptors of the world. New York: Houghton Mifflin Company.

ffrench, R. 1991. A guide to the birds of Trinidad & Tobago. Ithaca, NY: Cornell University Press.

Flikweert, M., T.G. Prins, J.A. de Freitas, and V. Nijman. 2007. Spatial variation in the diet of the Barn Owl *Tyto alba* in the Caribbean. Ardea 95:75-82.

Floyd, T. 2008. Smithsonian field guide to the birds of North America. New York: Harper Collins.

Forshaw, J.M. 2010. Parrots of the world. Princeton, NJ: Princeton University Press.

Fry, C.H., and K. Fry. 1992. Kingfishers, bee-eaters & rollers. Princeton, NJ: Princeton University Press.

Garrido, O.H., and A. Kirkconnell. 2000. Field guide to the birds of Cuba. Ithaca, NY: Cornell University Press.

Garrigues, R., and R. Dean. 2007. The birds of Costa Rica: A field guide. San Jose: Zone Tropical Press, and Ithaca, NY: Cornell University Press.

Geelhoed, S.C.V., A.O. Debrot, J.C. Ligon, H. Madden, J.P. Verdaat, S.R. Williams, and K. Wulf. 2013. Important bird areas in the Caribbean Netherlands. IMARES Wageningen UR, Netherlands.

Geelhoed, S.C.V., N. Janinhoff, J.P. Verdaat, R.S.A. van Bemmelen, and M. Scheidat. 2014. Aerial surveys of marine mammals and other fauna around Aruba, Curaçao and Bonaire, November 2013. IMARES Wageningen UR, Netherlands, Report number C012/14.

Gibbs, D., E. Barnes, and J. Cox. 2001. Pigeons and doves: A guide to the pigeons and doves of the world. New Haven, CT: Yale University Press.

Grant, P.J. 1986. Gulls: A guide to identification, 2nd edtion. Vermillion, SD: Buteo Books.

Goessling, J. M., W.I. Lutterschmidt, H.K. Reinert, L.M. Bushar, and RA. Odum. 2014. Multiyear sampling reveals an increased population density of an endemic lizard after the establishment of an Invasive snake on Aruba. Journal of Herpetology.

Goslinga, C.C. 1979. A short history of the Netherlands Antilles and Surinam. The Hague, Netherlands: Martinus Nijhoff.

Government of Aruba. 2009. National Report Aruba: Status quote implementation protocol concerning pollution from land-based sources and activities (LBS), June 5th, 2009. Orenjestad, Aruba: Government of Aruba.

Government of Curaçao. 2014. National Report of Curaçao for the Third International Conference on Small Island Developing States, Apia, Samoa, September 1-4, 2014. Willemstad, Curaçao: Ministry of Health, Environment and Nature, Government of Curaçao.

Gwynne, J.A., R.S. Ridgely, G. Tudor, and M. Argel. 2010. Wildlife Conservation Society birds of Brazil: The Pantanal & Cerrado of Central Brazil. Ithaca, NY: Cornell University Press.

Halcrow Limited. 2010. Strategic tourism master plan for the island of Curaçao 2010 – 2014. Dubai, United Arab Emirates: Halcrow International Partnership.

Harms, K.E., and J.R. Eberhard. 2003. Roosting behavior of the brown-throated parakeet (*Aratinga pertinax*) and roost locations on four southern Caribbean islands. Ornithol. Neotrop. 14: 79-89.

Harrison, P. 1987. Seabirds of the world: A photographic guide. Princeton, NJ: Princeton University Press.

Harrison, P. 1983. Seabirds: An identification guide. Boston: Houghton Mifflin Company.

Hartert, E. 1893. On the birds of the islands of Aruba, Curaçao, and Bonaire. Ibis 35:289-338.

Hartog, J. 2003. Aruba: short jistory. Oranjestad, Aruba: Dewitt and VanDorp Stores.

Hartog. J. 1978. A short history of Bonaire. N.V., Aruba: De Wit Stores.

Hartog, J. 1968. Curaçao: From colonial dependence to autonomy. Aruba, Netherlands Antilles: De Wit, Inc.

Hartog, J. 1961. Aruba past and present: From the time of the indians until today. Orenjestad, Aruba: D.J. De Wit.

Hayman, P., J. Marchant, and T. Prater. 1986. Shorebirds: An identification guide to the waders of the world. Boston: Houghton Mifflin Company.

Henkel, J. R., B.J. Sigel, and C.M. Taylor. 2012. Large-scale impacts of the Deepwater Horizon oil spill: Can local disturbance affect distant ecosystems through migratory shorebirds? BioScience 62:676-685.

Henriquez, P.C. 1962. Problems relating to hydrology, water conservation, erosion control, reforestation, and agriculture in Curaçao: questions and answers. Nicuwe West-Indische Gids 42:1-54.

Hilty, S.L. 2003. Birds of Venezuela, 2nd edition. Princeton, NJ: Princeton University Press.

Hilty, S.L., and W.L. Brown. 1986. A guide to the birds of Colombia. Princeton, NJ: Princeton University Press.

Hippolyte, J.C., P. Mann. 2011. Neogene-Quaternary tectonic evolution of the Leeward Antilles islands (Aruba, Bonaire, Curaçao) from fault kinematic analysis. Marine and Petroleum Geology 28:259-277.

Holian, P. 2012. Terns in paradise: Why five small islets off the coast of Aruba host more tern species than any other place on

Earth. Birdwatching Magazine: Feb 2012.

Holyoak, D.T. 2001. Nightjars and their allies: The caprimulgiformes. Oxford: Oxford University Press.

Howell, S.N.G. 2012. Petrels, albatrosses, and storm-petrels of North America: A photographic guide. Princeton, NJ: Princeton University Press.

Howell, S.N.G. 2010. Molt in North American birds. New York: Houghton Mifflin Company.

Howell, S.N.G., I. Lewington, and W. Russell. 2014. Rare Birds of North America. Princeton University Press, Princeton, NJ.

Howell, S.N.G., and S. Webb. 1995. A guide to the birds of Mexico and Northern Central America. Oxford: Oxford University Press.

Hulsman, H., R. Vonk, M. Aliabadian, A.O. Debrot, and V. Nijman. 2008. Effect of introduced species and habitat alteration on the occurrence and distribution of euryhaline fishes in fresh- and brackish-water habitats on Aruba, Bonaire and Curaçao (South Caribbean). Contributions to Zoology 77:45-52.

International Boreal Conservation Science Panel. 2013. Conserving the world's last great forest is possible: Here's how. Available at: http://borealscience.org/wp-content/uploads/2013/07/conserving-last-great-forests1.pdf (Accessed February 2015).

Isler, M.L., and P.R. Isler. 1999. The tanagers: Natural history, distribution, and identification. Washington, DC: Smithsonian Institution Press.

Jaramillo, A. 2003. Birds of Chile. Princeton, NJ: Princeton University Press.

Jaramillo, A., and P. Burke. 1999. New World blackbirds: The ccterids. Princeton, NJ: Princeton University Press.

Jones, H.L. 2003. Birds of Belize. Austin: University of Texas Press.

Juniper, T., and M. Parr. 1998. Parrots: A guide to parrots of the world. New Haven, CT: Yale University Press.

Kaufman, K. 2000. Birds of North America. New York: Houghton Mifflin.

Kenefick, M., R. Restall, and F. Hayes. 2007. Field guide to the birds of Trinidad & Tobago. Princeton, NJ: Princeton University Press.

Konig, C., F. Weick, and J-H. Becking. 1999. Owls: A guide to the owls of the world. New Haven, CT: Yale University Press.

Lace, M. J., and J. E. Mylroie. 2013. The biological and archaeological significance of coastal caves and karst features. Coastal Karst Landforms. Springer, Netherlands.

Latta, S., C. Rimmer, A. Keith, J. Wiley, H. Raffaele, K. McFarland, and E. Fernandez. 2006. Birds of the Dominican Republic & Haiti. Princeton, NJ: Princeton University Press.

Levesque, A., and P. Yeson. 2005. Occurrence and abundance of tubenoses (Procellariiformes) at Guadeloupe, Lesser Antilles, 2001-2004. North American Birds 59: 672-677.

Ligon, J. 2006. Annotated checklist of birds of Bonaire. Self-published.

LIGTT. 2014. 50 Breakthroughs: Critical scientific and technological advances needed for sustainable global development. Berkeley, CA: LBNL Institute for Globally Transformative Technologies.

Locke, H. 2013. Nature needs half: a necessary and hopeful new agenda for protected areas. Parks: The International Journal of Protected Areas and Conservation 19(2):13-22.

Luksenburg, J.A. 2014. Prevalence of external injuries in small cetaceans in Aruban waters, Southern Caribbean. Plos ONE 9(2): e88988. doi:10.1371/journal.pone.0088988.

Luksenburg, J.A. 2013. Cetaceans of Aruba, southern Caribbean. Journal of the Marine Biological Association of the United Kingdom: 1-14; doi:10.1017/S0025315413000337.

Luksenburg, J.A. 2011. Three new records of cetacean species for Aruba, Leeward Antilles, southern Caribbean. Journal of the Marine Biological Association of the United Kingdom 4:1-4; doi:10.1017/S175526721001193.

Luksenburg, J.A., and E.C.M. Parsons. 2013. Attitudes towards marine mammal conservation issues before the introduction of whale-watching: a case study in Aruba (southern Caribbean). Aquatic Conservation: Marine and Freshwater Ecosystems. doi:10.1002/aqc.2348

Luksenburg, J.A., and G. Sangster. 2013. New seabird records from Aruba, Southern Caribbean, including three pelagic species new for the island. Marine Ornithology 41:183-186.

Madge, S., and P. McGowan. 2002. Pheasants, partridges, and grouse: A guide to the pheasants, partridges. Quails, grouse, guneafowl, buttonquails and sandgrouse of the world. Princeton, NJ: Princeton University Press.

Madge, S., and H. Burn. 1999. Crows and jays. Princeton, NJ: Princeton University Press,.

Madge, S., and H. Burns. 1988. Waterfowl: An identification guide to the ducks, geese, and swans of the world. Boston: Houghton Mifflin Company.

McFarlane, D.A., and J. Lundberg. 2002. A middle Pleistocene age and biogeography for the extinct rodent *Megalomys curazensis* from Curaçao, Netherlands Antilles. Caribbean Journal of Science 38: 278–281.

Meyer de Schaeunsee, R., and W.H. Phelps, Jr. 1978. A guide to the birds of Venezuela. Princeton, NJ: Princeton University Press.

McMullan, M., T.D. Donegan, and A. Quevedo. 2010. Field guide to the birds of Colombia. Bogota, Colombia: ProAves Colombia.

Mearns, A. J., D.J. Reish, P.S. Oshida, T. Ginn, M.A. Rempel-Hester, C. Arthur, and N. Rutherford. 2013. Effects of pollution on marine organisms. Water Environment Research, 85:1828-1933.

Meesters, H.W.G., D.M.E. Slijkerman, M. De Graaf, and A.O. Debrot. 2010. Management plan for the natural resources of the EEZ of the Dutch Caribbean. IMARES Report Nr. C100/10. IMARES Wageningen UR, Netherlands.

Miller, J.Y., A.O. Debrot, and L.D. Miller. 2003. A survey of butterflies from Aruba and Bonaire and new records for Curaçao. Caribbean Journal of Science 39:170-175.

Mlodinow, S.G. 2004. First records of Little Egret, Green-winged Teal, Swallow-tailed Kite, Tennessee Warbler, and Red-breasted Blackbird from Aruba. North American Birds 57: 559–561.

Mlodinow, S.G. 2006. Five new species of birds for Aruba, with notes on other significant sightings. Journal of Caribbean Ornithology 19: 31–35.

Mlodinow, S.G. 2009. Two new bird species for Aruba, with notes on other significant sightings. Journal of Caribbean Ornithology 22: 103–107.

Mlodinow, S.G. 2015. Three new species for Aruba, with notes on other significant sightings. Journal of Caribbean Ornithology 28:1–5.

Morin, K.A. and N.M. Hutt. 1998. Remnant environmental effects from gold mining in Aruba after a century. Internet Case Study #8, http://www.mdag.com/case_studies/cs9-98.html (accessed 10 Jan., 2015).

Muhs, D. R., J.M. Pandolfi, K.R. Simmons, and R. R. Schumann. 2012. Sea-level history of past interglacial periods from uranium-series dating of corals, Curaçao, Leeward Antilles islands. Quaternary Research 78:157-169.

Murphy, W.L. 2000. Observations of pelagic seabirds wintering at sea in the southeastern Caribbean. Pp. 104-110 in Studies in Trinidad and Tobago Ornithology Honouring Richard ffrench (F. E. Hayes and S. A. Temple, Eds.), Dept. Life Sci., Univ. West Indies, St. Augustine, Occ. Pap. 11.

Nagelkerken, I.A., and A.O. Debrot. 1995. Mollusc communities of tropical rubble shores of Curaçao: long-term (7+ years) impacts of oil pollution. Marine Pollution Bulletin 30:592-598.

Nagelkerken, I.A., S. Kleijnen, T. Klop, R.A.C.J. van den Brand, E. C. de la Moriniere, and G. van der Velde. 2001. Dependence of Caribbean reef fishes on mangroves and seagrass beds as nursery habitats: a comparison of fish faunas between bays with and without mangroves/seagrass beds. Marine Ecology Progress Series 214:225-235.

Netherlands Antilles National Parks Foundation. 1983. Field guide: National Park Washington-Slagbaai, Bonaire. Netherlands Antilles National Parks Foundation, STINAPA No. 23, Bonaire.

Nijman, V., J. Booij, M. Flikweert, M. Allabadian, J.A. de Freitas, R. Vonk, and T.G. Prins. 2009. Habitat use of raptors in response to anthropogenic land use on Bonaire and Curaçao, Netherlands Antilles. Caribbean Journal of Science 45:25-26.

Nijman, V. 2010. The importance of small wetlands for the conservation of the endemic Caribbean coot *Fulica caribaea*. Caribbean Journal of Science 46(1): 112-115.

Nijman, V., M. Aliabadian, A.O. Debrot, J.A. de Freitas, L.G.L. Gomes, T.G. Prins, and R. Vonk. 2008. Conservation status of Caribbean coot *Fulica caribaea* in the Netherlands Antilles, and other parts of the Caribbean. Endangerd Species Research 4:241-246.

Nijman, V., J.T. Booij, M. Flikweert, M. Aliabadian, J.A. de Freitas, R. Vonk, and T.G. Prins. 2009. Habitat use of raptors in response to anthropogenic land use on Bonaire and Curaçao, Netherlands Antilles. Caribbean Journal of Science 45(1): 25-29.

Nijman, V., T.G. Prins, and J.H. Reuter. 2005. Timing and abundance of migrant raptors on Bonaire, Netherlands Antilles. Journal of Raptor Research 39(1): 94-97.

Noss, R.F., A.P. Dobson, R. Baldwin, P. Beier, C.R. Davis, D.A. Dellasala, J. Francis, H. Locke, K. Nowak, R. Lopez, C. Reining, S.C. Trombulak, and G. Tabor. 2012. Bolder thinking for conservation. Conservation Biology 26:1-4.

O'Brien, M., R. Crossley, and K. Karlson. 2006. The shorebird guide. New York: Houghton Mifflin Company.

O'Brien, M.J., J.B. Patteson, G.L. Armistead, and G.B. Pearce. 1999. Swinhoe's Storm-Petrel: First North American photographic record. North American Birds 53: 6-10.

O'Callahan, T. 2009. Parrots of the Caribbean. Audubon Magazine, July-August 2009.

Olsen, K.M., and H. Larsson. 2003. Gulls of North America, Europe, and Asia. Princeton, NJ: Princeton University Press.

Olsen, K.M., and H. Larrson. 1997. Skuas and jaegers: A guide to the skuas and jaegers of the world. New Haven, CT: Yale University Press.

Olsen, K.M., and H. Larsson. 1995. Terns of Europe and North America. Press, NJ: Princeton University.

Onley, D., and P. Scofield. 2007. Albatrosses, Petrels, and Shearwaters of the World. Princeton University Press, Princeton, NJ.

Pandolfi, J.M., and J.B.C Jackson. 2001. Community structure of Pleistocene coral reefs of Curaçao, Netherlands Antilles. Ecological Monographs 71:49-67.

Pandolfi, J.M., G. Llewellyn, and J.B.C. Jackson. 1999. Pleistocene reef environments, constituent grains, and coral community structure: Curaçao, Netherlands Antilles. Coral Reefs 18:107-122.

Paulson, D.R., C. de Haseth, and A.O. Debrot. 2014. Odonata of Curaçao, southern Caribbean, with an update to the fauna of ABC islands. International Journal of Odonatology 17:237-249.

Paulson, D. 2005. Shorebirds of North America: The photographic guide. Princeton, NJ: Princeton University Press.

Perlo, B.V. 2009. A field guide to the birds of Brazil. Oxford: Oxford University Press.

Peterson, G. 2016. Aruba Birdlife Conservation Checklist Version 2, ABC Birds of Aruba, September 5, 2016 - 254 Species. Aruba Birdlife Conservation, Aruba.

Petit, S. 2011. Effects of mixed-species pollen load on fruits, seeds, and seedlings of two sympatric columnar cactus species. Ecological research 26:461-469.

Petit, S. 2001. The reproductive phenology of three sympatric species of columnar cacti on Curaçao. Journal of Arid Environments 49:521-531.

Petit, S. 1995. The pollinators of two species of columnar cacti on Curaçao, Netherlands Antilles. Biotropica:538-541.

Petit, S., A. Rojer, and L. Pors. 2006. Surveying bats for conservation: the status of cave-dwelling bats on Curaçao from 1993 to 2003. Animal Conservation 9:207-217.

Petit, S., and L. Pors. 1996. Survey of Columnar Cacti and Carrying Capacity for Nectar-Feeding Bats on Curaçao. Conservation Biology 10:769-775.

Polido, A., E. Joao, and T.B. Ramos. 2014. Sustainability approaches and strategic environmental assessment in small islands: an integrative review. Ocean & Coastal Management 96:138-148.

Ponson, M. 1997. Sustainable development and tourism in Aruba. Islander Magazine July 4, 1997.

Pors, L. P. J. J., and I. A. Nagelkerken. 1998. Curaçao, Netherlands Antilles. Pp. 127-140. In: B. Kjerfve (ed), CARICOMP- Caribbean Coral Reef, Sea Grass and Mangrove Sites, Coastal Region and Small Island Papers 3, UNESCO, Paris.

Prins, T.G., and A.O. Debrot. 1996. First record of the Canada Warbler for Bonaire, Netherlands Antilles. Caribbean Journal of Science 32: 248-249.

Prins, T.G., J.A. de Freitas, and C.S. Roselaar. 2003. First specimen record of the barn owl Tyto alba in Bonaire, Netherlands Antilles. Caribbean Journal of Science 39(1): 144-147.

Prins, T.G., and V. Nijman. 2005. Historic changes in status of Caribbean coot in the Netherlands Antilles. Oryx 39:125-126.

Prins, T.G., J.H. Reuter, A.O. Debrot, J. Wattel, and V. Nijman. 2009. Checklist of the birds of Aruba, Curaçao and Bonaire. Ardea 97(2): 137-268.

Prins, T.G., K. Roselaar, and V. Nijman. 2005. Status and breeding of Caribbean coot in the Netherlands Antilles. Waterbirds 28:146-149.

Proosdij, A. S. J. van. 2012. Arnoldo's pocket flora: what grows and blooms in the wild on Aruba, Bonaire and Curaçao. Foundation for Scientific Research in the Caribbean Regions (Netherlands) 144.

Quick, J.S., H.K. Reinert, E.R. de Cuba, and R. A. Odum. 2005. Recent occurrence and dietary habits of *Boa constrictor* on Aruba, Dutch West Indies. Journal of Herpetology 39:304-307.

Radtke, U., G. Schellmann, A. Scheffers, D. Kelletat, B. Kromer, and H.U. Kasper/2003. Electron spin resonance and radiocarbon dating of coral deposited by Holocene tsunami events on Curaçao, Bonaire and Aruba (Netherlands Antilles). Quaternary Science Reviews 22:1309–1315.

Raffaele, H., J. Wiley, O. Garrido, A. Keith, and J. Raffaele. 1998. A guide to the birds of the West Indies. Princeton, NJ: Princeton University Press.

Reinert, H.K., L.M. Bushar, G.L. Rocco, M. Goode, and R. A. Odum. 2002. Distribution of the Aruba island rattlesnake, *Crotalus unicolor*, on Aruba, Dutch West Indies. Caribbean Journal of Science 38:126-128.

Restall, R., C. Rodner, and M. Lentino. 2006. Birds of Northern South America: An identification guide, Volume 1. New Haven, CT: Yale University Press.

Restall, R., C. Rodner, and M. Lentino. 2006. Birds of Northern South America: An identification guide, Volume 2: Plates and Maps. New Haven, CT: Yale University Press.

Reuter, J.H. 1999. Stern guide to the birds of Aruba. Oranjestad, Aruba: STERN NV.

Ridderstaat, J., R. Croes, and P. Nijkamp. 2013. Tourism and long-run economic growth in Aruba. International Journal of Tourism Research 16:472-487.

Ridgely, R.S., and P.J. Greenfield. 2001. The birds of Ecuador: Status, distribution, and taxonomy, Volume I. Ithaca, NY: Cornell University Press.

Ridgely, R.S., and P.J. Greenfield. 2001. The birds of Ecuador: Field guide, Volume II. Ithaca, NY: Cornell University Press.

Ridgely, R.S., and G. Tudor. 2009. Field guide to the songbirds of South America: The passerines. Austin: University of Texas Press.

Ridgely, R.S., and G. Tudor. 1994. The birds of South America, Volume II, the suboscine passerines. Austin: University of Texas Press.

Ridgely, R.S., and G. Tudor. 1989. The birds of South America, Volume I, the oscine passerines. Austin: University of Texas Press.

Rivera-Milán, F.F., F. Simal, and P. Bertuol. 2012. Population surveys of the yellow-shouldered parrot (*Amazona barbadensis rothschildi*) on Bonaire in March and October 2010–2012. Bonaire: STINAPA.

Rodner, C., M. Lentino, and R. Restall. 2000. Checklist of the birds of Northern South America. New Haven, CT: Yale University Press.

Rozemeijer, P. 2011. First record of sungrebe *Heliornis fulica* on Bonaire, Netherlands Antilles. Cotinga 33:123-124.

Rupert, L.M. 2012. Creolization and contraband: Curaçao in the early modern Atlantic world. Athens, GA: University of Georgia Press,.

Schellmann, G., U. Radtke, A. Scheffers, F. Whelan, and D. Kelletat. 2004. ESR dating of coral reef terraces on Curaçao (Netherlands Antilles) with estimates of younger pleistocene sea level elevations. Journal of Coastal Research 20:947–957.

Schreiber, E.A., and D.S. Lee. 2000. Status and conservation of West Indian seabirds. Society of Caribbean Ornithology, Ruston, LA.

Schulenberg, T.S., D.F. Stotz, D.F. Lane, J. P. O'Neill, and T.A. Parker III. 2007. Birds of Peru. Princeton, NJ: Princeton University Press,.

Sekeris, R. 2012. The yellow-shouldered Amazon: perspectives. ALTERRA, Wageningen UR, Netherlands.

Sibley, D. 2000. National Audubon Society The Sibley Guide to Birds. Alfred A. Knopf, New York, NY.

Simal, F., P. Holian, E. Albers, and E. Wolfs. 2010a. Monitoring program for water birds inhabiting the salt flats located on north-western Bonaire, Dutch Caribbean. Bonaire: STINAPA.

Simal, F., P. Holian, and E. Wolfs. 2010b. Brown booby (*Sula leucogaster*) monitoring program: year report 2010. Bonaire: STINAPA.

Simal, F., and F. Rivera-Milán. 2010. A manual for the landbird monitoring program of STINAPA Bonaire, Netherlands Antilles. Bonaire: STINAPA.

Slijkerman, D.M.E., R. de Leon, and P. de Vries. 2014. A baseline water quality assessment of the coastal reefs of Bonaire, Southern Caribbean. Marine Pollution Bulletin 86:523-529.

Slijkerman, D.M.E., P. de Vries, M.J.J. Kotterman, J. Cuperus, C.J.A.F. Kwadijk, and R. van Wijngaarden. 2013. Saliña Goto and reduced flamingo abundance since 2010: ecological and ecotoxicological research. IMARES Wageningen UR, Netherlands.

Slijkerman, D.M.E., R.B.J. Peachey, P.S. Hausmann, and H.W.G. Meesters. 2011. Eutrophication status of Lac, Bonaire, Dutch Caribbean including proposals for measures. IMARES Report Nr. C093/11.

Smith, S.R., W.J. van der Burg, A.O. Debrot, G. van Buurt, and J.A. de Freitas. 2014. Key elements towards a joint invasive alien species strategy for the Dutch Caribbean. IMARES Report Nr. C020/14. IMARES Wageningen UR, Netherlands.

Smith, S.R., N. Davaasuren, A.O. Debrot, F. Simal, and J.A. De Freitas. 2012. Preliminary inventory of key terrestrial nature values of Bonaire. IMARES Report Nr. C003/13. IMARES Wageningen UR, Netherlands.

Staring, M. 2003. Rondje Aruba: A walk around Aruba in 6 trails. Aruba: Editorial Charuba.

Steinstra, P. 1991. Sedimentary petrology, origin and mining history of the phosphate rocks of Klein Curaçao, Curaçao and Aruba, Netherlands West Indies. Uitgaven / Natuurwetenschapplijke Studiekring voor het Caraïbisch Gebied (Publications / Foundation for Scientific Research in the Caribbean Region), No 130, Amsterdam.

Stephenson, T., and S. Whittle. 2013. The warbler guide. Princeton, NJ; Princeton University Press.

Stiles, F.G., and A.F. Skutch. 1989. A guide to the birds of Costa Rica. Ithaca, NY: Cornell University Press.

Stoffers, A. L. 1956. The vegetation of the Netherlands Antilles. Studies on the Flora of Curaçao and Other Caribbean Islands 1, Natuurwetenschappelijke Studiekring Voor Suriname en de Nederlandse Antillen, Utrecht.

Sweerts-de Veer, R. van der Wal, and O. Byers, eds. 2008. WildAruba Conservation Planning Workshop Final Report. Apple Valley, MN: IUCN/SSC Conservation Breeding Specialist Group.

ter Bals, M. 2014. Demography of Curaçao, Census 2011. Bureau of Statistics, Willemstad, Curaçao.

Thielman, S. S. B. 2013. The effects of tourism developments on the built environment in small island developing states, scenarios, frameworks: The case study of Curaçao. Ph.D. dissertation, TU Delft, Delft University of Technology, Netherlands.

Turner, A., and C. Rose. 1989. Swallows & martins: An identification guide and handbook. Boston: Houghton Mifflin Company.

van Buurt, G. 2005. Field guide to the amphibians and reptiles of Aruba, Curaçao and Bonaire. Frankfurt, Germany: Edition Chimaira.

van Buurt, G., and A.O. Debrot. 2012. Exotic and invasive terrestrial and freshwater animal species in the Dutch Caribbean. IMARES Wageningen UR, Netherlands.

Van Der Lely, J.A.C., P. van Beukering, L. Muresan, D.Z. Cortes, E. Wolfs, and S. Schep. 2013. The total economic value of nature on Bonaire. IVM Institute for Environmental Studies, University of Amsterdam, Netherlands.

van Halewijn, R. 2009. The Netherlands Antilles II: Lago Reef, Aruba. Pp. 202-206 in (P.E. Bradley and R.L. Norton, eds.), An inventory of breeding seabirds of the Caribbean. University Press of Florida, Gainesville, FL.Vermeij, M.J.A. 2014. An overview of the ecological values of the Eastpoint area on Curaçao. CARMABI, Curaçao.

Voous, K. H. 1985. Additions to the avifauna of Aruba, Curaçao, and Bonaire, south Caribbean. Ornithological Monographs: 247-254.

Voous, K. H. 1984. Striated or green herons in the South Caribbean Islands? Annalen des Naturhistorischen Museums in Wien. Serie B für Botanik und Zoologie:101-106.

Voous, K.H. 1983. Birds of the Netherlands Antilles. Netherland: Foundation for Scientific Research in Surinam and the Netherlands Antilles, Utrechts.

Voous, K. H. 1957. The birds of Aruba, Curaçao, and Bonaire. Studies Fauna Curaçao and Other Caribbean Islands 29:1-250.

Voss, R.S., and M. Weksler. 2009. On the taxonomic status of *Oryzomys curasoae* McFarlane and Debrot, 2001, (Rodentia: Cricetidae: Sigmodontinae) with remarks on the phylogenetic relationships of O. *gorgasi* Hershkovitz, 1971. Caribbean Journal of Science 45:73-79.

Wells, J.V. 2007. Birder's conservation handbook: 100 North American birds at risk. Princeton, NJ: Princeton University Press.

Wells, J., D. Childs, F. Reid, K. Smith, M. Darveau, and V. Courtois. 2014. Boreal birds need half: Maintaining North America's bird nursery and why it matters. Boreal Songbird Initiative, Seattle, Washington, Ducks Unlimited Inc., Memphis, Tennessee, and Ducks Unlimited Canada, Stonewall, Manitoba.

Wells, J., F. Reid, M. Darveau, and D. Childs. 2013. Ten cool Canadian biodiversity hotspots: how a new understanding of biodiversity underscores the global significance of Canada's Boreal Forest. Boreal Songbird Initiative, Seattle, Washington, Ducks Unlimited Inc., Memphis, Tennessee, and Ducks Unlimited Canada, Stonewall, Manitoba.

Wells, J. V., and A. Debrot. 2008. Bonaire, in: Birdlife International (2008) Important bird areas in the Caribbean: key sites for conservation. Cambridge, UK: Birdlife International. (Birdlife Conservation Series no. 15).

Wells, J. S. Casey-Lefkowitz, G. Chavarria, and S. Dyer. 2008. Danger in the nursery: Impact on birds of tar sands oil development in Canada's Boreal Forest. Natural Resources Defense Council, Washington, DC.; Boreal Songbird Initiative, Seattle, WA; Pembina Institute, Calgary, Alberta.

Wells, J.V., and P. Blancher. 2011. Global role for sustaining bird populations. Chapter 2. Pp. 7-22. In: J.V. Wells (ed.), Boreal Birds of North America. Studies in Avian Biology (No. 41). Berkeley: University of California Press.

Wells, J.V., and A.C. Wells. 2006. The significance of Bonaire, Netherlands Antilles, as a breeding site for terns and plovers. J. Carib. Ornithol 19:21-26.

Wells, J.V., and A.M. Childs Wells. 2005. First record of Caribbean martin (Progne dominicensis) for Aruba. North American Birds 59:670-671.

Wells, J.V., and A. M. Childs Wells. 2004. Photographic documentation of Philadelphia vireo on Aruba. North American Birds 57:562-563.

Wells, J.V., and A. M. Childs Wells. 2002. Extreme extralimital summer record of western tanager (Piranga ludoviciana) from Bonaire, Netherlands Antilles. Cotinga 18:96-97.

Wells, J.V., and A. M. Childs Wells. 2001. First sight record of Philadelphia vireo (Vireo philadelphicus) for Curaçao, Netherlands Antilles, with notes on other migrant songbirds. El Pitirre 14:59-60.

Wells, J.V. 2002. For the island of the yellow-shouldered parrot. Birdscope 16(1).

Wells, P. G., J.N. Butler, and J.S. Hughes, eds. 1995. Exxon Valdez oil spill: fate and effects in Alaskan waters (No. 1219). ASTM International.

Wentink, C., and A. Wulfsen. 2011. Recreational and land use survey for Lac Bay Bonaire: A study towards mapping human activities in Lac Bay Bonaire and its catchment area and advising about the current management system. IMARES Wageningen UR, Netherlands.

Williams, S. 2003. The yellow-shouldered Amazon parrot (Amazona barbadensis): 2003 Field Study on Bonaire. Kralendijk, Bonaire: Unpublished report to the Bonaire Department of Planning and Resource Management (DROB).

Williams, S.R. 2009. Factors affecting the life history, abundance and distribution of the yellow-shouldered Amazon parrot (Amazona barbadensis) on Bonaire, Netherlands Antilles. Ph.D. Thesis. University of Sheffield, UK.

Woodard, C. 2008. Curaçao's crude legacy. Christian Science Monitor, November 28, 2008.

Wosten, J.H.M. 2013. Ecological rehabilitation of Lac Bonaire by wise management of water and sediments. IMARES Wageningen UR, Netherlands.

Zijlstra, J.S., P.A. Madern, and L.W. van den Hoek Ostende. 2010. New genus and two new species of Pleistocene oryzomyines (Cricetidae: Sigmodontinae) from Bonaire, Netherlands Antilles. Journal of Mammalogy 91:860-873.

Zijlstra, J.S., D.A. McFarlane, L.W. van den Hoek Ostende, and J. Lundberg. 2014. New rodents (Cricetidae) from the Neogene of Curaçao and Bonaire, Dutch Antilles. Palaeontology 57:895-908.

Photo Credits

Front cover: male (above) and female (below) Ruby-topaz Hummingbird (*Chrysolampis mosquitus*)
Back cover (top to bottom): Brown-throated Parakeet (*Eupsittula pertinax*), Burrowing Owl (*Athene cunicularia*), Bananaquit (*Coereba flaveola*), American Flamingo (*Phoenicopterus ruber*)

p. ii: Troupial (*Icterus icterus*), Robert Dean
p. vi: Troupial (*Icterus icterus*), John McKean
p. viii: Bananaquit (*Coereba flaveola*), Michiel Oversteegen
p. 4: American Flamingo (*Phoenicopterus ruber*), Glenn Bartley
p. 8: Jeff and Allison Wells
p. 10: Jeff and Allison Wells
p. 12: Suzie McCann
p. 14: Jeff and Allison Wells
p. 18: Jeff and Allison Wells
p. 20: Jeff and Allison Wells
p. 23: White-cheeked Pintail (*Anas bahamensis*), Michiel Oversteegen
p. 26: Jeff and Allison Wells
p. 31: Jeff and Allison Wells
p. 33: Jeff and Allison Wells
p. 35: Jeff and Allison Wells
p. 37: Henkjan Kievit/VanellusVanellus.com/DCNA
p. 40: Jeff and Allison Wells
p. 42: Ferdinand Kelkboom
p. 45: Jeff and Allison Wells
p. 47: Alex Nieuwmeyer
p. 49: Jeff and Allison Wells
p. 50: Glenn Bartley
p. 52: Ross Wauben
p. 55: Jeff and Allison Wells
p. 60: John McKean
p. 63: Sipke Stapert
p. 67: Lauren Schmaltz
p. 70: Jeff and Allison Wells
p. 73: Jeff and Allison Wells
p. 74: John McKean
p. 77: Glenn Bartley
p. 81: CARMABI
p. 84: Jeff and Allison Wells

Index of Common and Scientific Names

467

About the Authors and Illustrator

Jeffrey V. Wells and **Allison Childs Wells** began birding together when they met in college. After graduating from the University of Maine at Farmington, they went on to graduate programs at Cornell University in Ithaca, New York, where Jeff received his M.S. and Ph.D. and Allison, her M.F.A. Both stayed on at Cornell, Allison as Communications Director for the world renowned Cornell Lab of Ornithology, and Jeff, also at the Lab, as Director of Bird Conservation for National Audubon (first for New York State then for the US). They returned to Maine in 2004 to raise their child among family and Maine's spectacular natural environment. They have published hundreds of bird-related articles and have collaborated on many projects, including the popular books *Maine's Favorite Birds* and *Birds of Sapsucker Woods*, a weekly newspaper birding column, as co-contributors to the *Sibley Guide to Bird Life and Behavior*, and as creators and webmasters of the websites Arubabirds.com and Bonairebirds.com, which provide field identification and bird-finding information about the birds of these popular vacation islands. They have been studying the birds of Aruba, Bonaire, and Curaçao for more than twenty years and have published numerous academic and popular papers, reports, and articles on the birds of the islands. They have also amassed the largest collection of audio recordings of bird species from the three islands, a collection housed at the Macaulay Library of Natural Sounds at the Cornell Laboratory of Ornithology. Jeff is also author of *Birder's Conservation Handbook: 100 North American Birds at Risk,* published in October 2007 by Princeton University Press, and editor of *Boreal Birds of North America,* published in 2011 by University of California Press. Allison was coeditor of *Birder's Life List and Diary* (third edition) and contributed many bird family accounts to Scholastic's *New Book of Knowledge.* Both Allison and Jeff are also active blog authors, writing for a number of different blogs. Jeff is now Senior Scientist for the International Boreal Conservation Campaign and Boreal Songbird Initiative and is a Visiting Fellow at the Cornell Lab of Ornithology. Allison is Senior Director of Public Affairs for the Natural Resources Council of Maine. They live in Gardiner, Maine, with their son and two bird-watching indoor cats.

Robert Dean has been studying and painting Neotropical birds for more than fifteen years, during which time he has gone on birding tours—both as a guide and a participant—throughout the Americas. Born and raised in London, England, he was a successful professional guitarist and recording artist for eighteen years before settling in Costa Rica in the early 1990s, where he revitalized his childhood passions for wildlife and art. In addition to executing art commissions for the Costa Rican national park service, he has produced illustrations for Rainforest Publications pocket guides on the wildlife of Mexico, Belize, Guatemala, Costa Rica, Panama, Trinidad and Tobago, Peru, and the United States. Dean illustrated *The Birds of Costa Rica: A Field Guide* (Cornell University Press, 2014), *The Birds of Panama: A Field Guide* (Cornell University Press, 2010) and was a contributing artist to *The Wildlife of Costa Rica: A Field Guide* (Cornell University Press, 2010).

474